POND FISHERIES

F.G. MARTYSHEV

RUSSIAN TRANSLATIONS SERIES

4

CRC Press
Taylor & Francis Group
Boca Raton London New York

CRC Press is an imprint of the
Taylor & Francis Group, an **informa** business
A BALKEMA BOOK

RUSSIAN TRANSLATIONS SERIES

1. K.Ya. Kondrat'ev, Yu.I. Rabinovich & W. Nordberg (editors): *USSR/USA Bering Sea Experiment*
2. D.V. Nalivkin: *Hurricanes, Storms and Tornadoes*
3. V.M. Novikov (editor): *Handbook of Fishery Technology, Volume 1*
4. F.G. Martyshev: *Pond Fisheries*
5. R.N. Burukovskii: *Key to Shrimps and Lobsters*

CRC Press
Taylor & Francis Group
6000 Broken Sound Parkway NW, Suite 300
Boca Raton, FL 33487-2742

First issued in hardback 2017

Translation of:
Prudovoe Rybovodstvo.
Vysshaya Shkola Publishers, Moscow, 1973

Translator : Dr. B.R. Sharma
General Editor : Dr. V.S. Kothekar

ISBN-13: 978-90-6191-410-5 (pbk)
ISBN-13: 978-1-138-44071-5 (hbk)

Visit the Taylor & Francis Web site at
http://www.taylorandfrancis.com

and the CRC Press Web site at
http://www.crcpress.com

This book discusses problems related to the culture of valuable fish species as food for man.

Information is provided on the organization and construction of fish ponds, production processes in fish farms for warm-water carp and cold-water trout, the industrial method of incubation of fish eggs, rearing of phytophagous fish, polyculture, intensive carp and trout culture, mixed fish farming (rice and fish, carp and duck), development of pisciculture in peat quarries, irrigation networks, and reservoirs constructed for multipurpose exploitation. The book also discusses pure stock breeding of fish for pond culture and the transportation of live fish.

This edition of *Pond Fisheries* has been significantly supplemented with the latest information on commercial carp culture, biotechniques, culturing of phytophagous fish in ponds, and the cage method of fish rearing.

Sections dealing with hydrochemistry, hydrobiology, ichthyology, and other associated disciplines, included in the previous edition, have been omitted here. Students specializing in pond fisheries can obtain this information in pertinent courses and in courses on fishery hydroengineering, fish diseases, and lake and reservoir fisheries.

The present book, although intended mainly as a textbook for the specialized subject of pond fisheries, should prove useful to students of the Faculty of Animal Husbandry and other institutes where a general course on this discipline is taught.

As in the previous edition, relevant references have been listed at the end of each chapter, and even after individual sections in one chapter, covering both Soviet and foreign publications. Students will find these references useful in research work, while writing theses, and in broadening their knowledge of different aspects of pond fisheries. Literature on the theory and practice of pond fisheries has been presented at the end of the book, covering the more important Soviet and foreign publications on fisheries, ichthyology, hydrobiology, and fish physiology.

The author takes full responsibility for any shortcomings in this textbook, not only with reference to its scientific content, but also its pedagogic sequence, appreciating the fact that the material is not simply to be read but readily comprehended. How far I have succeeded in this attempt, only the reader can judge.

I am deeply grateful to my colleagues in the Department and Laboratory of Pond Fisheries of the Timiryazev Agricultural Academy for their valuable advice on reviewing this manuscript, especially those who suggested improvements in presentation. I am particularly indebted to Kamburova (Sofia Agricultural Institute) for providing data on pond fisheries in the Bulgarian Peoples' Republic.

Any constructive criticisms of this work are most welcome.

F.G. Martyshev

PREFACE

The present edition of *Pond Fisheries*, intended as a textbook for agricultural universities, differs from the previous one (1964) in that it concentrates on material mainly for students specializing in fish farming. Hence nearly every chapter in the book has been expanded to include recent advances in scientific knowledge and practical experience, both in the Soviet Union and abroad. Since the publication of the earlier edition the methods of management of fish farms have changed and new techniques have been introduced for a more efficient utilization of waters in intensive fish culture. The range of fish species used in pond culture has widened, necessitating an elaboration of the methods of fish rearing. New topics, not included in the earlier edition, are: (a) the industrial method of incubation of fish eggs and rearing of young carp; (b) production processes in the culture of phytophagous fish in ponds; (c) carp rearing in fallow rice fields flooded with water; (d) fish rearing in irrigation networks and multipurpose reservoirs; (e) rearing of carp and rainbow trout in cages; and (f) use of estuaries (ilmens*) and bays.

Additional information has been compiled in recent years on feeding of carp and trout and on methods of manuring fish ponds. The need for generalization and analysis of the data available demands that "intensive forms of pond fisheries" be considered a separate course, rather than a part of the general course on "pond fisheries." Such a separation would ensure a better assessment of our knowledge and experience on different forms and levels of intensification and a more rational utilization of the higher form, i.e., complex intensification. This would tend to speed up the development of different forms of intensification not only for carp and trout culture, but also for growing other types of valuable fish.

However, until such a course is offered, the relevant sections (Chapters XI, XII, and XIII), as in the earlier edition, constitute an abbreviated course on "pond fisheries." The present edition also includes a section on "Production processes in a warm-water and full-system carp farm with two-year rotation."

*Ilmens—delta lakes in the Volga River—General Editor.

CONTENTS

PART III

PART I

INTRODUCTION

Due to the steady rise in population throughout the world, considerable attention is being given to the problems of increasing protein resources and enhancing the biological value of different food products. Hence it is not by chance that pond fisheries have come under intensive study.

The area under fish ponds and fish yields are increasing, pisciculture technology is improving, and fish farming becoming accepted as a branch of agriculture.

Fish as a fairly valuable item of human nutrition is gaining greater recognition. The world catch of fish in 1969 was 57.5 million tons, with the specific weight of fish 45% higher, and the world production of meat of homoiothermal animals, including culled birds, tallow, and by-products in dressed weight was 127.7 million tons.

A significant part of a fish catch is subjected to one or another kind of processing, which reduces the nutritional value. From this point of view, the requirements of the consumer, which can be met to some extent by fresh, and to a greater extent by live fish, are assured by fresh-water resources, primarily fish ponds.

The increasing demand for proteinic food has made it necessary to exploit more completely and efficiently the waters available, especially those inland. Inland waters throughout the world comprise about 500 million hectares, yet the total production of this area is only 70 million quintals of fish or 15% of the total world catch, excluding catches of immature fish and miscellaneous angling.

The demand for fish as a nutritious food for man is increasing with each passing year. Simultaneously, inland resources in general and fish ponds in particular are being threatened by pollution from industrial effluents and toxic chemicals.

The widespread use of chemicals for plant and animal protection has led to a disruption in the biological balance in nature and a need has arisen for its restoration.

Often chemicals, especially hexachlorane, used for controlling harmful insects and in leaching waste water in ponds, have an adverse effect on the fish population. N.E. Kononiva, who studied the effect of chemical treatment of forests on wildlife in Ukrainian SSR, came to the conclusion that for trout, especially those living in man-controlled waters, exposure to high doses (200 kg/ha) of mineral-oil emulsion (DDT and hexachlorocyclohexane or HCCH) is dangerous. In the absence of water replacement this emulsion leads to mortality among trout of all age groups, while in running waters it leads to a total mortality among young fish and a partial mortality among older ones.

In the USA, in 15 tributaries of the Alabama and Tennessee Rivers all the fishes died due to chlorinated hydrocarbons running off of fields (DDT, etc.). A total mortality of salmon was recorded in four large Canadian rivers as a result of spraying forests with chemicals. The widespread use of DDT in the USA as an insecticide has more than once led to mass mortality of small birds, fishes, and other animals. The residual effects of DDT and HCCH, spread over areas surrounding water resources, were recorded not only in the body of fish which had died, but also in fish which had survived the treatment. The fish died not only as a result of direct treatment of the water resources, but also as a result of insecticides washed by precipitation from the plants entering the ground water. Oil emulsions (polychloropinene, DDT, HCCH) are very toxic in nature.

Pesticides and insecticides of the DDT type, and toxaphene, heptachlor, and others have been detected in most of the large surface waters of the USA. In the course of field treatment these chemicals enter rivers and other water resources. Cases of pollution with insecticides and pesticides of ground water have also been reported. These substances have a cumulative property, as a result of which they gradually accumulate in the tissues of various hydrobionts. There have been cases of human poisoning due to the consumption of fish caught from waters polluted with insecticides.

Thus the widespread and annually increasing use of poisonous chemicals, especially DDT and other chlorinated compounds of carbon, has led to environmental pollution and a reduction in the number of beneficial, sometimes endemic, organisms, posing a serious threat not only to animals but also to man. The increasing pollution of soil and water must therefore be taken into account in the exploitation of ponds and other waters for the purpose of fish farming.

Various types of growth stimulants should also be used with caution until their effects on the water resource and its various populations, in addition to fish, and on man in consuming the same, have been thoroughly investigated.

The General Secretary of the Central Committee of the Communist Party of the Soviet Union, L.I. Brezhnev, in one of his speeches said: "... an economic and rational use of natural resources, a concern for the earth, for the forests, for the rivers and clean air, and for the plant and animal kingdom—all these constitute a sincere communistic concern."[1] The IVth Session of the Supreme Soviet of the USSR (1972) was devoted exclusively to this problem.

The pond fisheries of the USSR have all the requisite conditions for large-scale development by collective and state enterprises. The main task before this branch of economy is to rear fish in specially adapted reservoirs, ponds, and small and floodland lakes to supply them to the people at various times for consumption in live or fresh form, since the fresher the fish the greater its nutritional value. From time immemorial fresh-water fish have been relished by the people. This need can be fully met with a wide assortment of fish if the problem of developing pond fisheries in the country is approached properly. With the passage of each year this problem has gained recognition, so much so that it is now an important State concern.

Essentially, the production processes of pond fisheries constitute a branch of agriculture. The methods of fish cultivation have much in common with the methods of rearing farm animals.

The development of pond fisheries as a branch of agriculture is determined by the fact that the foundation of socialistic agricultural exploitation of the land is based not only on the soil, but also on large expanses of water. It is known that water resources are extremely important for agricultural crops, especially in arid regions. Using such resources additionally for pisciculture is one way of increasing their profitability.

It is commonly stated that fish is an agricultural commodity; however, this statement is often disputed as it is generally considered that farm animals comprise cow, horse, pig, and so on. The affiliation with agriculture developed from the time man changed from simple fish catching to culturing fish in ponds and other waters used for agriculture. By constructing special ponds and providing proper conditions for growth, development, and feeding of fish, man treats them in the same manner as any other farm animal.

With each passing year the need for a wider development of pond fisheries is better appreciated. Gone are the days when the fish requirements of people could be met by angling alone, without the participation of man in the regeneration of fish stocks. The time has come for a greater preoccupation with fish cultivation, especially in those numerous water resources located within the boundaries of collective and state enterprises.

The XXIV Congress of the Communist Party of the Soviet Union in projecting the Five Year Plan for the development of the economy of the USSR during the period 1971 to 1975, made a provision for increasing fish production in the state fish farms by 2.7 times. Thus the development of pond fisheries in terms of agricultural production must be ensured.

[1]L.I. Brezhnev. *50 let velikikh pobed sotsializma* (Fifty Years of the Great Victory of Socialism). Gospolitizdat, 1967, page 33.

Fisheries occupy a major position in the national economy and their role is increasing year after year since fish is a source of the proteins essential to man. The meat of such widely distributed culturable fish as carp, trout, tench, sander, and vendace contain proteins to the extent of 18.0, 20.8, 17.5, 17.7, and 21.2% respectively. The proteinic matter of fresh fish is assimilated two to three times better by man than cattle meat.

Since fish oil contains vitamins and is easily assimilated, it too is a product of great importance. The meat of fish cultured in ponds contains oil ranging from 0.7% (pike) to 15.0% (peled).

The USSR is fairly rich in fresh-water inland resources: lakes comprise about 25 million hectares, reservoirs about 5.5 million hectares, and ponds more than 500,000 hectares. If rice fields and peat quarries are included, about one million hectares of waterspread become additionally available in which fish can be effectively raised. With the exploitation of rice fields for fish rearing, two harvests are obtained—rice and fish.

Many ponds and other water resources can be used for mixed carp-duck farming with larger profits for both poultry and fisheries. An appropriate quantitative ratio of duck to fish makes it possible to significantly reduce the cost of production of duck meat while augmenting fish production.

The area of peat quarries is increasing every year. These areas are largely inundated with water which, if not used rationally, could cause secondary swamping with all its attendant adverse effects. The organization of collective and state fish farms makes it possible not only to improve such areas through fish culture, but also to obtain a high quality product, namely, fish.

The efficacy of pond fisheries is confirmed by the following example. To obtain 75 quintals of meat from 100 hectares of arable land or 16 quintals of meat from 100 hectares of agricultural holdings a high expenditure is incurred; to obtain 200 quintals of fish (carp) from 100 hectares of pond area very little expenditure is required. This quantum of fish can be obtained because of the natural foods already available in the pond. With a moderate level of intensification and expenditure on additional feeding and fertilizers, it is possible to obtain 600 to 1,500 quintals of carp or even more from 100 hectares of pond area.

In the development of pond fisheries in the USSR efforts have been directed toward increasing productivity of the existing pond area, construction of new ones, and utilization of water in rice fields, irrigation canals, small and floodland lakes, estuaries, ilmens, bays, and shallows of reservoirs, peat quarries, waters released from thermal electric stations, sewage purification plants, and thermal waters. I recommend the construction of fish ponds in the beds of small rivulets also.

Special attention should be paid to cage fish culture, using cooling reservoirs and shallow waters of large reservoirs for this purpose.

Figure 1. Map showing distribution of pond fisheries in the prerevolution period (1914). *1*—Baltic trout farms; *2*—Baltic carp farms; *3*—western carp farms; *4*—northwestern carp farms; *5*—Belorussian carp farms; *6*—southwestern carp farms; *7*—isolated fish farms.

Figure 2. Andrei Timofeevich Bolotov (1738–1833).

Attention focused on fisheries as a branch of agriculture in Russia in 1869, although attempts at fish culture had been made even earlier. The famous ichthylogist F. Sudakevich, while describing fish culture in the USSR and elsewhere, mentioned that this branch of agriculture was neglected in Russia compared to other countries in the world. This remark succinctly reflects the conditions of fisheries in general and pond fisheries in particular during the prerevolution period (Figure 1). The progressive scientists of Prerevolutionary Russia clearly understood that given the depletion of fish stocks in major fisheries (marine, riverine, and lacustrine), the question of fish culture should be vigorously pursued, especially in resources of the pond type and in small lakes. Such talented scientists as A.T. Bolotov (Figure 2), V.P. Vrasskii (Figure 3), O.A. Grimm (Figure 4), and others advocated this policy but their suggestions did not elicit a sympathetic response, to say nothing about support, under the feudal system of the czar.

After the great October Revolution pond fisheries had essentially to be organized anew. From 1929 to 1940 pond fisheries developed rather intensively due to the collectivization of the Soviet village.

Figure 3. Vladimir Pavlovich Vrasskii (1829–1862).

During this period the total area increased six times compared to the prerevolution period. Pond fisheries were more numerous in the RSFSR and Ukrainian SSR (Figure 5).

Figure 4. Oskar Andreevich Grimm (1845–1921).

The area under fish ponds in the country, similar to areas in collective and state enterprises and specialized state farms, increased to 56,000 hectares in 1937 with a yield of about 97,000 quintals. The area by 1940 had increased to 99,500 hectares and production to 208,650 quintals.

After the Second World War and the commencement of the sixth Five Year Plan, pond fisheries covered an area of 156,000 hectares with 114,000 hectares in collective enterprises. Total production reached 307,000 to 344,000 quintals, of which collective enterprises contributed 153,000 quintals. At present, there are more than 200 state fish farms (taking into account those under construction), out of which 40 are trout farms (Figure 6).

In 1971 the production from pond fisheries in the country was about 830,000 quintals. The area under fish culture in collective and state enterprises was 74,000 hectares.

To ensure an increase in pond fisheries, it is essential to know not only the theoretical fundamentals but also the techniques of running a fish farm, whereby the proper conditions for the fish under culture are guaranteed. In this respect an important role is played by a number of factors: estimates of fish stocks, systematic control of their development, better utilization of natural food resources of the pond, requisite measures for drainage and intensification (feeding of fish, pond fertilizers, etc.). Controlling these factors is not possible in natural resources (seas, lakes, rivers, reservoirs), since the quantity of different fish species, their ratio to food resources, etc. are difficult to determine. In natural resources it is not possible to observe inputs on the basis of estimation and control to the same extent as in the case of fish ponds. Hence fish productivity in natural resources is comparatively less. For example, the yield from lakes is 20 to 30 kg of fish per hectare, with only a few more productive lakes yielding 50 to 80 kg per hectare, whereas from one hectare under pond pisciculture even with extensive fish farming it is possible to harvest 2 to 3 quintals of fish, and

Figure 5. Distribution of pond fisheries in the USSR as of January 1, 1941.

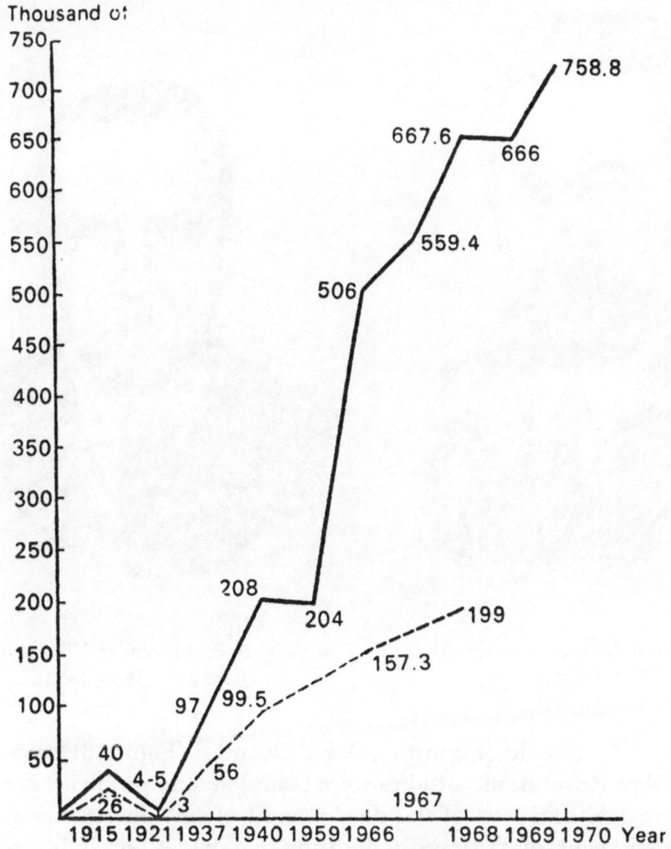

Figure 6. Growth of pond fisheries in the USSR.
—— fish in thousand quintals; ---- area of ponds in thousand hectares.

under intensive fish farming 15 to 20 quintals or more.

A rational development of pond fisheries is based on culturing those valuable species and varieties of fish for food which, over a short period, provide a high quality product. Locating the farms and fish ponds in close proximity to well-populated and industrial centers makes it feasible to supply live fish to the consumer without resorting to long-distance transportation. Because of the ease in supply and higher productivity, pond fishery has an advantage over lacustrine, riverine, and even marine fisheries.

Pond fisheries constitute an aspect of biology which must not be overlooked. A knowledge of the requirements of culturable fish at different stages of their life cycle makes it possible to regulate their developmental processes in the right direction.

In the development of indigenous fisheries a very significant role was

Figure 7. Ivan Nikolaevich Arnold (1868–1942).

Figure 8. Alexander Nikolaevich Eleonskii (1886–1952).

also played by I.N. Arnold (Figure 7), A.N. Eleonskii (Figure 8), and others after the October Revolution. Studies were conducted on problems pertinent to production: determination of standard weight of carp fingerlings (underyearlings), assessment of fattiness (condition factor) as an index to cold resistance, development of norms for mixed and additional stocking in a two-year rotation of carp farming, mixed carp-duck farming, methods of raising marketable one-summer carp in the southern regions, determination of fundamentals for two-year rotation of carp pond fisheries, determination of conditions for transportation of live fish without water by air, introduction of new species in pond culture (grass carp, silver carp, bighead, etc.), development of methods for intensification of pond fisheries, including formulation of theories on fish feeding and manuring of ponds, methods of fish rearing in peat ponds, establishment of work norms in pond fisheries and assessment in terms of man-days, investigation of problems related to breeding and advancing carp culture in the north, cage culture of carp, trout, and other species of fish, evaluation of mixed forms of pond fisheries (carp-duck farming, rice-fish farming), and others.

Placing inputs on a scientific basis in pond culture made it possible to increase the fish productivity of both state and collective enterprises. However, many problems of pond fisheries have yet to be resolved and hence certain recommendations for application in the field cannot be formulated. The most important of these problems pertain to mechanization of labor, economics, organization, methods of rearing juvenile fish, fish breeding, factors associated with fish feeding and pond manuring, and the introduction of valuable new fish species for food in pond culture. In the course of resolving the theoretical aspects research work on the development and application of respective methods has assumed great significance. As Academician I.P. Pavlov has written: "Science advances intermittently, depending on the success made by the techniques of scientific research. With each step of advanced methodology, so to speak, we rise one step higher and broaden the horizon of previously unknown fields."[2]

Pond pisciculture has already proved a profitable venture and in a number of places has expanded rapidly. For example, in Krasnodar pond fisheries have been established in more than 250 collective farms and 67 state farms, and in Altai 75 collective and state farms. In these and other farms not only are production areas under construction, but also fish nurseries; in addition to carp, trout and phytophagous species are being cultured and incubation sections for obtaining fish fingerlings under construction. The construction of fish ponds has been undertaken as a measure to prevent water and air erosion of cultivated lands in Stavropol.

These examples of the organization of pond fisheries have been provided by progressive collective and state enterprises which have mastered the techniques of creating fish ponds and fish farms. They are also becoming important as pig farms, poultry farms, and so on. This is a relatively younger branch of agriculture but the number of genuine enthusiasts, experts, and amateur fish farmers is steadily rising.

The problem now is to convert pond fisheries into a highly productive sector of collective and state farms on the basis of scientific principles. This can be done by educating cadres in zoo-technology in the three centers of higher agricultural education in the country (Timiryazev Agriculture Academy, and the Novosibirsk and Kherson Institutes of Agriculture).

[2]I.P. Pavlov. 1957. Collected works, vol. 2, no. 2, p. 22.

Chapter 1

TYPES, SYSTEMS, ROTATIONS, AND FORMS OF POND FISHERIES

Modern pond fisheries are of two types: warm-water and cold-water. This division is based on the biological peculiarities of the fish cultured in the ponds, the foremost being their behavior toward external environmental factors, mainly temperature and hydrochemical conditions.

In warm-water fisheries, in addition to common carp (the mainstay of fish culture), other fish such as silver carp, bighead, grass carp, and black amur are also raised, followed by tench, Crucian carp, goldfish, sander, pike, largemouth bass, trout-perch, starlet, etc. In cold-water fisheries, mainly trout, especially the rainbow, is cultured, together with some species of whitefish, cisco, vendace, peled, etc.

In terms of organization fish farms are divided into full-system or complete and partial-system or incomplete. In full-system fish farms fish are raised from eggs to the marketable stage (for human consumption) (Figure 9). In the partial-system only one of the two stages of production is undertaken: only stock material or seed fish reared, or foraging farms in which the seed fish are fattened to marketable weight and size. Farms of the first type are called fish nurseries and those of the second type one-year foraging (production) farms.

The selection of one of these two systems depends on the biological, technical, organizational, and economic conditions which prevail. Most of the fish ponds in the agricultural collective and state enterprises constitute foraging areas, which receive stock material (carp yearlings) from full-system fish farms or fish nurseries. State owned specialized fish farms are usually full-system farms, only rarely fish nurseries (Figure 10), and more rarely one-year production farms.

In organizing fish culture in the collective and state enterprises in each region, territory, and autonomous republic, one should first of all consider the geographic distribution of the existing foraging ponds and those under construction, second determine the requirements for stock material under one or the other system of intensification, and third identify the most convenient places for the construction of both full-system farms and one-year foraging farms and proper approach roads.

Pond fisheries are organized on different rotation schemes. Rotation in pond fisheries means the length of time required for raising fish from eggs up to the

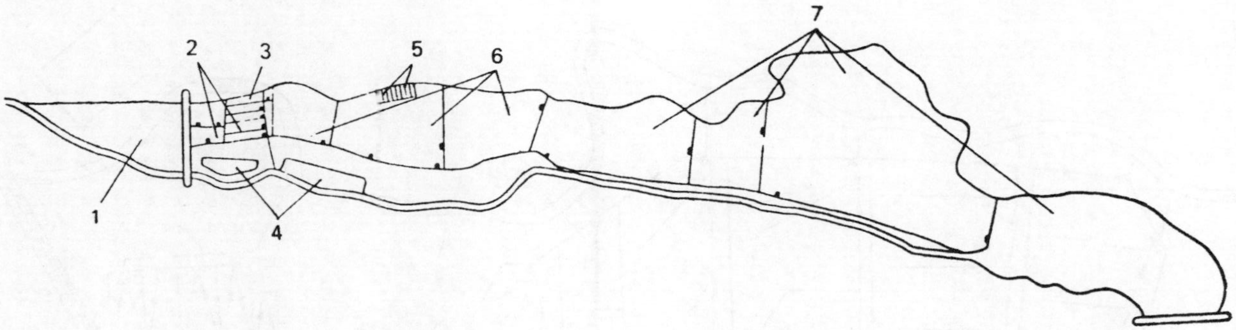

Figure 9. Sketch of full-system carp farm.

1—head pond (water supply pond); *2*—hibernation (wintering) ponds; *3*—transplantation pond; *4*—stock ponds; *5*—spawning (hatchery) ponds; *6*—rearing ponds; *7*—production (foraging or fattening) ponds.

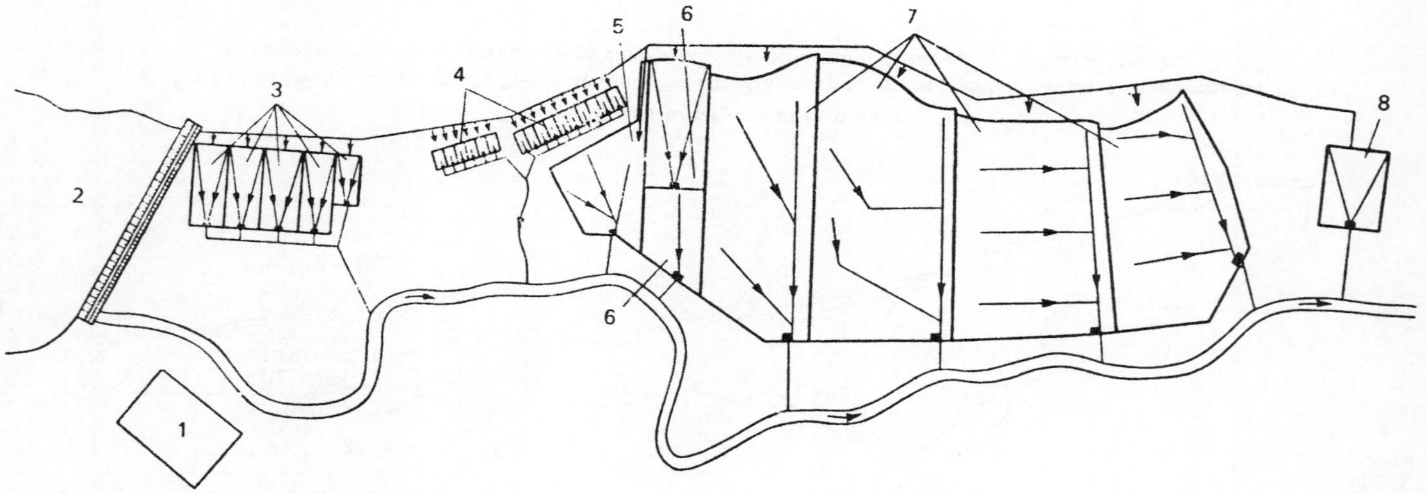

Figure 10. Sketch of a carp nursery.

1—administrative center; *2*—head pond; *3*—wintering ponds; *4*—hatchery ponds; *5*—nursery pond; *6*—stock ponds; *7*—rearing ponds; *8*—quarantine ponds.

marketable stage. This period varies according to the type of pond fishery and the culture of individual species of pond fish in different climatic conditions.

In warm-water carp culture one-year, two-year, and three-year rotations are common; three-year rotation is more prevalent in Western Europe.

In the one-year rotation scheme carp is raised to marketable size for 5 to 6 months (fish of one summer), in the two-year rotation for 16 to 17 months (fish of two summers), and in the three-year rotation marketed after 28 to 29 months (fish of three summers). In the USSR two-year rotation currently predominates in warm-water carp culture. In cold-water trout culture (with rainbow trout), two-year rotation has been adopted in the USSR and in most other countries.

The adoption of one or the other type of rotation in pond fisheries depends on the biological peculiarities of the fish, climatic conditions, level of fishery engineering, demand of consumers for certain size of carp, and so on.

Finally, pond fisheries are categorized on the basis of their operational organization as extensive, semi-intensive, and intensive. Usually the extensive type entails raising fish only on the natural foods available in the pond. However, in the present work this concept of extensive fish farming should be considered conditional, since even in the so-called extensive pond fisheries elements of intensification have already been included. This is to say that man interacts with nature by constructing special ponds for each age group, rearing fish from eggs up to marketable size, selecting brood stock carefully, improving the breeds of culturable species, draining of ponds, and so on. All these features impart properties to an extensive fish farm which reflect a more organized or cultured type. In other words, the so-called extensive fish farm is really the simplest form of intensification of pond fishery. The production of extensive fish farms is generally less but can be increased by rational use of the pond area.

As the base of extensive fish farms, the presently existing intensive fish farms should be universally extended since these units make it possible to significantly increase the production of marketable fish. In this type of management the inputs (feeding of fish, pond fertilizers, etc.) are intensified not randomly, but systematically. As a result of these measures, either coordinated or isolated, it is possible to increase the yield of fish per unit area of pond several times compared to yields obtained under extensive fish farms.

The semi-intensive type of fish farming is a transitional phase between extensive and intensive. In such farming, in addition to natural foods, the fish are periodically given artificial feed and the ponds partial fertilizer applications. The quantity of fish stocked for rearing is slightly higher in accordance with the quantity of periodic inputs—food and fertilizers. The purpose behind the periodic introduction of these inputs is not only to increase the quantity of fish stocked but also to increase the individual weight of the fish reared.

Chapter II

BIOLOGICAL AND TECHNICAL BASIS OF WARM-WATER CARP CULTURE

BIOLOGICAL CHARACTERISTICS OF CARP CULTURE. The biological peculiarities of culturable fish determine the methods of management and the nature of the ponds in which they are reared.

Carp are heat-loving fish which because they feed intensively during the vegetative or growth period are maintained in ponds rich in nutrients, i.e., which exhibit the characters of eutrophy.[1] Carp ponds should consist of stagnant or mildly flowing shallow waters which permit better warming up.

A small layer of soft silt, rich in organic and readily mineralized substances promotes a significant development of food organisms, increasing the food base of the pond and making it highly productive. No coarse plants should rise above the pond surface (reed, bullrush, cattail, sweet flag, etc.); if any are present they should be destroyed. The development of soft submerged plants (pondweed, hornwort, water thyme, water milfoil, etc.) over a small area of the pond is beneficial since they enhance the water with oxygen, afford places of shade for the carp during high temperatures, create a concentration of food organisms, and enrich the soil with nutrients when the plants perish and decompose.

Late in autumn, when the temperature drops, carp stop feeding and enter a state of low activity. During this time a sufficient quantity of oxygen and water for respiration must be provided. Ponds for the winter maintenance of carp are deeper and provided with running water. They are constructed on nonfertile soils to avoid losses of oxygen through oxidative processes.

CATEGORIES OF PONDS FOR CARP CULTURE. Since the requirements of carp vary according to its life stages and at different times of the year, ponds are constructed to accommodate changes in size and production requirements. Fish culture is usually carried out in ponds which can store water in summer and winter as well as in specially constructed ponds.

Summer ponds include spawning ponds, ponds for fry or nurseries (transplanting ponds), rearing ponds, and production or foraging ponds. These ponds serve as places for the growth and development of carp at

[1]A eutrophic pond is a water resource rich in nutritional substances.

different stages of its life. Winter ponds are only used for fish storage (hibernation) during winter.

Special ponds include stock ponds (summer and winter), quarantine ponds, and isolation ponds.

The head pond should be constructed near a perennial source of water that can adequately meet the requirements of the entire farm, taking into account losses through seepage and evaporation.

Every pond of the above-listed categories should have a sloping bottom with an outlet for drainage.

The distinguishing features in terms of biological and technical characteristics for the different categories of ponds are given below.

Spawning (hatching) ponds are used for breeding fish. They are small water bodies with stagnant water covering an area of 0.2 ha (preferably) or 0.1 to 0.3 ha (permissible range). Their bottom is covered with soft meadow vegetation serving as a suitable substratum for the attachment of common carp eggs. The ponds should be so situated as to be well-heated by the sun, protected from northern and north-eastern winds, and away from roads and mass movement of people and cattle since noise disturbs spawning. Hatching ponds should never be constructed on marshy soils with a high acidic reaction since this is lethal for the eggs and fry. The average depth of spawning ponds varies from 0.6 to 0.8 m, but may include a deeper (up to 1.25 to 1.50 m) and shallower (15 cm) section. Such a distribution of depth is attained by constructing a channel with a slope toward the dike. The shallow water region (up to 30 cm deep) should comprise 40 to 50% of the total pond area.

In low-lying areas with a high water table and in marshy regions, Samoilov (1967) suggests that spawning ponds be constructed not with an uneven bottom, as is usually done, but in the form of truncated pyramids (Figure 11, A and B). Such a construction, with a network of water-collecting channels in the shape of a Christmas tree, ensures good movement of the surface and bottom waters and makes the catching of young fry easier. In providing such modifications in the bottom relief of the spawning pond different depths are provided in the peripheral as well as in the middle parts of the pond. A shallow water zone is also beneficial for spawning and the development of zooplankton, required as food for different stages of postembryonic development. In the deeper zones carp fry can shelter from unfavorable weather conditions, especially sharp reductions in temperatures.

Transplantation ponds (nurseries) are usually 0.25 to 1.0 to 1.5 ha with an average depth of 0.5 m. Young fry about three to five days old are transferred from spawning ponds to nurseries where they remain for about 40 days. The nurseries are constructed near the spawning and rearing ponds on fertile soils redolent with food organisms and so located as to be well warmed by the sun. The main objective of these ponds is to create suitable conditions for food availability and growth of fry, which will later be trans-

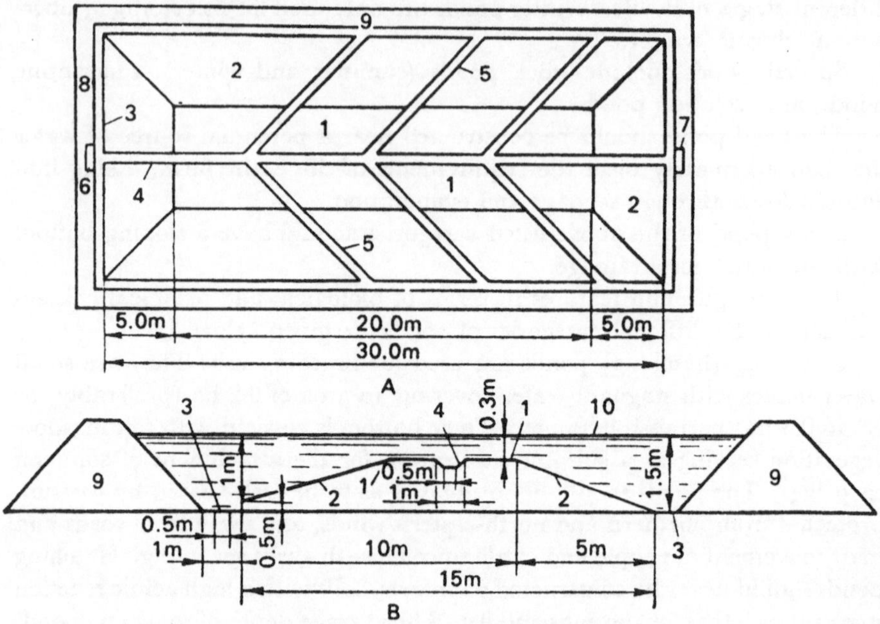

Figure 11. Plan of spawning ponds in lowlands with a high water table and in a marshy region (designed by M.D. Samoilov).

A—view from top; *B*—longitudinal section.

1—crest as spawning area 20 m long, 5 m wide, and total area 100 m²; *2*—sloping sides of spawning area about 350 m²; *3* and *4*—main water collection channels along perimeter and in middle of pond (width at top up to 1 m, at bottom about 0.5 m, and depth 40 to 50 cm) slanting toward monk; *5*—additional lateral draining ditches or channels up to 0.5 m wide and 30 to 40 cm deep; *6*—water-supply monk; *7*—water-drainage monk; *8* and *9*—cross and longitudinal dikes or embankments of pond; *10*—water surface (line of water surface of spawning pond filled).

ferred to rearing ponds, while attaining high fish productivity. Transferring young fry directly from spawning ponds to rearing ponds poses great hazards for them in struggling against predators and the mechanical effect of the waves formed in larger sheets of water. Nevertheless, at collective and state farms, where the area under rearing ponds in most cases is not very large, nurseries are usually not provided; the young fry are transplanted directly from the spawning ponds to the rearing ponds and thus extra stocking avoided. With proper technical management it is possible to obtain fingerlings of the standard weight and in good condition toward autumn, which are capable of wintering and attaining a sufficiently high growth in the second year of their life.

Rearing ponds are used for raising young fish (under-yearlings or fingerlings) up to the standard weight and condition established for a given zone.

Young fry are transplanted to these ponds from spawning ponds or nurseries (if the latter exist). The desirable area for a rearing pond is 10 to 15 ha (upper permissible limit 20 ha) with an average depth of 0.5 to 0.8 m (minimum depth 20 to 30 cm and maximum 1.5 m). Such ponds are located near spawning and nursery ponds.

In a three-year rotation of carp culture it is essential to have rearing ponds of the first and second order. The rearing pond described above is for a two-year rotation farm or a pond of the first order for a three-year rotation scheme. Rearing ponds of the second order are deeper than ponds of the first order (maximum depth 2 m, average 1 m), and used for raising two-year-old fish which, in a three-year rotation farm, are still not marketable. These fish become ready for marketing after three summers of feeding.

Production (foraging) ponds are the largest in area, ranging from 0.25 to several hundreds or even thousands of hectares (Figure 12), and are used for feeding marketable fish. These ponds are deeper than the rearing ponds, 0.5 to 2.0 m (average 1.0 m). At greater depths, 3.0 to 4.0 m, water warming is slower and light penetration poor, creating unfavorable conditions for the development of food organisms. Such depths are permissible only near the lower parts of the flush weir.

Wintering (hibernation) ponds are used to maintain under-yearlings or fingerlings during winter and fish of an older age left at the farm. Depending on the total acreage of the farm, each pond of this category may range in area from 0.5 to 1.0 ha, and in larger farms even up to 1.5 ha. The depth of the wintering pond depends on climatic conditions. In the central regions

Figure 12. Production pond at collective farm "Gigant" in Altai. Area, 30 ha.

of the USSR, where the winters are long and severe, the depth of the wintering pond is usually 1.8 to 2.0 m, but with the condition that the water layer left unfrozen be not less than 0.8 to 1.3 m. In the northern and southwestern regions of the USSR the depth of the wintering pond near the outlet is 2.5 to 3.0 m and in the upper part 1.5 m. Here, it is essential that the bottom of the pond uniformly slope from the upper part to the outlet.

In the southern regions, where the winter is not so severe and prolonged, a depth of 1.5 m suffices. The bottom of the wintering pond need not slope and the soil, contrary to that in the summer pond, should be compact—neither silty nor marshy. Preferably the soil should consist of sand or loam.

Stock ponds serve for brood fish and young fish during summer (summer stock ponds) and winter (winter stock ponds). Summer stock ponds are built according to the specifications for production ponds, and winter stock ponds in accordance with those for wintering ponds. The area of the stock pond depends on the quantity of brood and young fish required at the farm, which is determined on the basis of stock norms for the growth period or wintering.

Quarantine ponds are used for fish received from other ponds in order to examine their health. The area of the quarantine pond is 0.10 to 0.25 ha. It must meet the specifications for production ponds with a depth ranging from 0.8 to 1.5 m. The water supply arrangements for this pond should be autonomous and the drainage outlet separate so that the contaminated water does not enter the other ponds of the farm. Hence quarantine ponds are located at the end of the farm (along the flow of the stream) at a distance not less than 20 m from the production ponds. The ponds of this category should not be constructed on peaty and marshy grounds. The bottom of such ponds should be flat to facilitate disinfecting. Quarantine ponds should never be used for any other purpose.

If the fish held in the quarantine ponds are found to be disease-free, the water of the pond may be drained into the stream without prior disinfection. If, however, some infection is found, the fish are caught in seines or any other suitable apparatus and the water disinfected prior to drainage. The fishing gear and other accessories should also be disinfected.

Isolation ponds are reserved for fish with either confirmed or suspected disease. Their construction is such that the diseased fish (if not destroyed) can remain in them and also hibernate. Therefore, about 60% of the area of the pond has a depth of not less than 1.5 m. The norms for stocking fish depend on the time of year and are the same for spring and summer as for production ponds, and in autumn and in winter correspond to the minimum norms of stocking in wintering ponds. The water released from the isolation pond after the retention of diseased fish must necessarily be chlorinated.

The different categories of ponds, typically provided in different systems of pond pisciculture, are given in Table 1.

Table 1. Categories of fish ponds at fish farms of different systems

Full-system farm: Three-year rotation	Full-system farm: Two-year rotation	Fish hatchery	Partial one-year production farm
	One-year		
Spawning ponds	Spawning ponds	Spawning ponds	Production ponds
Nurseries (transplantation)	Nurseries (transplantation)	Nurseries (transplantation)	—
Rearing ponds of first order	Rearing ponds	Rearing ponds	—
Wintering ponds	Wintering ponds	Wintering ponds	—
	Two-year		
Rearing ponds of second order	Production ponds	—	—
Wintering ponds	—	—	—
	Three-year		
Production ponds	—	—	—

The area under ponds in each category in a full-system carp farm with a two-year rotation and in a fish hatchery must be laid out in a strictly defined ratio. For example, in an extensive full-system carp farm with a two-year rotation, where stock material is raised only for the farm's requirements, the ratio of ponds of different categories should be approximately: spawning ponds—0.1%, rearing ponds—5.9%, production ponds—93.8%, and wintering ponds—0.2% of the total pond area. This ratio may change, depending on the problems and assignments of the farm (for example, increasing the quantity of seed fish as stock not only for the farm but to sell, etc.).

The ratio of the ponds of different categories in a fish hatchery should be as follows: spawning ponds—2%, rearing ponds—94 to 95%, and wintering ponds—3 to 4% of the total pond area.

The area under special ponds (stock ponds, quarantine ponds, and isolation ponds) does not depend on the ratio of the ponds of different categories. The area under special ponds is calculated in accordance with the norms specified above. An approximate ratio of the ponds of different categories in a three-year rotation farm is given in Chapter VI.

The main task of a partial-system farm is to grow marketable fish. For a successful management of such a farm the fish hatchery from which the farm receives seed fish must be carefully assessed. The growth rate at collective and state fish farms in the country depends on the quality and quantity of the seed fish received.

Concept of Natural Fish Productivity and the Factors Determining it. The extensive type of fish farming is very close to an extensive pasture well maintained for farm animals, since in such cases the fish feed only on natural foods consisting of plankton and benthos. The only difference is that most of the culturable species of fish do not feed during winter.

In extensive fish farms a definite number of fish must be fed in the summer ponds. Fish of various species and age are fed until they reach the desired weight and size. Hence these ponds must contain a definite quantity of natural foods of that quality which will ensure the magnitude of growth desired or the natural fish productivity expected.

In pond fisheries natural fish productivity is evaluated in weight units of growth rate (increase in weight) of fish per unit area over one vegetative period. If, for example, the natural fish productivity calculated on the basis of the natural food resources of a pond is only equal to 300 kg per ha, this means that the increase in growth of fish over one vegetative period per hectare of area is 300 kg. The magnitude of this index is determined by several factors: quality of water; type of soil; climatic and meteorological conditions (intensity of light and heat); culturable state of water resource; quality of water; quality, quantity, and age of fish; etc.

Depending on the conditions of existence, fish attain a certain magnitude of growth, which determines the magnitude of natural fish productivity. The quality of the soil of the pond exerts a considerable influence on the natural fish productivity. Soils with high fertility and rich in readily soluble substances are best for carp culture; sandy, stony, and pebbled bottoms are not desirable. A sandy bottom has a high rate of seepage and is less productive; stony and pebbled bottoms contain insoluble salts, which lower natural fish productivity. Soil fertility is important not only in carp ponds, but also in trout ponds, especially when natural foods constitute the main source of nutrition. Peat soils tend to sink here and there, complicating hydrotechnical works; often floating peat beds appear which hinder fish culture.

The large water bodies (peat quarries) left after peat recovery have not been exploited for the purpose of fish culture; however, from the standpoint of fisheries peat quarries of low-lying sedge bogs yield a sufficiently high fish productivity, especially when reclaimed.

In constructing ponds for the summer retention of fish (spawning ponds, rearing ponds, production ponds, and summer stock ponds), the most productive soil layer must be retained; its removal renders the bottom infertile and considerably reduces the natural productivity of fish. For ponds for the winter retention of fish (hibernation ponds) the upper productive layer of the soil should be removed, since decomposed organic matter reduces the oxygen content of the water. It is well known that warm-loving fish such as carp do not feed in wintering ponds and the creation of the best oxygen regime for their normal respiration is extremely important.

The magnitude of natural fish productivity also depends on the quality of the water-collecting (catchment) area, since nutrient-laden water can considerably influence fish development. This factor should be taken into account not only when assessing the quality of the water-heading area in general, but when the ponds are filled with flood waters in spring. Ponds are better filled at the end of the floods, since the water is richest in nutrients at that time.

Water and soil are the two most important elements in carp culture in ponds. These two factors, in conjunction with others obviously, exert the greatest influence on natural fish productivity.

Natural fish productivity does not remain at a constant level; it may fluctuate with the passage of time and decrease in the absence of systematic check-ups on technical processes. It depends on a number of conditions, ultimately under the direct influence of man. Productivity is usually higher in ponds of young fish (for example, rearing ponds), since they utilize food resources more profitably and grow more rapidly than older fish. The growth rate of fish is generally higher in their first year of life. Carp exhibits a significant growth rate in the second (20 to 30 times) and third year. Growth slows down in subsequent years while the consumption of natural foods per unit of growth rate increases; food now is utilized more for sustenance than for growth. Thus it is understandable that carp fingerlings utilize natural food resources more effectively in rearing ponds than, for example, the two-year and three-year-old carp in production ponds. Given similar conditions, fish productivity in rearing ponds is 20 to 25% and, according to V. Shäperclaus (German Democratic Republic), even 40% higher than in foraging ponds.

Natural fish productivity in nurseries is generally good in spite of the fact that the fish remain in such ponds for only a short time (25 to 40 days), reaching 300 to 500 kg/ha.

Of the different climatic and meteorological factors influencing the natural food base, the development of carp, and consequently, the natural fish productivity of a pond, temperature plays the most important role.

Natural fish productivity also depends on density of carp stocking. High density leads to food shortages and a decrease in natural fish productivity. In such a situation the regeneration of food organisms may also be disturbed.

Obviously, the wide variations in climatic and soil conditions in the USSR preclude a single recommendation with regard to natural fish productivity in all the areas of the country. A more detailed analysis of this very important problem, both theoretical and practical, is still needed.

When fish culture is undertaken in exploited ponds put to use again without their natural productivity having been established, then stocking must be calculated on the basis of approximate data (Table 2).

Table 2. Natural fish productivity of carp foraging ponds located in different zones

Nature of catchment area	Administrative regions	Natural fish productivity of fishing ponds (kg/ha)
	Northern zone	
Forests and marshy, partially ploughed lands with podsolic soils	Pskov, Leningrad, Kalinin, Kostrom, Volgograd, Yaroslavl', and Kirov districts; northern part of Gorkii district; Perm; Sverdlovsk; Chelyabinsk district; Udmurt Autonomous SSR; southern part of Estonian SSR	125–175
	Siberian zone	
Partly forest lands	Kurgan, Tyumen, Omsk, Kemorev, Irkutsk, and Chita districts; Altai territory*; Buryat ASSR	150–200
	Northwestern zone	
Low fertility, podsolic and sandy soils, partly forests and marshy lands	Kaliningrad, Kalinin, Baranovich, Polots, Velikolusk, Minsk, Gomel', and Vitebsk districts; Lithuanian and Latvian SSR	150–200
	Central region and central belt	
Mixed, forest and ploughed fields with podsolic soil	Smolensk, Bryansk, Moscow, Kaluga, Tula, Ivanov, and Vladimirsk districts; southern part of Gorkii district; Chuvask, Marian, and Bashkirian ASSR (except for southern part)	175–300
	Eastern arid zone	
Cultivated fields and steppes	Ryazan; Penza, Kuibyshev, and Orenburg districts; southern part of Krasnoyarsk territory; Mordov; Tatar; southern part of Bashkirian ASSR; Buryat ASSR	200–225
	Central black soil zone	
Cultivated and manured chernozem soils	Tambov, Voronezh, Orlov, Kursk, Chernigov, Kharkov, Kiev, and Vinnits districts	225–275 and more
	Southwestern zone	
Manured chernozem fields	Kamenets-Podolsk, Poltava, Kirovograd, Dnepropetrosk, and Voroshilovograd districts	275–300

Nature of catchment area	Administrative regions	Natural fish productivity of fishing ponds (kg/ha)
	Southern regions	
Manured fields with chernozem and chestnut soils	Moldavia; Armenia; Georgia; Azerbaidzhan; Kirgizia;** Kazakh* (except southern part); Uzbek SSR; Dagestan ASSR; Krasnodar territory; southern part of Rostovo district; southern part of Volgograd district; Odessa, Nikolaev, Zaporozh, and Crimean districts***	300–350 and more

*Range from 150 to 300 kg/ha.

**For Kirgizia, Uzbek, and Kazakh—not more than 300 kg/ha.

***For Ukrainian SSR the average natural fish productivity according to Shpet (1962) is as follows: steppe regions—400 kg/ha, forest-steppes—320 kg/ha, and forest areas—150 kg/ha. For Trans-Carpathia the average is 230 kg/ha (range 180 to 370 kg/ha).

Natural fish productivity for valleys with gray soils and in Tadzhik SSR has been estimated as 150 to 200 kg/ha.

The figures projected in Table 2 are based on ponds originally stocked and subsequently exploited again. In autumn, when netting has been completed, the quantity of fish caught is estimated and, on the basis of growth rate obtained over the vegetative season converted into growth per hectare; the true natural fish productivity of a given pond is calculated on the basis of its natural food sources. This value is used in calculating the fish stock required in the subsequent year. Greater reliability is ensured by using the average values from two vegetative periods.

References

Karzinkin, G.S. 1951. O biologicheskoi produktivnosti vodoemov (Biological productivity of reservoirs). *Zool. Zhurn.*, vol. 30, no. 4.

Karzinkin, G.S. 1967. Razvitie problemy biologicheskoi productivnosti vodoemov za 50 let Sovetskoi vlasti (Progress in biological productivity of reservoirs during 50 years of Soviet rule). *Vopr. Ikhtiol.*, no. 5, p. 46.

Samoilov, M.D. 1967. Klassifikatsiya karpovykh nerestovykh prudov (Classification of carp spawning ponds). Sbornik *Rybnoe Khoz-vo.* Urozhai, Kiev, no. 4.

Shpet, G.I. 1962. Ryboproduktivnost' prudov Ukrainskoi SSR (Productivity of Fish Ponds in Ukrainian SSR). Gossel'khozizdat UkrSSR.

Chapter III

CONSTRUCTION OF FISH PONDS

In the construction and organization of a fish farm a number of hydrotechnical structures are essential (dikes, inlet and outlet channels, superstructures, bottom drains, siphons, weirs, sluices, water collection channels, fish pits, fishing devices, aerators, and so on). For all these items the following requirements must be met:

(a) Supply of good quality water in sufficient quantity;

(b) Filling of ponds with water through a system of water supply channels and troughs;

(c) Provision for complete draining of water from ponds;

(d) Transportation facilities (animal carts, automobiles, narrow gauge rails) inside the farm between ponds; accordingly, some dikes should be appropriately wide and strong;

(e) Complete and rational catchment of fish ensured.

Selection of Area for Construction of Fish Farms. Before selecting the site for the construction of a fish farm, one should first of all determine the nature of its future exploitation—whether simply for fish rearing or this in combination with other projects. The site and total area required, the quantity of water necessary, the waterworks, etc. are then considered. Special attention should be paid to the water used for fish culture as well as for the other requirements of the farm from the viewpoint of sanitation and hygiene, since here the most important aspect is the source of water supply. Places such as cemeteries, cattle-burying areas, and dumping grounds, where the soil is apt to be polluted with the decomposed products of organic matter or which may be infected with pathogens dangerous to man and animals, should never be considered for the construction of fish farms. The water required at the farm for domestic purposes should be clean and safe from pollution; the ponds for this purpose ought to be located along the water course above polluted places. A zone of sanitary maintenance should be created around such ponds, especially around the sources for the living quarters.

The blueprint for any fish farm should be approved by the State Sanitation Department.

Water analysis should be done twice a year to determine the quality of the water—once in summer and once in winter (in the central belt of the USSR preferably in February–March).

The approximate optimum chemical parameters characterizing the suitability of water for carp and trout farms are given in Table 3.

Table 3. Chemical parameters characterizing the suitability of water for carp and trout farms

Parameters	Optimum values for carp farm	Optimum values for trout farm	Remarks
Chromaticity, degrees:			
Summer ponds	Up to 30	Less than 30	
Winter ponds	Up to 30–50		
Transparency	Transparent, mild turbidity	Transparent	
Oxygen content, ml/l	More than 4	6–8	For incubation in apparatus
		8–10	For rearing fry
Free carbon dioxide, ml/l:			
Summer ponds	Up to 10	Up to 10	
Winter ponds	Up to 20	Up to 10	
Hydrogen sulphide, mg/l	0.0	0.0	
pH	7.0	7.0–7.5	
Alkalinity, kg/eq	1.8–3.5	1.5	
Total hardness, degrees	5.0–8.0	10–20	
Oxidizability, mg O_2/l:			
Summer ponds	Up to 30	Up to 8	
Winter ponds	Up to 10	Up to 8	
Nitrogen in the form of albumin-nitrogen, mg N/l:			
Summer ponds	Up to 1.5	Up to 0.08	
Winter ponds	Up to 0.5	Up to 0.08	
Ammonia salts, mg N/l:			
Summer ponds	Up to 1.0	Up to 0.5	
Winter ponds	Up to 0.5	Up to 0.5	
Nitrites, mg N/l:			
Summer ponds	Up to 0.2	Hundredth fraction of mg	
Winter ponds	Thousandth fraction of mg	Hundredth fraction of mg	
Nitrates, mg N/l:			
Summer ponds	Up to 2.0	Up to 0.08–0.70	
Winter ponds	Up to 1.0	Up to 0.08–0.70	
Phosphates, mg P_2O_5/l:			
Summer ponds	Up to 0.5	Up to 0.01–0.02	
Winter ponds	Up to 0.2	Up to 0.01–0.02	
Total iron, mg/Fe/l:	From 1/10th of mg up to 1–2	0.1–0.3	

(Contd.)

Parameters	Optimum values for carp farm	Optimum values for trout farm	Remarks
Chloride, mg Cl/l:			
Winter ponds in central belt of European part of the USSR	Up to 25	3.4–7.0	
Foraging ponds in southern regions located on saline soils	Up to 5,000	3.4–7.0	
Sulfates, mg SO_4/l:			
Winter ponds in central belt of European part of the USSR (except for regions with soils rich in gypsum)	Up to 25		
Salinity of water, g/l:	Up to 1.0	Fresh water	

The quantity of water required depends on the type and system of culture, size of the farm, area and depth of the ponds, water consumption in summer ponds taking into account percolation and evaporation, and maintenance of water flow in wintering ponds. The water consumption in summer carp ponds is approximately 1 l/sec/ha, although in practice this norm is reduced to 0.5 l/sec/ha and sometimes even lower. Consumption depends on the quality of the soil, climatic conditions, quantity of summer precipitation, and magnitude of losses due to evaporation and percolation.

In selecting the area for construction it is essential to estimate the discharge or output of the source of water supply to ascertain whether it will adequately fill the ponds in each category to the planned level within the approximate time established for such (Table 4). Recently, artesian wells have been adopted for supplying water to fish ponds.

The periods for filling and draining water from ponds given in Table 4 mainly relate to medium and large-sized farms. Filling and draining small ponds goes much more rapidly if their construction is technically correct.

The periods of filling of foraging ponds located in estuaries and river beds are estimated on one pond and may vary from 25 to 40 days. The time taken for draining such ponds depends mostly on the correctness of the pond's gradient and the draining capacity of the water outlet. In wintering ponds, where the main requirement is a continuous flow of water, water replacement (rate of flow) is closely related to a hydrochemical analysis. It is useless to increase the rate of flow to such an extent that replacement takes place in less than 10 days, since the stronger inflow of water has a disturbing effect on wintering fish and may not necessarily increase the oxygen content of the

Table 4. Periods of filling and draining water from ponds (in days)*

Parameters	Stock ponds	Spawning ponds	Nurseries	Rearing ponds	Wintering ponds: One-year old fish	Wintering ponds: Brood fish	Quarantine ponds
Period of filling all ponds with water	2	2	7	30	Up to 10	Up to 1	—
Period of filling each pond:							
Desirable	0.5	0.2	1	10–20	0.5–1.0	—	0.3
Permissible	1.0	0.3	2.0	30	1.5	—	0.5
Period for draining water from each pond:							
Desirable	0.2	0.1	0.5	3.0–5.0	1.0–1.5	Up to 0.5	0.2
Permissible	0.5	0.2	0.8	10	2	Up to 0.5	0.3

*Periods of filling and draining water for foraging ponds are determined on the basis of their area. In large ponds these periods may reach 20 days and more.

water (but may be done in the case of oxygen shortage in the water supply of the wintering pond or in the case of poor aeration).

Recently, due to a general deficit in fresh-water fish whereby the consumer demand notably exceeded supply, large-scale plans for the development of pond fisheries have led to studies for determining the norms of water consumption, and in particular the water requirement per fish biomass in unit area.

According to experiments conducted at fish farms in the Ukraine (forest-steppes, steppes, and forest areas), preliminary data indicate that to raise 1 kg of marketable carp in foraging ponds 10 to 15 m^3 of water is required (Shpet, Kharitonova, Kubyshkin, and Stebel'skii, 1967; Shpet, Kharitonova, and Kubyshkin, 1969). On the average the same quantity of water is consumed in wetting the bottom of the pond, percolation, and evaporation. Thus, according to these authors, the water consumption for raising 1 kg of table fish is about 30 m^3. To raise 1 kg of carp fingerlings consumption is about 10 m^3 with about 15 m^3 lost in soaking the bottom, percolation, and evaporation, i.e., a total consumption of around 25 m^3. For 1 kg of carp fingerlings held in wintering ponds, consumption is about 3 m^3 water to fill the ponds and an equal amount for losses, for a total consumption of about 6 m^3.

These findings should not be used as a basis for calculating the water requirements in all the zones of the USSR, not even in the Ukraine. Ponds of different types, with different conditions of water supply, methods of pond management, and nature and degree of intensification would obviously vary in this respect. Furthermore it should be kept in mind that water is an important commodity not only for fish culture but also for irrigation, domestic purposes, cultivation, and other needs and, therefore, in assessing the consumption of water for fish culture its other uses should be taken into account and included in estimates of water costs for the fish farm. It is important to note here also that increasing the productivity of fish should gradually lead to a reduction in water consumption per 1 kg of fish cultured.

In selecting a site attention should be given to the topography of the locality, as the cost of construction varies according to the terrain. Uneven terrain raises costs while a wide smooth floodplain reduces them. In most cases carp farms are constructed in the shallow valleys of small rivers and on lowlands which can be easily flooded and drained.

The type of soil on the bottom of a fish pond is another important consideration. Fertile soil is more productive for fish culture, but it is generally not possible to spare land suitable for the cultivation of valuable agricultural crops for pisciculture. Hence so-called "unsuitable" lands—marshes and reclaimed low-fertility areas—are generally converted into highly productive fish culture units.

Sandy soils and pebbled bottoms are not suitable for fish ponds as they have a high rate of percolation leading to greater water losses. Highly marshy soils with a thick layer of poorly decomposed peat are also not suitable.

Loamy and sandy-loamy soils (layer 1 to 2 m thick) provide a better bed for wintering ponds; here charging by ground waters is not permitted. In other categories of ponds the ground water table should desirably be at a depth of not less than 0.5 to 1.0 m.

Arrangement of Ponds in a Fish Farm. In the general scheme of a farm the ponds of each category should be laid out according to the course of production to avoid unnecessary transportation of fish during stocking and transplantation.

Wintering ponds with running water should be located closer to the source of water supply (head pond) to facilitate clearing the trough and inlet channels of ice and snow, which is particularly important in winters of heavy snowfalls. Spawning, rearing, wintering, and stock ponds should be arranged in a compact block to facilitate the process of fish transplantation. Stock ponds are also located closer to the source of water to ensure a high natural fish productivity. The arrangement of foraging (production) ponds is generally less compact; nevertheless it is highly desirable that they be located close to the nursery and rearing ponds (if possible at the planning stage of the area). The administrative center, residences, and service blocks should be located near the wintering ponds.

Layout of Bottom of a Fish Pond. An essential condition for a well-managed fish farm is to make complete water drainage possible for each pond. An uneven bottom provides depressions where fish can shelter, making their catchment difficult. The bottom should be so devised therefore as to render netting fish quick, easy, and complete. This is achieved by providing a slope toward the drain, constructing a deeper central channel which also slopes toward the drain, and less deep lateral collecting channels which slope toward the central one (Figure 13). The width of the channels should not be more than 1 m at the bottom and the depth 40 to 50 cm. Near the drain the central water channel should be broadened to a fish collection pit equal to the channel with its bottom at the same level as the threshold of the drain. This area depends on the pond and varies from 15 to 25 m^2.

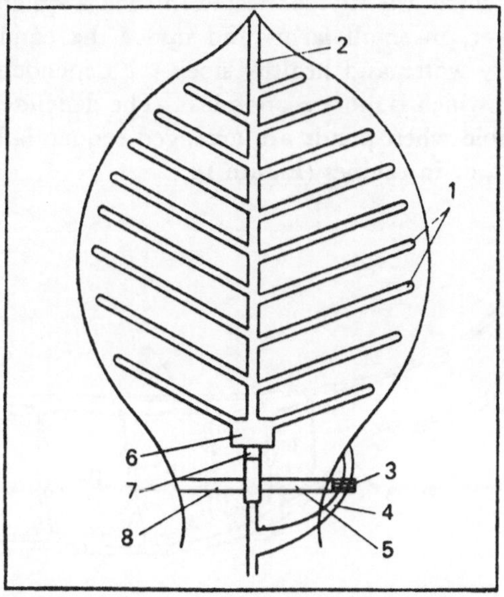

Figure 13. Layout of bottom of a fish pond.

1—lateral channels; *2*—central water collection channel; *3*—bridge; *4*—spillway; *5*—dike; *6*—fish pit (resting place of fish); *7*—vertical (raised) drain; *8*—horizontal water discharge channel.

The number of lateral channels depends on the bottom relief of the pond. The drainage system should consist of channels arranged in the form of divergent rays, similar to a herring bone, taking into account of course the bottom relief. Such a design—a central channel with smaller lateral channels—makes it possible to drain the pond completely. Fish which congregate in the lateral and central channels as well as in the fish pit are readily caught.

Systems of Water Supply to Fish Ponds and their Types. Independent and dependent types of water supply are found in pond fisheries. In the independent system each pond is filled with water from the head pond or some other source through channels and troughs which are not interconnected with another pond. The independent system of water supply is of paramount importance for wintering ponds where the replacement of water under high stock density in certain periods is an essential technical requirement. It is also suitable for other categories of ponds since it facilitates filling and draining. In the dependent system water flows directly from the head pond or other source only into those ponds in close proximity and from them to ponds located farther away. The inconvenience in the dependent system of water supply is the greater length of time it takes to fill the ponds. Often it is not possible to dry them either. The risk of spreading infection in ponds which are interconnected in such a system should not be overlooked. However, in small farms and under the conditions of maintaining good quality water and healthy stock the dependent type of water supply can be used, which is more economical. The dependent type of water supply is unavoidable when ponds are arranged sequentially along a river bed, a small rivulet, or in ravines (Figure 14).

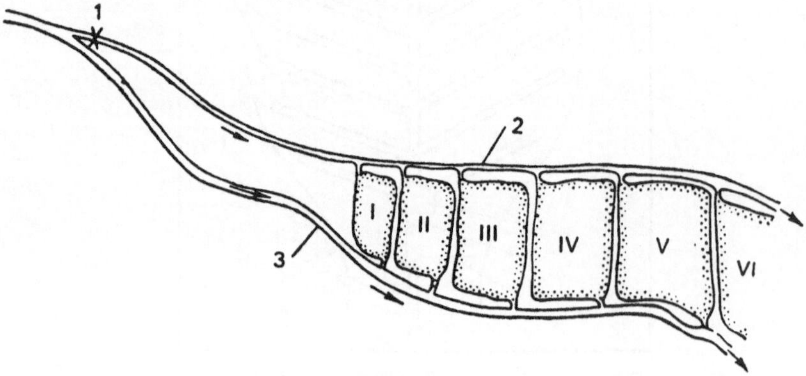

Figure 14. One system of water supply to ponds fed by streams and rivulets. Water supply from inlet channel and drainage of water into stream.
1—sluice; *2*—inlet channel; *3*—stream.
I to *VI*—ponds.

Water sources include lakes, reservoirs, small rivers, and streams. Water is supplied to the fish ponds by gravity or pumping.

Fish ponds may differ in shape, the contours depending on the topography of the area and the production requirements. Wintering ponds should preferably be long straight rectangles to ensure complete water replacement.

Requirements for Individual Hydroengineering Installations at Fish Ponds. The largest and costliest structure in the system of fish ponds is the head pond. This pond serves the purpose of storing water and should have its source on a perennial water course. The water stored in this pond must meet the entire requirements of the fish farm, taking into account the perennial water supply which is essential for replenishing the losses caused through evaporation and percolation.

A spillway is constructed at the head dam for discharging the excess water flowing into the pond from the source of water supply as well as flood waters. At the place of connection between the head pond and the source of water supply (river, rivulet, etc.) a screen is provided, which checks the ingress of undesirable fish and the egress of fish held in the head pond when it is used as a foraging pond. Even if the head pond is not used as a foraging pond, its overcrowding with trash or predatory fish is not advisable since these may enter the lower foraging ponds in the course of their filling and destroy the conditions of fish raising or become a source for disease.

The upper part acts as a barrier and is in the form of a simple sluice (Figure 15), consisting of netted screens installed in a series of grooves. It is essential to provide two sets of screens for alternate cleaning especially at the time of floods. Before catching fish from the pond the screens are replaced by plates which temporarily block the entry of water. The simplest kind of barrier installation is made of wattle or netting.

In recent years mechanical and electrical installations have been developed which restrict the movement of fish in undesirable directions. Studies are being done on the use of ultrasonics for this purpose. Experiments with vobla, bleak, perch, sander, and white bream in the Volga-Caspian regions have shown that these fish respond from a distance of 25 to 50 cm to ultrasonic waves produced by a magnetic-striction transformer. After entering the marked zone (area of influence), the fish move less rapidly and soon shift toward the opposite direction. The same response in fish was observed in stationary waters and in waters moving at a speed of 0.6 to 0.7 m/sec. The maximum sensitivity to ultrasonic waves was recorded in sander and bleak.

It is not always desirable to construct costly head dams for water supply to fish ponds. If there is a lake or river near the fish farm with sufficient gradient and output of water (without constructing the spillway block), the water from these sources is obtained by gravity and the need for constructing a head pond disappears.

In arid regions and places where it is not possible to use a cheaper method the water supply must be pumped, irrespective of the difference in the levels of the water source and the level of the fish ponds.

In a number of places it is more appropriate to use the reservoirs of hydroelectric power stations as the main source of water; all the categories of fish ponds are constructed below such large reservoirs.

Water Outlets. These constitute an important structure which varies according to the nature of the fish being reared. Outlets regulate the water level in the pond and ensure complete drainage. Those constructed in head ponds and foraging ponds laid out in river beds should be of the shaft type and spillways (sluices) constructed to increase the water discharge. In other categories of fish ponds a tubular bottom outlet of the monk type should be installed. This consists of two leakproof tubes connected at a right angle.

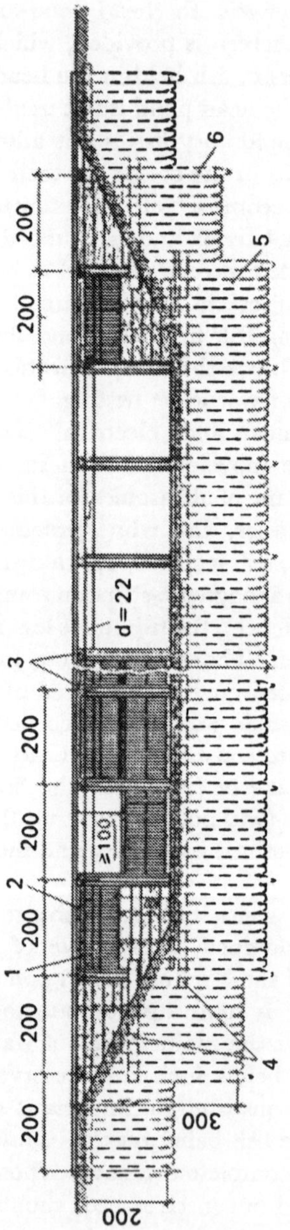

Figure 15. Top barrier.

1—metal screen; *2*—plank covering; *3*—filling (packing); *4*—directing logs (d=22); *5*—log pile; *6*—main pile (d=22) (dimensions in cm).

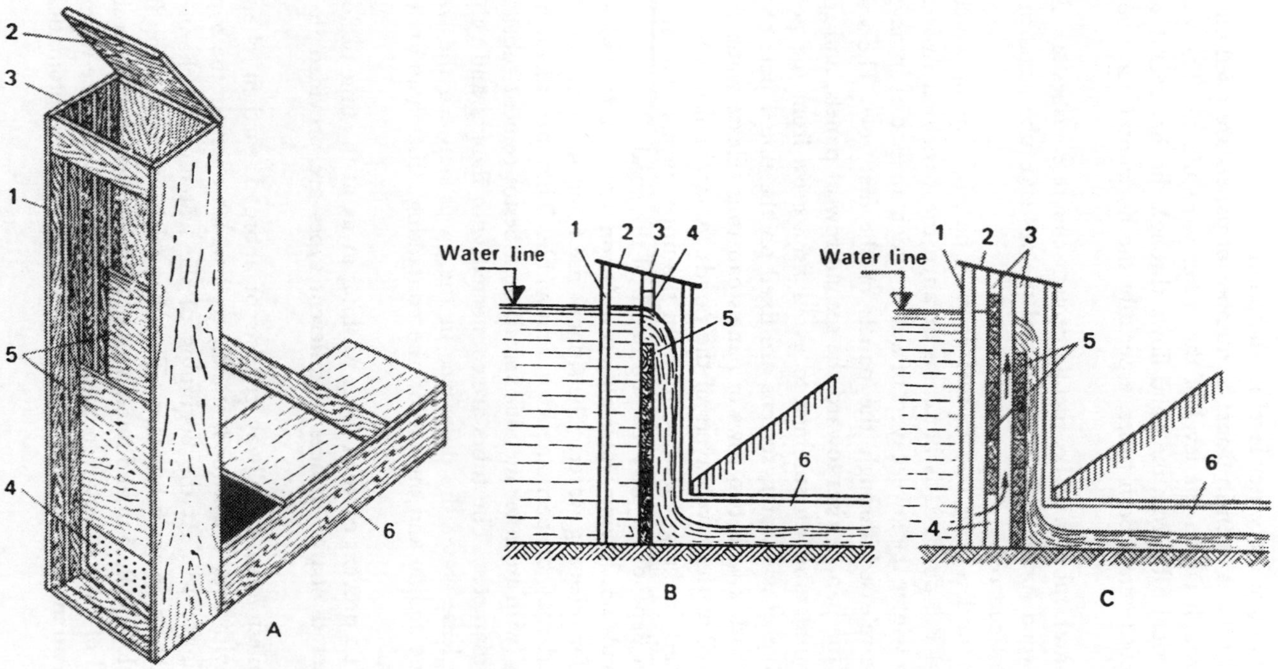

Figure 16. *A*—general view of bottom outlet; *B*—with one row of panels; *C*—with two rows of panels. *1*—vertical tube; *2*—lid; *3*—grooves for panels; *4*—frame with screen; *5*—shutters; *6*—horizontal tube; *7*—water line.

The horizontal tube is placed at the bottom of the pond at its deepest point or even 10 to 20 cm below the level of its pit. The vertical tube (monk) is situated slightly above the water level in the pond.

The vertical tube is open in front; a number of panels are fixed on the lateral sides through which the level of the water is regulated. The water entering the vertical tube from the pond flows through the horizontal tube and out its opposite end. Both tubes, especially the horizontal one, should be leakproof.

The horizontal tube may be metal, cement-concrete, asbestos, clay, brickwork, or wood (better oak or pine), and the vertical tube wood, iron, brick, or cement-concrete.

The bottom outlets have one or two rows of panels,[1] one of which is directed toward the pond. It is better to arrange the panels in two rows (Figure 16). In winter the lower layers of water, which are poorer in oxygen content, can be released through the panels of the first row. The water passes through the panels, rises toward the second row of panels, and spills into the horizontal tube. In summer to avoid fish egress from the ponds while discharging excess water, screens are fixed on the upper part of the panels. Water outlets with two rows of panels can thus release water from the upper as well as the lower layers of the ponds. A mechanism with just one row of panels discharges water only from the upper layers. The size of the water outlet depends on the area and depth of the pond.

In small ponds not having water outlets a siphon type water discharge system is used for releasing water. The siphons are fitted through the dike (Figure 17) and may be permanent or removable. They are made from metallic-rubber (with an inner metallic spiral) or asbestos-cement tubes 200 to 300 mm in diameter. The tubes are connected with flanges and rubber washers. To release water the difference in the levels between the water 0.10 to receiving siphon and the end of the contacting tube should not be less than 0.25 m.

Spillway. To prevent earthen dikes washing away at the time of floods and rains, water discharge devices of different types are provided in the ponds.

In the simplest form a spillway consists of a broad earthen channel (strengthened with sod, brushwood, stones, etc.) which opens in the bypass of the dike (Figure 18). At the beginning of the water outlet channel a screen is placed to prevent fish escape. The channel is constructed on firm ground with a slope of 0.05 m and lined with stones to protect it from erosion. The outlet level of the spillway is located at the normal level of the water in the pond; with an increase in water above this level, the excess automatically drains out.

[1]Height of panels, 20 cm.

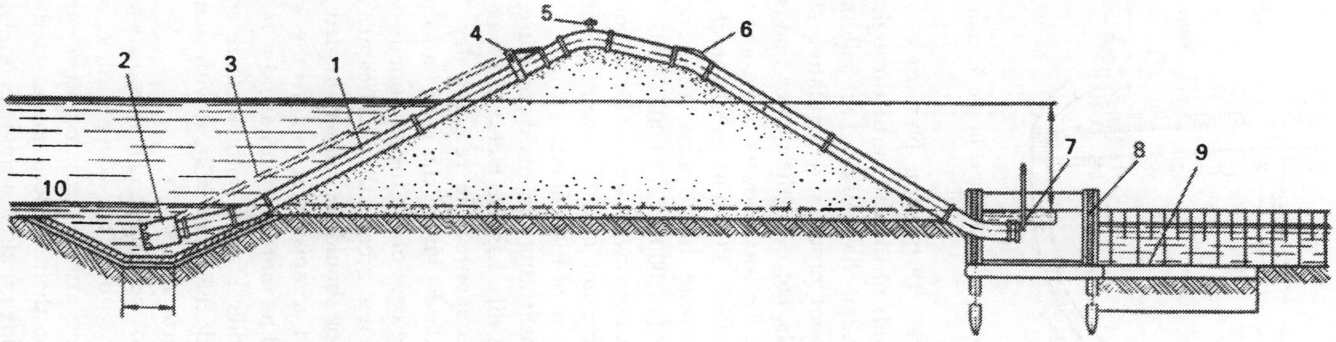

Figure 17. Diagram of a dismountable siphon.

1—metal tube; *2*—entry gate of siphon; *3*—rope or cable; *4*—windlass for regulating entry gate; *5*—protruded tube for filling siphon with water; *6*—crest of dam; *7*—outlet gate of siphon; *8*—fault pit; *9*—supportive material; *10*—bottom of pond.

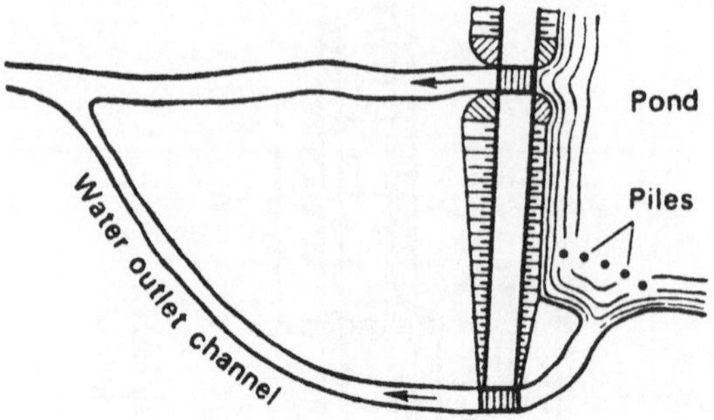

Figure 18. Layout of water discharge channel in the form of a spillway.

WATER SUPPLY AND WATER DISCHARGE NETWORK. Water from the source of supply enters the ponds of the fish farm through a special water supply system. It consists of earthen channels or, if the soil has high seepage, wooden channels, asbestos-cement pipes, and regulating structures (sluice gates, partitioning arrangements, etc.). In the winter ponds and sometimes in the summer ponds as well (if the water source has a low oxygen content), an aeration system is also installed. The bottom of the channel should be above the maximum level of the water in the pond; the slope depends on the relief of the area but is usually 1:1,000 to 1:3,000.

The channels carrying the water away are similar in construction to the water supply channels and serve not only for draining the ponds but for releasing excess water at the time of high floods.

EARTHEN DIKES. Dikes, constituting the main structure, are erected to create artificial water bodies. A dike has two faces: the wet side is directed toward the water, while the opposite side remains dry. The slopes for inside and outside depend on the height of the dike and the quality of the soil from which it is made. The base of the inside slope is twice, and in the case of head ponds even three times, greater than its height. For the summer category of ponds the inside slope should be gentler (angled less) to create a shallow water zone rich in food organisms for fish. In the wintering ponds, on the contrary, this side should be more acute to avoid a reduction in pond area (Figure 19). To prevent dike erosion both sides are covered with sod and grasses sown. In large ponds the inside is lined with stones held in place with wattle mats, wattle supports, and so on.

Planting trees on dikes is not permitted. Their roots would weaken the dike, their crowns create shadows over the water surface, and their leaves pollute the pond. Moreover, trees attract birds and other enemies of fish.

Earthen dikes, depending on their purpose, are erected along the outer boundaries of a group of adjoining ponds or a single pond, while partitioning dikes are built between neighboring ponds.

The service life of hydrotechnical installations is considerably enhanced with proper and systematic maintenance.

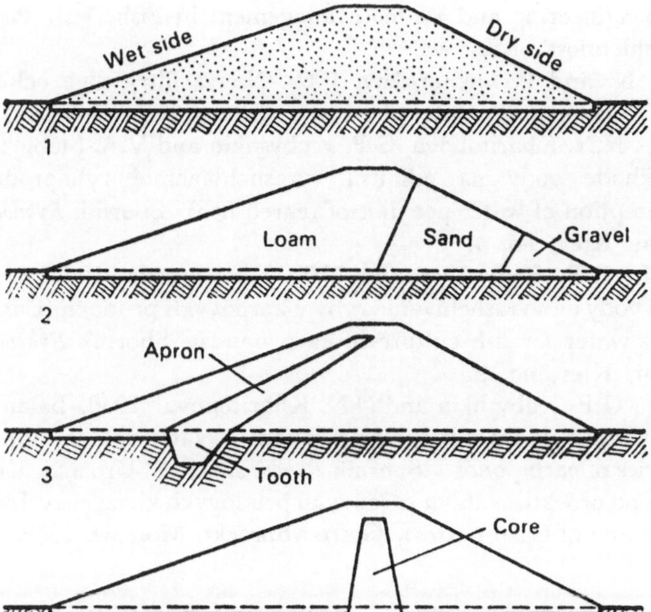

Figure 19. Dikes made of loamy and sandy-loamy soils.

1—dike of homogeneous mineral soil; *2*—dike of heterogeneous mineral soil; *3*—dike with apron; *4*—dike with core; *5*—dike made with peat; *6*—dike with foreapron; *7*—dam with apron and foreapron.

References

Orlova, Z.P. 1969. Rybokhozyaistvennaya gidrotekhnika i melioratsiya (Hydroengineering and Water Management in Fisheries). Pishchevaya Promyshlennost', Moscow.

Pivovarov, N. and P. Sukhorukov, 1969. Novye filtry dlya ochisthi vody (New filters for purification of water). *Rybovodstvo i Rybolovstvo,* no. 5.

Shpet, G.I., N.N. Kharitonova, G.P. Kubyshkin and V.A. Stebel'skii. 1967. O raskhode vody na edinitsu vyrashchivaemoi ryboproduktivnosti (Consumption of water per unit of reared fish). Sbornik *Rybnoe Khoz-vo.* Urozhai, Kiev, no. 4.

Shpet, G.I., G.P. Kubyshkin and N.N. Kharitonova. 1969. O raskhode presnoi vody na vyrashchivanie ryby v karpovykh prudakh (Consumption of fresh water for fish culture in carp ponds). Sbornik *Rybnoe Khoz-vo.* Urozhai, Kiev, no. 9.

Shpet, G.I., G.P. Kubyshkin and N.N. Kharitonova. 1969. Balans vody po osnovnym kategoriyam karpovykh prudov (Water balance for the main categories of carp ponds). Sbornik *Rybnoe Khoz-vo.* Urozhai, Kiev, no. 9.

Ukazaniya po proektirovaniyu karpovykh prudovykh khozyaistv (Instructions on Planning of Carp Farms). Gidrorybproekt, Moscow, 1968.

Chapter IV

PRODUCTION PROCESSES IN A WARM-WATER AND FULL-SYSTEM CARP FARM WITH TWO-YEAR ROTATION

In the USSR the fish generally cultured in pond fisheries is the common carp (*Cyprinus carpio*). Economically, this is a very convenient species because of its quick growth and less demanding conditions for life. The common carp has a good flesh content and its meat is of a high quality (8.6% fat and 16.5% proteins). The current commercially valuable breeds of this fish have been developed from sazan (wild carp) by fish breeders during the last five centuries. Aside from its external appearance, carp is distinguished from sazan by a more rapid growth and better utilization of food. The biological peculiarities of sazan and carp are, for the most part, similar. Both are warmth-loving fishes in which the main life functions are performed at higher water temperatures. Spawning usually occurs at 17 to 20°C, although cases are known of spawning at 14 to 15°C; Taparavan sazan (Taparavan Island) spawn at 12 to 14°C, sazan in the waters of Tatar ASSR spawn at 15°C, and Amur sazan spawns at 16 to 20°C. The common carp attains a better growth at 20 to 28°C. With a temperature reduction to 13 to 14°C food intake and digestion are poorer. A further reduction of water temperature to 7 to 8°C leads to an even lower consumption of food. Thus, a reduction in temperature is closely associated with a drop in intensity of feeding and consequently in growth rate. At a temperature of 1 to 2°C carp enters a state of little activity, lying in the deeper parts of the wintering ponds, and refusing to feed at all. During this period it consumes the stored matter (fat) accumulated over the vegetative (feeding) period. A prolonged increase in water temperature up to 30°C and above is not only undesirable but even harmful to carp, which is adapted to temperate and moderate climates. For carp acclimatized to tropical conditions, the upper limits of temperature would, probably, be higher.

Carp growth depends on environmental conditions: feeding and chemical composition and temperature of water. According to these conditions the weight of under-yearlings may fluctuate considerably, from 15 to 500 g and more, the weight of two-year fish from 150 to 1,000 g, three-year fish from 350 to 2,000 g, and four-year fish from 1,000 to 3,000 g. In the fish farms of the central belt of the European part of the USSR the following norms of cultured carp are usually taken as standard: under-yearlings—25 to 30 g, two-year fish—500 to 800; three-year fish—1,200 to 2,000 g, and four-year fish—2,000 to 3,000 g.

Carp is considered an omnivorous fish, feeding on plant as well as animal organisms. Many authors include it in the group of nonpredatory carnivorous fish. Juveniles consume plankton and crustaceans found in thickets of vegetation, showing a preference for cladocerans, e.g., *Chydorus sphaericus, Daphnia* sp., *Bosmina* sp., *Alona* sp., *Sida crystallina, Eurycercus lamellatus,* etc. Copepods also constitute food, primarily species of *Cyclops.* Rotifers play a significant role in the nutrition of fingerlings, affording a high rate of growth. Benthic larvae, especially those of the family Chironomidae, are important for carp growth. Finally, carp consume the larvae of *Chaoborus,* Ephemeridae, *Sialis* and other worms (*Tubifex*), small mollusks (*Limnaea auricularia, Valvata,* etc.), waterbugs (*Corixa*), etc.

Carp have a high fecundity rate. A female of 1 kg weight produces about 180,000 small yellow adhesive eggs (about 1.5 mm in diameter). The age of the spawners, condition of environment, nutrition, and maintenance influence the fecundity of the fish and the quality of its offspring. Carp attain sexual maturity in the central belt in the fourth or fifth year of life, and earlier in the southern regions (in the third or even second year); the Kuban carp matures in the sixth to the eighth month of life. Under conditions in Karelian ASSR females attain sexual maturity at the age of six years and males at four to five years, i.e., one year later than in the Leningrad district. The average fecundity of the female is less than in the southern regions, up to about 130,000 eggs.

Condition of Carp Brood Fish. In carp culture attention should first be paid to high quality brood fish and the selection of a good breed, taking into consideration the climatic conditions of a given zone. Unlike other sectors of animal husbandry, in pond fisheries the spawners are sexually mature individuals of both sexes (male and female). Whatever may be the breed of carp taken for culture, sufficient attention should be given during selection to the fact that the spawners are well fed; they should have a good external appearance, growth, and breeding potential and exhibit no symptoms of weakness or degeneration. They must be able to withstand the given environmental conditions and be resistant to various types of disease. The physical condition of a fish is a very important index not only for spawners but also for fingerlings, since it reflects winter-hardiness, a factor determined on the basis of external appearance, data from chemical analyses, and calculations according to Fulton's formula.

The external characteristics of a well-fed fish include a healthy appearance, a correct ratio between body parts and shape, good meat-filling, and a relatively small-sized head. The color of a well-fed carp spawner is yellowish-golden and of a poorly fed or lean carp gray with a dull silver tinge.

Fulton's formula for determining the condition factor of fish is as follows:

$$K = \frac{g \times 100}{L^3},$$

where K = coefficient of condition;
g = weight of fish, g;
L = total length of fish in cm (greater length taken from tip of snout to vertical line passing through tip of bigger lobe of caudal fin).

Fulton's formula is applicable only to fish specimens of the same sex and similar in age.

The coefficients of condition for selection of brood fish of different carp breeds are given in Chapter XVI; the coefficient for tench is 1.3, for rainbow trout—1.1, and for three-year-old Dnieper sander—1.5 on the average.

Fulton's formula has been modified by the All-Union Research Institute of Pond Fisheries as follows:

$$K_1 = \frac{v \times 100}{d^3},$$

where v = weight of fish, g;
d = lesser length measured from tip of snout to end of scale covering in cm (Figure 20).

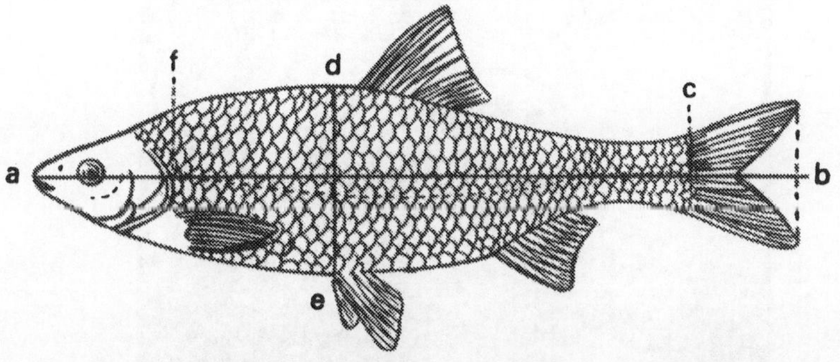

Figure 20. Scheme of recording length measurements of fish.
a to *b*—total length of body (L);
a to *c*—standard length of body without caudal fin (l);
d to *e*—maximum height (H); *a* to *f*—head length (C).

The coefficient of condition in this formula is designated by K_1. This value reflects the condition of the carp more correctly.

The coefficient of condition is not a constant value. It depends on the age of the fish, its growth, species, breed, and stage of gonadal maturation. The estimation of the coefficient of condition by the above formula should be treated as incomplete, since a precise picture can only be formulated by a chemical analysis. However, in day-to-day practice in collective and state fish farms it is not possible to conduct a chemical analysis; moreover, for this purpose it is necessary to kill some brood fish which, in the case of a deficiency of fish, is not desirable. Therefore, until some new improved method is devised for estimating the condition factor of fish, one must be guided by the formula proposed by the Institute of Pond Fisheries.

On the basis of external characters the breeds of carp reared in ponds can be divided into two groups: those with a high back and those with a broad back. This division is based on morphological differences—the ratio of the body height to its length (H:l). In high back carp the ratio of H:l applicable for characterization of external features is 1:2.05 to 1:2.60, and in broad back carp 1:2.6 to 1:3.0. The higher fluctuations in the ratio reflect carp closer to wild forms.

Carp in the age group of 5 to 10/11 years are considered better spawners. Fish older than 11 years are usually replaced by younger ones from the replenishing stock. It is difficult to determine the sex of carp and, in the case of young fish and sexually immature individuals, impossible to do so on the basis of external appearance. Only at the beginning of spawning can the sexes be differentiated on the basis of the following characters:

(a) In females the abdominal vent enlarges, bulges slightly, and turns reddish; the profile is fuller due to the development of ripe ovaries (Figure 21);

Figure 21. Female carp ready for spawning (from I. Fišer and S. Stochl).

(b) In males the vent elongates inwardly in the form of a narrow pale-colored slit;

(c) Small hard tubercles appear on the head and opercula of males before the spawning period (although poorly expressed); these constitute a unique feature of the spawning livery of the male.

D.V. Shaskol'skii has proposed a method of tagging the sex of spawners (based not only on the characters prior to spawning, listed above) so that the sex of a fish can be ascertained at any time of the year. This method consists of singeing the fish with a red-hot iron wire or rod 60 cm long and 4 to 6 mm thick (Figure 22) with a hook at the end. To tag mirror carp a place without scales should be selected. The following rules must be observed during tagging:

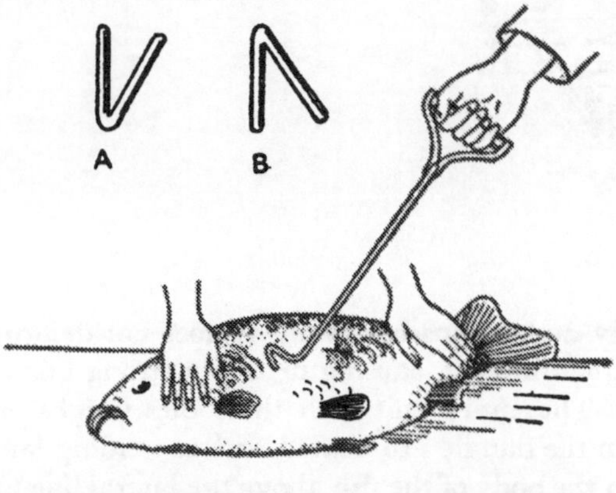

Figure 22. Tagging of carp spawners by singeing.
A—female mark; *B*—male mark.

(1) Tagging is done on spawners soon after spawning;

(2) Fish are carefully cleaned prior to tagging to remove slime;

(3) Singeing is done quickly with firm pressure and the fish immediately returned to the pond;

(4) The last figure of the year in which the fish hatched from eggs is placed on the right and the serial or individual mark of the fish on the left. For example, if a particular fish hatched in 1971, the figure 1 is marked;

(5) For sex identification the distinguishing mark of male or female is placed on the left in the form of two lines which narrow downward (the first two lines of the letter H, which indicates female) or the narrow point kept upward (the first two lines of the letter M, which signifies male).

This method of tagging fish has invited a number of objections: in autumn the marks are poorly discernible, the spawners suffer pain for a long time after singeing and even die, and the process is time-consuming. To speed up this process and reduce its harmful effect Korovin (1968) devised a special apparatus (Figure 23). The die consists of a steel sheet 2 mm thick. It can

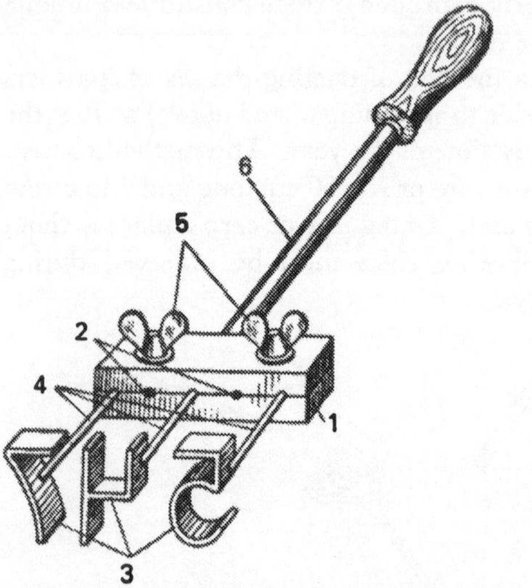

Figure 23. Equipment for tagging fish
1—holder; *2*—openings for fixing of dies; *3*—dies; *4*—stems; *5*—screws with wing nuts; *6*—handle with distal end inserted in lower segment of holder.

be heated quickly and retains heat well; it does not deform during heating and leaves a distinct imprint, causing neither a strong burn nor a loss of the adjoining scales. The entire unit with three dies weighs only 500 to 600 g. The dies fitted in the handle are heated with a welding lamp until red-hot, then pressed into the body of the fish above the lateral line for 1 to 2 seconds, after which the fish is immediately released into water. This method has the following advantages:

(a) Reduces time of fish out of water;

(b) Reduces risk of heat shock, if singeing done by a few dies;

(c) Does not affect a wide portion and provides a better mark since the surface of the die touching the body of the fish is not very large.

Trial tagging by the above-described method was done during spring up to spawning at the Oyashinskii State Farm in the Novosibirsk district and yielded good results: a week after tagging the wounds had healed, the spawners were normal, there was no disease, there were no losses, and all the females spawned.

Maintenance of Brood Fish and Replenishing Stocks. Considering the special significance of proper maintenance of brood fish and replenishing stocks and their care in summer and winter, a yearly plan for maintenance in different seasons (winter and summer) is prepared in the respective ponds of the farm. Brood fish are kept in different ponds: the stock required in the current year; reserve stock in case of the spawners in the current year do not spawn; and replenishing stocks (if special ponds do not exist for replenishment), selected on the basis of desirable qualities for replacement of rejected spawners.

A fish farm should have a minimum of two brood fish stock ponds (in addition to the pond for replenishment), so that the males and females can be kept apart before spawning. These ponds, as far as possible, should be fully filled with water so that the temperature in them remains fairly low, preventing premature ripening of the gonads. Carp are the most sensitive fish with reference to heat in the last 12 hours of gonadal maturation. If there are no stock ponds, the brood fish are isolated before spawning in ponds of other categories, usually in the wintering ponds. After spawning the males and females may be left together. It is essential that all the carp spawners held at the farm, including reserve spawners as well, be given an opportunity to spawn. An unusual delay in spawning has a harmful effect on the quality of spawners; it debilitates them and, probably, leads to the output of a weak progeny and intervals in spawning in the subsequent year or years. Spawning carp in batches (two to three) may cause a large accumulation of young fish in the stock ponds where active spawners as well as reserve brood fish are held after the spawning campaign. Young fish not required at the farm are released in the nearest natural water resource or used as forage for small pike and yearlings of sander and largemouth black bass or rainbow trout which may be reared in the stock ponds (at the stocking rate of 150 fish/ha of small pike and yearlings of largemouth black bass). These predatory fishes feed on young carp and as a result provide an additional yield of valuable meat in the autumn.

If possible, it is better to keep males and females separated after spawning in summer stock ponds, and the replenishing stock separated by age groups in specially constructed ponds.

Mixing brood fish and different groups of replenishing stocks disrupts their normal maintenance and growth checks (on the basis of years), leads to a loss of distinction between different age groups which need different environmental conditions, and makes it impossible to approach their maintenance and rearing in a differential manner. Hence special ponds should be planned for the maintenance of different age groups of young fish stocks. The area of these is calculated on the basis of normal stocking density of each age group of the replenishing stock. As an exception brood fish may be kept together with two-year fish in foraging ponds; the different age groups of the replenishing stocks should be kept in the best foraging or rearing ponds. Yearlings are better stocked in rearing ponds together with the fingerlings of the replenishing stocks, but this should be done only if they are perfectly healthy; diseased fish of older age groups cause the infection of fingerlings. In the case of combined maintenance of the main stocking material with the replenishing stocks the difference in age should not be less than two years.

Exceptionally, yearlings and fry with an age difference of one year only may be combined.

Mixed holding is allowed only when the density of the main stocking material does not exceed the density of the other groups by more than double. The number of brood fish and quantity of replenishing stock to be held depend on the magnitude of the natural fish productivity and the established growth rate over the growth period of the respective groups.

In winter brood fish and different age groups of replenishing stock are better kept separated in the wintering ponds. If these ponds are insufficient, mixed holding is permissible on the condition that the difference in age is not less than two years. If the brood fish and replenishing stock have age tags, such a difference in age is not necessary.

Estimation of Required Number of Brood Fish. In each full system fish farm a definite number of brood fish must be kept, estimated on the projected production plan. To estimate the required number of brood fish the following example should prove useful. A well-fed female carp weighing 7 kg can produce 1,200,000 eggs. Only one-third of this quantity under natural conditions in the spawning ponds develop up to the larval stage, yielding about 400,000 fry. However, at this stage of development losses in the spawning ponds can be quite significant due to unfavorable weather conditions, various kinds of disease, and predation by enemies, reaching 70% or even higher. Consequently, about 120,000 young fish remain alive. Of these, about 80% can be expected from hatching ponds and transplanted to rearing ponds. Thus, one can project a total of about 100,000 juvenile carp from a single female weighing 7 kg.

According to the existing standards of the Ministry of Fisheries of the USSR, the output of four- to eight-day-old fry from the hatching ponds per female fluctuates from 50,000 to 100,000 depending on the zone.[1]

The output of carp fingerlings (fish of the current year group caught from wintering ponds and stocked in rearing ponds developing up to the spring) from rearing ponds toward autumn is taken as 70% for regions north of 51° and 80% for regions south of 50° N. Thus, 70,000 young fish survive for stocking in the wintering ponds. The output of yearlings (age group obtained in spring, comprising the catch of autumn-stocked fingerlings in wintering ponds) from wintering ponds and ponds adopted for this purpose fluctuates from 60 to 85% in different zones. Taking the output from wintering ponds as 75%, the spring catch would comprise 70,000 fish planted in the wintering ponds in autumn, and 52,500 yearlings. Two-year-old fish after winter have a weight of 500 g.

Approximate estimations make it possible to work out the number of brood fish required for pond fisheries with a certain production capacity.

[1]Progressive farms obtain a notably higher number of fry.

F.M. Sukhoverkhov has proposed the following formula (modified by D.V. Shaskol'skii) for calculating the requirements of brood fish, depending on concrete conditions:

$$I = 2\left(\frac{G \times P \times 100 \times N}{V \times r \times M} + \frac{H}{M}\right),$$

where I = number of females required at farm, which is equal to number of sets or redds[2];
2 = coefficient of reserve stock (equivalent to 100%);
G = area of rearing ponds (ha);
P = average natural fish productivity of rearing ponds (kg/ha);
100 = calculated constant coefficient;
N = coefficient of stocking density of young fish taking into account food of fish (see Chapter XI);
V = average weight of fingerlings (g) toward autumn in accordance with established norms of pond fisheries;
r = output of yearlings from rearing ponds according to established norms (% of stocking);
M = output of young fish from a single female according to established norms (thousands);
H = total quantity of fry required for mixed stocking in production ponds (thousands) (see Chapter VII).

Given concrete conditions, the quantity of brood fish required can be readily estimated by the foregoing formula.

Preparations for Carp Breeding in Spawning Ponds. Preparations for breeding work in the central belt of the USSR are initiated in the middle or at the end of April with the catching of fish from wintering ponds and the transportation of brood fish—males and females—to different ponds. Before transportation the brood fish are carefully examined. The transfer and breeding of spawners with mature gonads should not be delayed, since overmatured eggs become excessively sensitive, the viability of the hatchlings reduces, losses increase, and there are more males in the progeny. Furthermore, such eggs lose the capacity for fertilization. Immature eggs, on the other hand, are also unsuitable since they are less receptive than ripe eggs to fertilization and thus cause losses in fish production.

Simultaneous with the examination of brood fish spawning ponds are prepared. All the dead vegetation is removed from these ponds and they are thoroughly cleansed of debris (with the help of iron rakes). The water courses and fish pits are also cleaned and dikes, troughs, and water outlets repaired. The latter are equipped with a small-mesh screen to prevent the egress of young fry from the ponds. If the difference in the level permits, it is advisable

[2]A single redd consists of one female and two males.

to create a fall of water near the water supply trough and to install debris eliminators. In the process of preparing spawning ponds for breeding work the bottom should be harrowed and treated with unslaked lime (at the rate of 50 g or more per 1 m^2 and in the water channels up to 80 g per 1 m running length), the vegetation of the previous year (weeds and moss) removed, and the tanks supplied with decomposed organic matter (humus). The vegetation of the previous year should be collected in a heap and burnt and the ash used as manure for the ponds. This holds true not only for spawning ponds, but also for rearing and foraging ponds.

The depressed area (pit) of the spawning pond is planted with tender meadow plants, which not only serve as a substratum for egg-laying by carp, but also afford favorable hydrochemical conditions (especially the gas regime) in the ponds and a natural food base for the fry. However, not all species of tender meadow plants are equally resistant to water; those which are less resistant will decompose, causing a deterioration in the hydrochemical status, which leads to a situation unfavorable for carp to spawn. Hence water-resistant species of meadow grasses should be cultivated in the spawning ponds. For the forest steppes and steppe zones of Ukrainian SSR, Samoilov (1968) recommends common slough grass, reed canary grass, meadow foxtail, fowl meadow grass, Kentucky bluegrass, fiorin, crawling couch grass, meadow timothy, and other water-resistant grasses. After the ponds are filled with water this vegetation enriches it with oxygen in the process of photosynthesis. Other species of tender meadow grasses (orchard grass, white clover, red clover, alsike clover, military grass, drooping brome, perennial ryegrass, etc.) decompose quickly after the ponds are filled. The subsequent deterioration of the hydrochemical regime has an adverse effect on the developing eggs and carp fry leading to mass mortality. Carp avoid egg-laying on coarse acidic grasses (dense overgrowth of sedge, rush, horsetail, etc.), while their favorite spawning substratum is a thicket of freshly flooded tender meadow grasses.

A rare or sparse grassy substratum causes considerable losses in eggs (up to 25.3 to 32.6%).

Care must be exercised therefore in the selection of plants as a substratum for carp spawning, since the process of spawning, egg incubation, and level of young fry production depend on this vegetation.

The studies initiated by Samoilov should be continued in other zones of the USSR to establish suitable plants which would improve the quality of spawning ponds and eventually fish productivity. Here it is necessary to consider their moisture index; plants suitable for arid and moderately moist zones are not recommended for moderate and highly humid areas.[3] Preference should be given to those species of perennial grasses which under certain

[3]Marshy areas should not be used for spawning ponds.

conditions of soil moistening are able to conserve their viability over a longer period of flooding. One should also check the growth of water-susceptible plant species which decompose in water on the 3rd to the 4th and 7th to the 10th day.

If vegetation is not provided, the bottom of the spawning ponds should be covered with sod or an artificial spawning medium. In making such arrangements mops are usually prepared from juniper or spruce. The branches of deciduous plants are not suitable since they wilt quickly and decompose in water. Mops are grouped in bundles and affixed at different points on the bottom with the help of stones or pegs in such a way that they are fully covered with water.

One or two days before the spawners are introduced into the ponds (better on the day of stocking) they are filled with water and its temperature raised to 15 to 16°C, but not higher, since at 17 to 18°C spawning commences.

The ponds should not be filled with water much in advance since this facilitates a greater development of the enemies of young fry (larvae of European diving beetle, water bugs, etc.). Water beetles of the family Hydrophilidae, most of the members of the family Dytiscidae, and the water bugs *Notonecta glauca* and *Corixa dentipes* are great enemies and competitors for the food of young fry. Cladocerans and young fry of carp are the favorite foods for three species of beetles: *Craphoderes bilineatus, Colimbetes striatus,* and *Hydaticus transversalis,* in addition to the above-mentioned species of water bugs.

While filling the spawning ponds with water, screens must be placed which will prevent the ingress of fish. This should also be done when filling ponds of other categories (rearing ponds, wintering ponds, production ponds, summer stock ponds). Simultaneous with the preparation of the spawning ponds the brood fish are examined a second time, weighed, and the best of the whole stock selected, excluding those with well-developed gonads. This is done on a wet tarpaulin or bast. To avoid deterioration in the results of spawning, the spawners should be handled as gently as possible so as not to cause the slightest damage or scar; and properly surrounded with water. A special type of carrier (Figure 24) in which replaceable tarpaulin trays are installed is best. The length of each tray should preferably not exceed 60 to 65 cm, width 20 to 25 cm, and depth 25 to 30 cm.

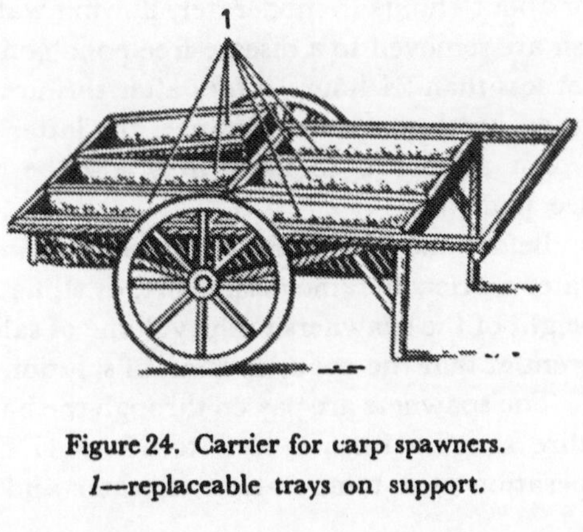

Figure 24. Carrier for carp spawners.
1—replaceable trays on support.

The introduction of fish for spawning is preferably done toward evening. The carefully selected and conditioned breeders should spawn the next morning. The fish are examined once again before introduction and those breeders selected which are suitable for spawning in the current year.

The brood fish selected for spawning are released in the spawning ponds in redds (1 female and two males). In the case of a deficit of males or when breeding work is being done (where it is necessary to know the precise origin of the progeny), spawners are released in pairs (1 male and 1 female). For large-scale production and with a sufficient quantity of males plantation is preferably done in the form of redds to ensure a more complete fertilization of the eggs. Spawning of a single female with a larger number of males (when they are available) guarantees successful spawning and an increased output of young fry.

Before the breeders are introduced for spawning they are given a saline bath containing a 5% solution of common salt; this is done twice for 5 minutes each time with an interval of 5 to 7 days. A salt solution of such a concentration is harmless for carp. It is essential to maintain a constant salt concentration in the solution, which means it should be checked regularly with a hydrometer. After the saline bath the fish are kept in fresh water.

The purpose of the saline treatment is the destruction of microparasites (on the skin and gills), especially the trematode *Dactylogyrus vastator,* which is extremely dangerous for young fry.

The salt solution is prepared in wooden or tarpaulin containers. Containers made of zinc or zinc-coated iron are not suitable since sodium chloride forms compounds with zinc which are poisonous to fish. Tarpaulin containers are therefore used, consisting of a wooden frame with a planked bottom 8 to 10 cm above the ground level. The length of the container is 100 cm, width 60 cm, and height 60 cm. The tarpaulin container is attached to its upper borders by means of planks or ropes. The spawners are passed through the saline treatment, then quickly taken to the rinsing container, and kept there for about 2 hours in moderately flowing water. After washing and grading the fish are removed to a disease-free pond containing clean water for a period of not less than 24 hours. Only after this are the fish transported. If transportation involves a short distance, the latter operation may not be necessary.

An ammonia solution may also be used as a prophylactic measure (see page 76).

Before the spawners are given a saline bath they are placed in freshwater carriers to remove silt, dirt, or slime from their body. The ratio of the weight of the spawners to the volume of saline solution should be 1 : 10. It is essential that the concentration of solution remain constant (5%).

The spawners are passed through the bath at a water and solution temperature ranging from 6 to 7 to 15 to 17°C. It is not advisable to do this operation at a temperature of water and solution higher than 19°C. The used saline solution is thrown out at a place where infection cannot be carried back into the ponds and other water bodies.

The above prophylactic treatment of fish in saline baths (as well as ammonia) under the conditions of large-scale breeding is highly laborius and time-consuming. To simplify this process the Laboratory of Ichthyopathology of the All-Union Research Institute of Pond Fisheries has proposed an antiparasite treatment of fish directly in the wintering ponds. For this purpose synthetic organic dyes are used—basic bright green and basic violet "K". These are capable of killing and suspending the development of pathogenic bacteria, fungi, and unicellular organisms (protozoa).

Experiments on the use of these dyes have shown that they have no adverse effect on fish and their productivity and that the losses sustained, compared to control ponds in which the fish were not treated with dyes, were notably less. Treatment was done in autumn and spring and considerably reduced the intensity of infection or totally destroyed the parasites.

For 1 m^3 of area, 0.15 to 0.20 g of chemical are dissolved in one bucket of hot water. The solution is poured into a portable, suspended type of spraying machine, designed by S.Z. Papov, and 400 l of pond water added. The machine is then fixed on the chassis of an SSH-20 tractor and wintering ponds sprayed at the rate of 0.5 to 1.0 hectare per hour. Treatment of fish with dyes is obviously much quicker and cheaper. Furthermore, traumatization, always a possibility in saline or ammonia baths, is totally excluded.

Spawning of Carp. Carp usually spawn in the morning hours. At the time of spawning the fish moves vigorously and swiftly, making a noise in the shallow water areas of the spawning ponds. The agitated activities of spawners at the time of a bout (run) is immediately detectable from splashes on the water surface (Figure 25). The eggs laid by the females are immediately fertilized by the sperm of the males. Since the eggs are adhesive they attach to the water plants where development takes place. One can estimate the extent of spawning on the basis of the eggs attached to plant branches. The percentage of fertilized eggs is determined by examining 100 eggs under a lens or microscope; these are collected 3 hours after spawning from each of the hatching ponds. Fertilization of 80 to 85% of the eggs is considered normal. Spawning is usually finished in the course of one day.

The selection of spawners for a redd is important. In some redds spawning takes place when it should and in others is considerably delayed. In the case of unusual delay in spawning the spawners (either male or female) should be replaced or even the entire redd changed. During breeding the oxygen regime of the water of the spawning ponds is important. To increase the oxygen content at the time of development of the fertilized eggs it is advisable that aeration of the water supplied to the ponds be increased, especially in the period between midnight and two to three hours after sunrise.

When spawning is finished, the breeders are removed from the ponds the next day, taking extreme care not to damage the eggs, and transplanted to summer stock ponds. This is essential since the spent carp could be vectors

Figure 25. Spawning activity of carp.

of different diseases for the young fry; moreover they cause mechanical damage to the eggs, and sometimes even prey on them. The spent fish are caught in a hoop or scoop net; during the process the water level should be lowered in the pond, but the spawning media with attached fertilized eggs never exposed.

The duration of development of fertilized carp eggs up to the emergence of embryos depends, first of all, on temperature conditions (naturally with a favorable oxygen regime) and secondly on favorable hydrochemical conditions. For the complete development of eggs up to hatching of the embryos a certain total quantity of heat hours is required, usually 60 to 80° per day. At a water temperature of 17 to 20°C, the development of the eggs continues for 3 to 6 days; if the water temperature drops to 8 to 12°C, the emergence of prolarvae (free embryos) may take place on the 10th to the 12th day after fertilization. Given the climatic conditions of the southern zone of western Siberia and other similar conditions where persistent frost prevails in spring up to May 20 to 25, and in some years even up to June 10 to 15, spawning can only be guaranteed in the second half of June. Studies by Tatarko (1965, 1966) have shown that favorable temperatures for the embryonic development of carp range from 16 to 24°C. At 27°C and in temperature fluctuations from 20 to 30°C embryonic development is advanced, but the hatching of prolarvae suffers. A temperature of 30°C is not favorable and a reduction in oxygen content to 3.6 mg/l sublethal. Although a still higher temperature has a greater effect on the morphogenesis of fish than on their growth, nevertheless it exerts considerable influence on the growth potential.

The size of the eggs is also important. Large eggs contain more yolk and are characterized by higher viability: these eggs develop more rapidly and a greater proportion of embryos are obtained from them than from medium or small eggs. The conditions of maintenance and the nutrition of brood fish considerably influence the quality of eggs. In the case of brood fish maintained under favorable conditions the fecundity and quantity of large eggs decrease and consequently the quality of the progeny obtained suffers.

Prolarvae after hatching (free embryos) are 4 to 6 mm long and contain a small yolk sac, the contents of which are consumed in the process of growth. On the first day, immediately after hatching, they enter a quiescent state, after darting about to hang onto objects with their tails downward. They are equipped with glands or cementing organs for attachment to underwater objects.

Toward the end of the second day and the beginning of the third, even though the yolk sac is still not fully absorbed, the larvae commence active feeding, i.e., by this time the prolarvae have become fully formed larvae. By the fifth to the sixth day, when the yolk sac of the larvae disappears, they are called young fry.

Hence the following stages are distinguishable in the development of fertilized eggs of carp and the first postembryonic period in the life of young fish:

(1) Embryonic period—from fertilization of the eggs up to the appearance of free embryos;

(2) Early postembryonic period—from emergence of free embryos up to changeover to active feeding;

(3) Larval period—from the beginning of active feeding up to total absorption of the yolk sac;

(4) Early fry—from total consumption of the yolk sac up to resemblance to adult fish.

Even with a high fecundity and a high percentage of egg fertilization the survival rate of fry and fingerlings may be low due first to the loss of fertilized eggs through diseases and predation, and second the fact that free-swimming larvae are extremely vulnerable to various kinds of predatory organisms. These are the main reasons why the survival rate of fingerlings during the first vegetative period is very low. Predatory invertebrates (European diving beetles and their larvae, water bugs), frogs, and large tadpoles weighing 6 to 10 g and more destroy young fry. For example, in experiments conducted by Samoilov (1968) lake frog consumed in one day 17 of the 20 three-day-old larvae of carp placed in a glass container, and large tadpoles consumed 5 to

8 fry each. The larvae of European diving beetle destroyed 7 to 9 larvae in one day, while water bugs (laserwort, ranatra, water scorpion) consumed 4 to 8 larvae each.

These findings reveal that under simulated natural conditions in spawning ponds the destruction of carp fry may be fairly high. Thus the main task of fisheries is to organize that type of preparation of spawning/hatching ponds, nurseries, and rearing ponds which will ensure an overall reduction in the number of predators of fish eggs, larvae, and young fry.

To protect fry in hatching and rearing ponds further against predation, additional measures are necessary:

(1) Frog eggs should be removed from the pond;

(2) Tadpoles are caught in a net fitted on a frame which is lowered onto a feeding trough. When the tadpoles have accumulated in sufficient quantity, the net trap is lifted. The young fry pass through the mesh but the tadpoles do not. Feed can also be placed in the trap (Figure 26), in addition to the feeding trough, as a lure to the tadpoles. The trapped tadpoles and collected frog eggs can be fed to older fish (see Chapter XI).

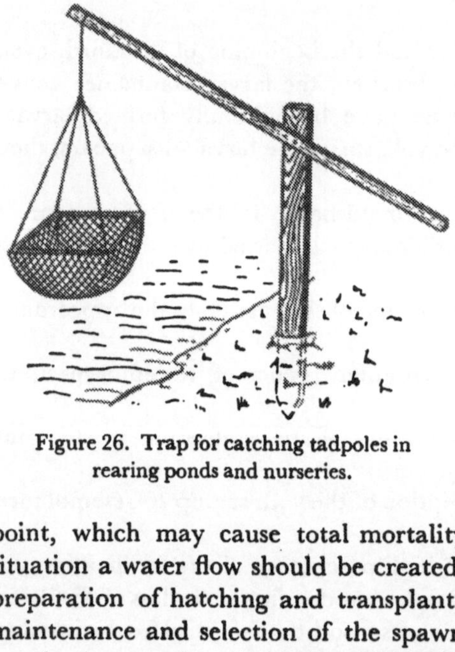

Figure 26. Trap for catching tadpoles in rearing ponds and nurseries.

(3) Beetles and bugs are destroyed by draining the water from the ponds and treating the bottom with lime; catching devices are also used.

(4) Notostracans are destroyed by draining the water from the ponds; larvae exposed to desiccation eventually die.

In spawning (hatching) ponds rich in vegetation the oxygen content in the morning hours often decreases to a lethal point, which may cause total mortality among young fry. To avoid this situation a water flow should be created throughout the night. With good preparation of hatching and transplantation or nursery ponds and proper maintenance and selection of the spawners the output of fry from one set can be increased to 150,000 to 200,000 and an experienced pisciculturist can even obtain from 500,000 to 800,000.

Feeding of Young Carp. A sufficient quantity of food organisms is very important for young fry since feed determines favorable progress in their growth in hatching, transplantation, and rearing ponds.

The nature of the food consumed by carp in different stages of its life varies and care must be exercised in the selection of food organisms during the short periods of development of the fry. Many aspects of this complex problem of rearing fry have not been studied well, especially in pond fisheries.

Lacking proper information about the nutrition and requirements of young fish from the moment of their changeover to active feeding, leads to shortcomings in the production activity of fish farms. Studies in this field must proceed in the direction of understanding the peculiarities of the nutritional regime of carp fry in rearing ponds, ratio of natural foods and supplementary feed, maximum permissible repetition of stocking, and daily feeding conditions in different soil-climatic zones. Such studies must be based not only on generalizations from practical experience, but also on physiological and biochemical data, as such data provide a theoretical basis for proposed recommendations.

During the period of changeover to active (external) feeding large-scale losses of larvae are likely. Natural foods must be present in the pond at that time which are essential for the larvae at a given stage of their development, namely, small forms of zooplankton. The food components at different stages of postembryonic development change quickly. This should be taken into account in the process of evolving the natural food base. In the central belt of the European part of the USSR, larvae dissected on the day of changeover to feeding on zooplankton were found to contain rotifers and small forms of cladocerans in large quantities. On the second day, in addition to the foregoing, small forms of copepods were also found in the gut contents. The composition of the gut contents usually reflects the species composition of the zooplankton in a pond. Small forms of chironomids appeared in the food contents during the third to the ninth day after the changeover of young fry to active feeding, while from the first day algae were often present.

Studies conducted by a number of authors have established that most larvae changeover to active feeding before the yolk sac is fully absorbed. Already on the second day of emergence from eggs prolarvae begin to feed actively. According to Chaplina (1948) young carp fry up to the age of 10 days (from 4.5 to 20 mm long) feed on small crustaceans (*Alona rectangulata, Bosmina longirostris, Chydorus sphaericus, Pleuroxus,* nauplii of copepods, etc.). Rotatoria and plants are insignificant as food at this age. Older carp fry consume larger food organisms. At the age of two weeks (from 25 to 35 mm) the fry can easily consume large crustaceans (*Daphnia magna, D. pulex, Moina rectirostris, Paradiaptomus, Acanthocyclops viridis*) as well as small chironomids (*Criscotopus, Microchironomus,* etc.). At the age of four weeks (60 to 70 mm) the juvenile fish changeover to feeding mostly on oligochaetes, chironomids, and larvae of other insects.

A knowledge of the differential feeding of carp fry makes it possible to produce live foods which, in species composition, must conform to the requirements of the fry in the different periods of their development. Feeding without discrimination is all too often the practice of fish farms.

More detailed studies on the feeding habits of carp fry (and other fish species used in pond culture) would broaden our knowledge of their food regime and lead to a more rational organiztion of fish culture.

The first period after hatching of the prolarvae is the most crucial since at this time one or the other zooplankton species determines the conditions for obtaining viable fry. The usual measures for increasing the development of zooplankton (filling the hatching ponds with water taking into account the degree of multiplication of zooplankton up to the time of changeover of the larvae to active feeding, application of feeding, application of manures and chemical fertilizers) do not ensure a stable and sufficient food regime. Starvation and a high mortality rate among young fry result because they have not yet become hardy. The early transfer of such fry to transplantation or rearing ponds also increases losses.

A more frequent application of organic manure for intensive reproduction of zooplankton is also incorrect since overloading hatchery ponds with decomposed dung has an adverse effect on the hydrochemical regime. During the short period of fry development an increase in the natural food base in spawning or hatching ponds can be achieved by adding chemical fertilizers.

Often the cause for a high mortality rate and a low survival of young fry from a single female can be traced not only to a deficiency in food, but also to faulty selection and rearing of replenishing stocks and maintenance of brood fish. The conditions for maintenance of brood fish exert a considerable influence on the development of oocytes in the ovary, ovulation, and the development of fertilized eggs.

Despite the food value of zooplankton, there are certain species, in particular cyclopoid copepods, which are enemies of fish larvae. It has been found that preying on fish larvae by these copepods probably influences the species composition of the zooplankton in general, and the species afflicting the fish larvae in particular.

According to Bogoslovskii (1955) the shell-less rotifer *Proales daphnicola* spends the greater part of its life in daphnids, feeding on the green algae located in their shells. This rotifer causes mortalities among carp larvae (as well as other species) by ensnaring them in its thin catching filaments. Mostly it ensnares the head and the pectoral fins of the larvae, i.e., those parts of the body which are more readily entangled. Since the rearing of daphnids as live food for young fry is practiced widely, these organisms should be carefully checked before releasing them into the ponds. If rotifers are observed, the daphnids should not be introduced.

The intensity of carp feeding varies throughout the day. In Moldavia the curve of the fullness index of the intestine of fry under conditions of a satisfactory oxygen regime shows three peaks: at 10:00 a.m., 4:00 p.m. and midnight.

In southern Krasnodar the daily rhythm of feeding of carp fry is also clearly expressed; most intensive from 10:00 a.m. to midday with a maximum peak in the second half of the day (toward 6:00 p. m.), gradually decreasing to a complete cessation at night. The daily feeding rhythm in different months of a single vegetative growth period, as well as in one and the same periods of different years, also varies. The rhythm depends on the availability of food, meteorological conditions, weather, and age and physiological condition of the fish. The daily feeding rhythm has its own peculiarities in ponds in different soil-climatic zones. All these factors must be taken into account when planning the food regime of carp fry (see Chapter XI).

COMMERCIAL METHOD OF FERTILIZATION OF CARP EGGS.[4] Besides the natural spawning of carp in ponds specially prepared for this purpose, the commercial method of artificial breeding has recently expanded. It is based on obtaining mature sex products with the help of hypophysation or pituitary hormone injections,[5] enzymatic detachment of fertilized eggs and their incubation, and the retention of larvae under artificial conditions.

The first method for the detachment of carp eggs was proposed by the Hungarian scientist Z. Woynarovich, who used solutions of common salt and urea. In view of the prolonged nature of the detachment process GosNIORKh has developed and recommended other methods (see pages 64–65).

The commercial method is particularly useful in regions where frosts are possible through the first ten days of June, as a result of which the period of rearing fingerlings is reduced to 70 to 75 days. In this short period the fingerlings are not able to reach the standard weight and condition necessary for ensuring winter-hardiness. Furthermore their retention in the wintering ponds is prolonged, 7.0 to 7.5 months.

The advantages of the commercial method of egg fertilization are as follows:

1. Spawners and developing young need not be placed together; hence the larvae obtained by this method are free of pathogens causing infectious diseases.

2. If in the course of incubation water is used which contains no infection, a more healthy progeny can be obtained from spawners which might have been suffering from infectious diseases.

3. The stock of brood fish can be reduced since fewer males are required. In this method 25 males suffice for 100 females, while in natural spawning

[4]This section is based on a set of instructions for obtaining larvae of carp and sazan by commercial methods prepared by GosNIORKh (State Research Institute of Lake and River Fisheries) and approved by the Ministry of Fisheries of the USSR (1969). Only the highlights of these instructions are presented here. For more detailed information the reader should consult the publication mentioned.

[5]The hypophysis is a gland in the brain which regulates the growth and development of an organism. The secretions (hormones) produced by different sections of the hypophysis pass into the blood and stimulate the process of sexual maturation and ovulation.

200 males are required. The quantity of reserve stock of brood fish is also reduced to 20 to 30% of the operating stock from 100% (minimum 50%) in natural spawning.

4. The processes for conditioning spawners, extrusion, fertilization, and incubation of the eggs, as well as temperature conditions, are fully under control and unaffected by unfavorable hydrometeorological conditions.

All these features make it possible to obtain fry almost one month earlier and thus fish stocking in the rearing (or transplantation) ponds also occurs earlier. This creates more favorable conditions for fry and increases the average daily weight gains of the fingerlings toward autumn.[6] Success with the commercial method of fertilization of carp eggs depends on the facilities and equipment available in the incubation section and the training and experience of the pisciculturists.

Biotechniques of Artificial Fertilization of Carp Eggs. At a water temperature of 16 to 18°C the females are ready for egg extrusion after a pituitary injection. For an earlier production of larvae the carp spawners are caught from the wintering ponds and transplanted to ponds or tanks with regulated temperatures. In the course of the first day the water temperature in the tank is gradually increased to 17°C. After this the females are kept four to five days at 17 to 19°C, the pituitary injection given, and the eggs extruded.

A prolonged retention of fish before giving the pituitary injection at temperatures suitable for spawning deteriorates the quality of the eggs, and at a temperature higher than 19°C renders the eggs useless because they become overripe. Male carp mature after the injection without prior retention in warm water. If for some reason the larvae must be obtained early, the males may be maintained before the injection for one to two days in tanks gradually heated up to 16 to 18°C. Males and females must be kept separated.

The hypophysis of sazan carp, bream, and Crucian carp is used in the hormone injection. The hypophysis is treated with acetone and a water suspension made according to the method of Professor N.L. Gerbil'skii.

Intact pituitary glands which have retained their form and are white or light brown in color are used. Fungal, damaged, and dark hypophyses are not suitable. The hypophyses are stored in dry, airtight tubes or empty phials of penicillin and the water suspension prepared immediately since the gland loses its requisite properties within a matter of a few hours.

The suspension is prepared in the same manner as that for phytophagous fish (see Chapter IX). The hypophyses are first broken in a porcelain mortar into coarse particles to which a few drops of distilled water are added and the

[6] An earlier production of carp fry by the commercial method is possible by using warm industrial waters Spawning temperatures in this case are achieved four to seven weeks earlier than in the natural reservoirs of the same zone.

contents ground into a homogeneous paste. Subsequently water is added to the homogeneous paste at the following rates: for females 1 ml and for males 0.5 ml.

A physiological solution (isotonic) (6.5 g of sodium chloride per liter of water) is most suitable for preparing the suspension for injection, although in extreme cases boiled, filtered, and cooled water may also be used.

The dose of pituitary extract depends on the age and weight of the spawner. To obtain fish eggs in the period of usual natural maturation, four to five pituitary glands of sazan are required to make enough extract for a female weighing 3 to 5 kg. Better results are obtained if females are administered an injection of 0.5 l hypophysis 12 to 20 hours before the basic dose.

The dosage is reduced by two times for males and the injection given only once. Males and females are injected simultaneously. The injection is given with a "Rekord" type syringe as described in Chapter IX. The fish are held either in tarpaulin carriers or kept on a specially prepared table with a soft top. The suspension is shaken before filling the syringe since particles of the hypophysis settle to the bottom. After administering the injection the males and females are kept in separate cages during the period of maturation. The injection is so timed that the work of egg collection can be carried out during the day.

The duration of ripening of the females after the injection depends on water temperature and weather conditions. The approximate periods of gonadal maturation of spawners after the injection in relation to the water temperature are as follows:

Water temperature, °C	Duration of ripening (hours)
15–16	23–28
17–18	20–23
19–20	18–20
20–22	14–18

The first check of the females is done 2.0 to 3.0 hours before the expected time of egg maturation and the second check 1.5 to 2.0 hours later. At the time of checking the female is held with the abdomen upward and the flanks pressed gently. If the eggs are easily extruded in the form of grains, the female is considered ripe; if the eggs are not extruded as grains, the process of ripening is considered incomplete.

Before collecting the sex products it is recommended that the spawners be cleaned with a napkin soaked in a 3% solution of chloramine. All the work and operations related to the collection of eggs are best done in a covered space in the incubation section or under a canopy to protect the sex products from the lethal effect of direct sunlight.

The containers used for the collection of milt and eggs should be thoroughly cleansed and dried. Water touching the extracted eggs and milt may prematurely activate them and obstruct fertilization. The eggs are collected in enameled basins of one liter capacity and stripping continued until small globules of eggs and blood clots start to appear (Figure 27).

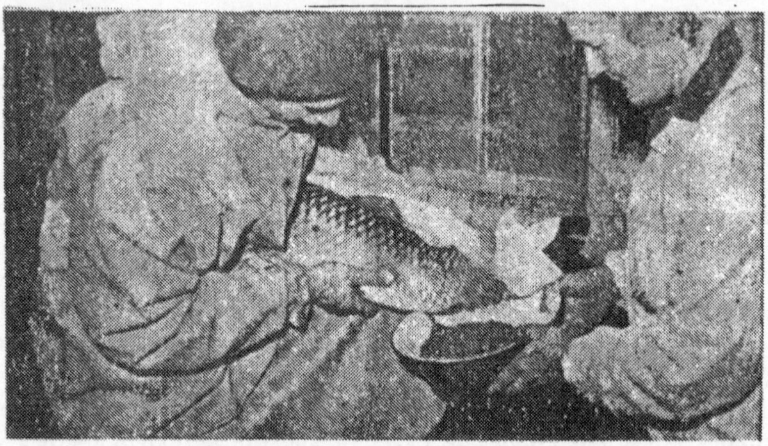

Figure 27. Extruding eggs from a female carp.

The milt is collected in wide dry tubes or sampling bottles and should be extracted after the eggs have been collected from the female. Eggs kept in containers without water and covered with a few layers of muslin cloth do not lose their fertility for 30 to 45 minutes.

The eggs are mixed with the milt from not less than three males. The following scale is used to assess the quality of the spermatozoa:

All sperm motile and movement progressive....................5
All sperm motile, but some perform sinuous movements..........4
All sperm motile, but larger fraction perform sinuous movements....3
Main bulk of sperm motile, movement predominantly sinuous.....2
Major part of sperm nonmotile..............................1

Spermatozoa graded 5 and 4 are suitable for artificial fertilization. A good quality of milt has the appearance and consistency of concentrated milk and a large number of the spermatozoa retain motility for one minute. Here it should be kept in mind that the motility time of the spermatozoa depends on the ratio of the quantity of water to the spermatozoa for fertilization. This ratio should be 10:1 of water and spermatozoa.

Under commercial conditions the spermatozoa retain their capacity for fertilization for about 90 minutes and their quality should be checked every 30 minutes.

The eggs are fertilized by mixing 3 to 5 ml of milt in each liter of eggs measured without water. The two are carefully and thoroughly mixed using a feather and the mixture then treated (against adhesion) with degumming solution No. 1 at the rate of 200 ml solution per liter of eggs. The degumming solution is a preparation of acetone-treated testes (PAS-G).[7] Hyaluronidase in the form of a powder prepared from acetone-treated calf testes was found to be less active than preparations made from pig testes.

Fish eggs obtained from fish reared at a water temperature below 16°C cannot be properly degummed with the help of hyaluronidase.

For this treatment 50 g of PAS-G powder are mixed in each liter of isotonic solution (8.5 g of sodium chloride or iodineless common salt per liter of distilled water) and the solution allowed to stand at room temperature for not less than three hours. The original (stock) solution can be stored in a refrigerator for five to seven days with no loss in efficacy.

To prepare solution No. 1 the stock solution of PAS-G is diluted ten times with clean water and the fish eggs placed in enameled containers immediately before degumming.

One more preparation is used in the process of degumming, solution No. 2—a solution of tannin (100 to 150 mg tannin per liter of water). The stock solution of tannin is prepared by dissolving 10 g of tannin powder in one liter of warm distilled water. It can be stored in glass containers throughout the entire period of spawning work. For solution No. 2, 100 to 150 cc of stock solution is mixed with one bucket of pond water. This solution is prepared immediately before utilization as it cannot be stored. When degumming solution No. 1 is used at the rate of 200 cc of solution per liter of fish eggs, it is thoroughly mixed and the eggs transferred to another enamel-coated tub of 5- to 6-l capacity. As the eggs swell, and to prevent the mixture from becoming too dense, fresh portions of the solution are added to this container. The layer of the solution above the eggs should not be more than 0.5 to 1.0 cm.

Degumming with solution No. 1 continues for about 15 minutes after which solution No. 2 is gradually added to the tub and thoroughly mixed. After about 30 to 35 minutes from the beginning of the degumming process the eggs stop swelling but solution No. 2 is nonetheless added right up to the termination of the degumming process. The layer of the solution above the eggs at this time is about 3 cm.

To assess the extent of degumming, about 40 minutes after the process has been initiated a sample of the eggs is placed in a petri dish containing clean water. If the eggs do not adhere to the undisturbed petri dish within five minutes, the process of degumming can be considered complete and the eggs placed for incubation without prior washing in the usual Weiss apparatus of 6 to 8 l capacity (see Chapter IX). If the eggs adhere to the glass degumming must be continued.

[7]Solution for detaching (degumming) of fish eggs is produced by the Rigma Meat Canning Factory

Given the normal conditions of incubation, after degumming each batch of eggs is placed in a separate container and water supplied at the rate of 2 to 3 l/min to prevent local stagnation of the eggs and mass mortality. Before submersion of the eggs in the apparatus the flow of the water is adjusted to 0.5 l/min, and only after the transfer of the eggs has been completed is the water supply gradually and carefully increased to ensure a slow mixing of the entire egg mass.

During the period of incubation of the fish eggs it is particularly important that the water supply be carefully maintained and controlled. Dead eggs are systematically siphoned out from the apparatus starting from the second day of incubation. Live eggs coming into the siphon together with dead ones are carefully removed to the reserve (control) Weiss apparatus (Figure 28).

Figure 28. Weiss apparatus at a fish farm.

Since fish eggs develop more rapidly at a water temperature above 20°C than the fungus *Saprolegnia,* the possibility of saprolegniosis is either totally excluded or almost nil. Egg treatment with malachite green is therefore not essential. If the temperature is below 20°C and egg development delayed, the possibility of saprolegniosis infection rises and hence this dye should be used once every two days. In this case the water supply is stopped and three minutes after the eggs have settled half of the water layer above them removed from the apparatus. For each liter content of the apparatus 10 ml of 0.05% solution of malachite green is added. After carefully stirring the contents of the apparatus with a feather the eggs are allowed to stand for 20 to 30 minutes. The water supply is subsequently renewed and the dye solution gradually removed by the water flow.

Table 5. Duration of incubation of carp eggs, depending on temperature

Incubation temperature, °C	Duration of incubation days
22	2.5–3.0
20	3.5–4.0
19	4.5–4.0
17	7.0–7.5
below 16	more than 8

The relationship between the duration of incubation and temperature is given in Table 5.

The hatching of larvae under conditions of normal water flow (2 to 3 l/min) takes place in 10 to 15 hours. To speed up this process in the beginning of mass hatching, it is recommended that the flow of water be reduced to 0.2 to 0.5 l/min. If the time has been correctly selected complete hatching is over within 20 to 40 minutes. However, if hatching is not complete in this period, then the normal water flow is restored and the attempt made again after two to three hours. It is thought that the acceleration of hatching with a reduction in water current results from the accumulation of the enzyme in the water which stimulates the hatching process.

After hatching the larvae are transferred to a trough or bucket and transported to cages for maintenance. They should not be allowed to stand in the apparatus after hatching since they die very rapidly in overcrowded conditions.

The larvae are kept in cages made of gauze No. 17 netting since they can pass through a net of large mesh, and the water exchange is inadequate with nets of a smaller mesh. The size of the cage should be 90 cm × 60 cm × 45 cm or 45 cm × 45 cm × 45 cm. Each cage can hold 250,000 and 150,000 larvae respectively. The cages are installed on a floating frame so that the bottom of the cage is even and without wrinkles, and the upper margin projects 10 cm above the water level. The cages are placed in those areas in the pond protected from wind and the impact of waves or kept in large concrete tanks. Since the larvae are very sensitive to oxygen deficiency, at the time of placement of the cages in the tanks the consumption of water should not be less than 1 m^3/hr per one million larvae. To improve water exchange the supply should come through a pipe or a nozzle opening at a distance of 15 cm from the bottom of the cage.

The larvae are kept in the cages for two to four days, depending on temperature, and are not fed during this period. Their changeover to external feeding coincides with the moment of appearance of air bubbles in the swim bladder and not later than one day after the bubbles appear. To avoid losses during the process of further rearing, the larvae should be transferred to the pond from the cages, a continuous 24-hour process. They should be scooped out together with water and never collected in muslin nets. The density of stocking should not exceed 500,000 larvae per hectare and the difference in their age during stocking in a single pond should not be more than one day. If it is not possible to prevent the entry of trash and predatory fish in the rearing ponds, then stocking is done in the nursery (transplantation) ponds. To calculate stocking the final weight is taken as 100 to 200 mg with a fish productivity of 50 to 100 kg/ha. The growth rate is checked after every two to three days.

Holding larvae in the nursery (transplantation) ponds yields more favorable results.

Temporary bionorms for the commercial breeding of carp

Rate of females to males	5:1
Reserve stock of brood fish	20–30%
Ripening of females after injection	80%
Relative working fecundity of females	50–60 thousand per 1 kg of weight
Yield of active larvae from collected eggs	50%
Yield of young fry from larvae	50%
Yield of fry from nursery ponds	70%
Yield of fingerlings from fry	80%

All the foregoing is related to the corresponding estimates obtained from a unified system of primary documentation consisting of the following material:

1. Temperature register completed by the man on duty.
2. Register for administering injections to spawners and collection of eggs maintained by the fishery expert responsible for these duties.

The number of females used in a particular spawning season are entered in this proforma. The fish are weighed after egg extraction, thus avoiding possible damage before undertaking this task. To estimate the quantity of eggs in 1.0 g of each sample of inseminated eggs, a sample of 0.5 to 1.0 g is taken (on VR=1 balance) and fixed in a 4% formalin solution. When the fixed eggs have lost their adhesiveness, they are counted with the help of a binocular lens or under a microscope in a Bogorov chamber.

3. Register for incubation of eggs completed by the man on duty.

The volume of fish eggs is determined in an incubation apparatus marked on a scale of 0.5 l for which water in the apparatus is stopped for three minutes. After three minutes the volume of the eggs is measured and the flow of water subsequently restored.

The percentage and number of fertilized and developing eggs per ml are determined by removing 2.5 to 3.0 ml of eggs from the apparatus the next day. Using a glass tube with rounded ends the eggs are transferred to a measuring cylinder of 10 ml capacity. When the eggs have settled in the cylinder, their volume is measured and counted in a Bogorov chamber, taking a separate count of live and dead eggs and calculating the percentage of developing ones. Batches of eggs which are similar in quality are pooled in one or several apparatus as required. In the case of removal of a large quantity of dead eggs and a sharp decline in their number the remaining portions are combined again and the selection and mixing of samples carried out in the main apparatus as well as in the control and the data recorded in the incubation register.

Only eggs collected on one and the same day should be pooled; eggs collected on different days should never be pooled.

For a uniform distribution of young fry in the cages the egg count is done before hatching for the purpose of maintenance and calculation of the area required in the ponds for the fry. Pond filling commences from the moment of hatching.

4. Register for maintenance of larvae completed by the fishery expert responsible for the maintenance and supply of larvae.

The number in the column of the register headed "quantity of transplanted larvae" is equal to the number of live eggs before hatching. Larvae supplied to other farms are counted in the same way as done during the transplantation of larvae or young fry from spawning ponds to rearing (or nursery) ponds.

The data compiled in the primary documentation make it possible to study spawners and identify their peculiarities.

5. Register for recording the results of the incubation section completed for the entire spawning period by the person in charge of the incubation section or the chief pisciculturist.

In this register the sum total of the results of egg extraction for each day are entered. The degree hours are calculated from the entries of the temperature register and the average temperature obtained for the period of maturation of the spawners and incubation of the eggs.

The five registers detailed above are always maintained for the commercial method of fertilization of carp eggs.

The commercial method of incubation of fish is carried out in a special incubation section (or division) where Weiss apparatus are installed together with a water collection trough. The apparatus are equipped with water supply etc. The incubation of the eggs proceeds normally at 70 to 90% water saturation with oxygen; oversaturation is not permissible.

At present treatment against adhesion is widely practiced in the incubation of eggs under commercial conditions. However, experiments have recently been done (Emel'yanov et al., 1969) on the incubation of sazan and carp eggs in a sticky state in the trough type incubator designed by Sadov-Kokhanskaya. These experiments need to be continued and methods developed.

Transplantation of Young Fry from Spawning Ponds to Rearing Ponds. Methods of Catch Estimation and Grading. Young fry no more than six to seven days old (or slightly older if the abundance of natural foods permits) are transplanted in accordance with calculated norms to nurseries or, if such are absent, directly to rearing ponds.[8] A deficiency of food leads to delay in growth, weakening of the fry, development of diseases, and large-scale mortality. The fry may be transplanted from the hatching ponds earlier as long as precautions are taken to see that neither injury nor weakening occurs.

Starvation of fry in the early stages of postembryonic development must be guarded against since the internal organs, especially the digestive, are in the process of formation. It is essential that the conditions of the natural food base be carefully observed during this period since starvation inevitably leads to significant losses and a deterioration in the quality of the progeny raised. The ponds should be saturated with live food and properly managed, prepared, and manured.

The transplantation of fry commences with netting. The technique of transplantation is not complex but rather laborious. Several methods have been developed for netting fry from spawning or hatching ponds. On a warm clear day when the fry concentrate on the surface layer of the water they can be easily caught with the usual muslin net of 0.5 m diameter or a small muslin drag seine. The fish are gradually released into the rearing ponds and forced to concentrate in the deepest places—canals and fish pockets. The water outlet is covered with a small mesh (1.0 to 1.5 mm) to prevent fry escape.

The tendency of fry to swim against the water current can be utilized. If a water current is created along the water conducting channels, the fry immediately respond to it. By moving against the current they form shoals at the place of water supply from the canal and can be caught with the help of dip nets or muslin seines.

To catch fish trapping boxes 1.5 m long, 0.9 m wide, and 0.6 m high are also used. The trapping boxes are attached by means of rings or straps to the horizontal tube of the monk (Figure 29). Four muslin or tarpaulin walls are fixed inside the trapping box at an angle of 30 to 40° in relation to the water surface. When the trap is ready the water is released. Fry fall into the box with the flow of water and are collected with the help of a muslin dip net. By this method 200,000 fry can be collected in one day. Losses are two to three times less than in the case of catching with a muslin drag seine. Experience has shown that with careful handling of the trapping boxes the fry sustain no damage whatsoever.

[8] In the case of spawning ponds with a very small area (100 to 200 m^2) the collection of young fry and their transplantation are carried out even earlier, on the fourth and fifth day.

A

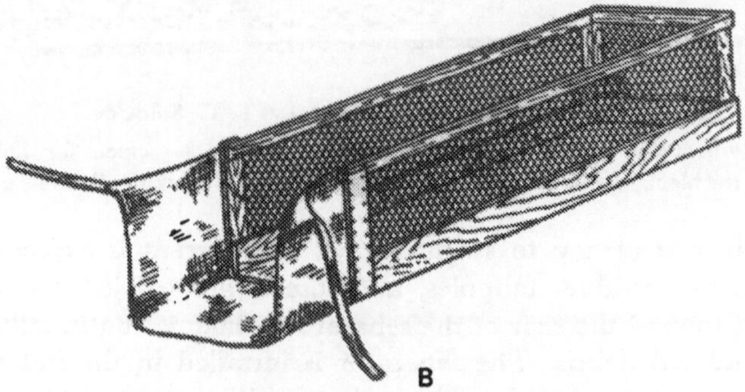

B

Figure 29. Trap for fry.

A—trap box under end of horizontal tube of monk in water collection channel; *B*—tarpaulin projection at end of trap for connecting it with end of horizontal tube of monk.

T.T. Solov'ev has proposed a special kind of trap for catching fry, which capitalizes on their tendency to move against the current toward the inflow (Figure 30). Nets of different mesh are fixed in the trap: the front net has a larger mesh and the rear one (immediately behind the outlet) a small mesh. The fry can be graded with the help of such a trap since the larger or stronger, healthier ones accumulate at the inflow. All kinds of admixtures, plant residues, tadpoles, larvae of insects, etc. are carried away by the current. This type of trap is also used for catching fingerlings and even marketable fish (Figure 30).

The length of time taken to transplant fry has great significance. In practice this operation is sometimes stretched over several days and even a week. A difference in growth rate between fry transplanted on the first day and those on the last must obviously be expected.

Regardless of the type of trap used larvae of insects, tadpoles, plant residues, and other admixtures drift into it with the fry. To separate the fry

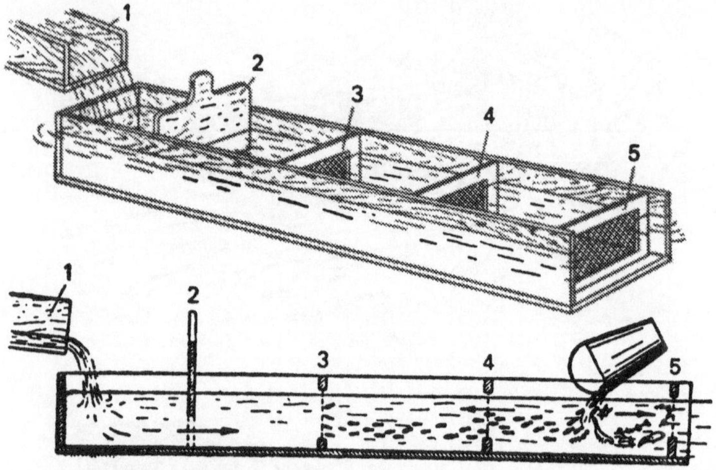

Figure 30. Grader for fry designed by T. T. Solov'ev.
1—water supply trough; *2*—shutter for regulating current; *3*—screen for holding fry; *4*—screen for blocking insects and tadpoles; *5*—screen for blocking dirt and pests.

from these their tendency to swim against the current is exploited. The admixtures, plant residue, tadpoles, and insects are carried away by the water current toward the rear of the separator, which is continually cleared of the accumulated debris. The separator is installed in the inflow of the water and current speed regulated by a shutter. The separated fry accumulate near the front wall and are collected preferably with muslin spoons or in scoops with a brail of wire netting (Figure 31) and placed in a counting apparatus (Figure 32).

The fry caught by one or the other method are kept in containers previously situated at convenient places (tarpaulin vats, cans, barrels, or wooden

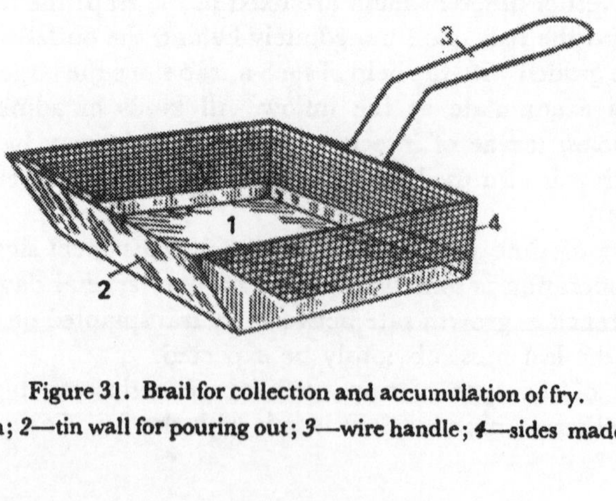

Figure 31. Brail for collection and accumulation of fry.
1—tin bottom; *2*—tin wall for pouring out; *3*—wire handle; *4*—sides made of metal net.

Figure 32. Estimation of number of fry by pouring them together with water through a spout.

boxes filled with water), from which they are transferred to the nurseries or rearing ponds. Throughout the procedure the fry must be protected from direct sunlight. When all the fry are caught, those obtained from each pond from a given redd or pair of spawners are counted. This makes it possible to assess the quality of the spawners. After this the spawning ponds are drained (see Chapter X) and left dry until the next spawning campaign.

The fry are stocked in rearing ponds in accordance with the calculations for natural productivity, which ensure well-fed fingerlings of a standard weight. The following formula is used in such calculations:

$$A = \frac{G \times P \times 100}{V \times r},$$

where A = rate of stocking of fry (no.);
G = area of rearing pond (ha);
P = natural fish productivity (kg of growth—addition in weight of fingerlings from 1 ha area in vegetative period);
100 = multiplication factor for conversion in percentage;
V = final per piece weight of fingerlings toward autumn (kg);
r = percentage of output of fingerlings toward autumn.

The quantity of fry to be stocked in each rearing pond should be calculated separately.

Assuming the availability of a sufficient number and area of rearing ponds, to determine the quality of spawners the fry obtained from one redd or pair of spawners should be isolated. The fry are measured by different methods: visual observations, volume, or total numbers. Eye estimates are made with the help of a trace, spoons for fry, etc. but they are less precise. Accuracy of count depends on the counter's experience and requires a fixed concentration that often causes eye strain. The fry are counted in a muslin cloth spoon-measure or by pouring them through a spout (Figure 32). The latter method is mainly used in research work where a high degree of accuracy is needed during stocking. Both methods are fairly difficult and not recommended for commercial purposes.

Of the different methods for estimating quantity of fry with or without water the easiest is with water in one of the apparatus designed by T.T. Solov'ev.

One Solov'ev apparatus (Figure 33) operates on the principle of water displacement by the fry. It consists of two parts: a container for estimation

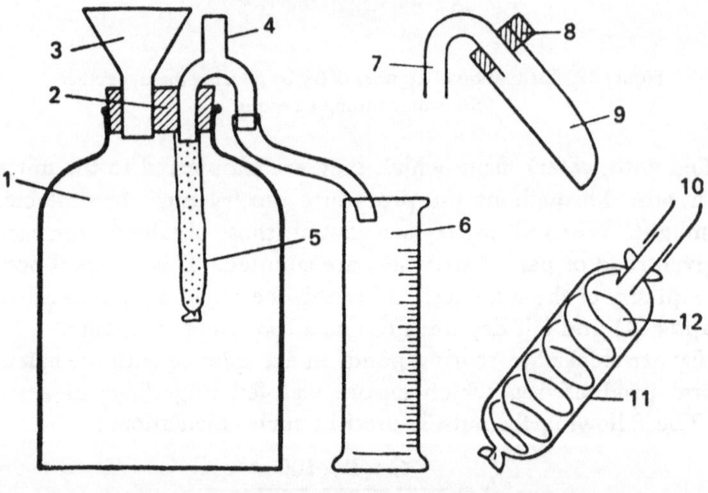

Figure 33. Apparatus designed by T.T. Solov'ev for estimating the quantity of fry.
1—bottle; *2*—cork; *3*—funnel; *4*—Y-piece; *5*—filter made of rubber tubing; *6*—measuring cylinder; *7*—outlet tube (in place of T-piece); *8*—cork; *9*—steel spout; *10*—tube of filter (another kind); *11*—muslin cloth; *12*—wire.

and glass cylinders for measuring. The estimation container is made of glass, galvanized iron, or painted steel sheet. It is filled with water taking care that any excess flows through the T-piece (through the outlet). The measuring cylinder is placed under it and the previously counted fish introduced into the container. The fry introduced displace some water, which flows out through the tube into the measuring cylinder, while the water level in the cylinder rises by a certain number of divisions. The number of fry per each division of the measuring cylinder is then calculated. For greater accuracy this procedure is repeated two to three times. Subsequently the number of fry to be stocked is determined on the basis of the volume of water displaced in the measuring cylinder.

Another apparatus designed by T.T. Solov'ev (Figure 34) for estimating the quantity of fry also operates on the principle of water displacement. Solov'ev describes the procedure as follows: "A scoop or ladle is placed in the empty container without fry in such a way that the lower margin of the scoop

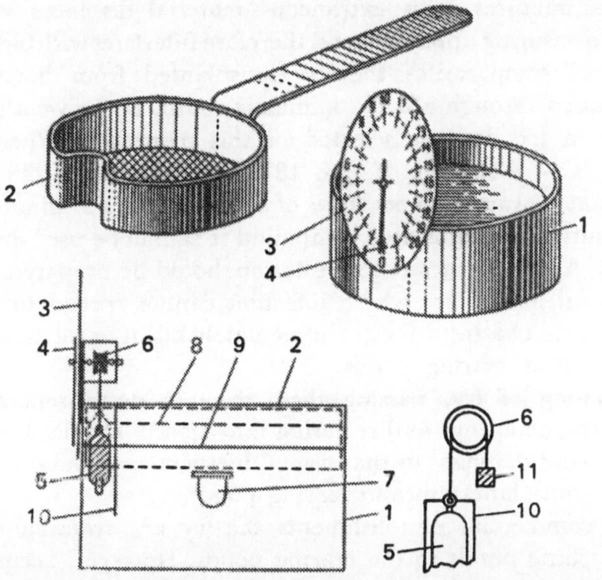

Figure 34. Apparatus designed by T.T. Solov'ev for estimating the quantity of fry.
1—container (tank); *2*—inserted scoop; *3*—scale; *4*—needle; *5*—float; *6*—spindle; *7*—support for scoop; *8*—level of water with fry; *9*—level of water without fry; *10*—tube; *11*—weight.

rests on the supporting plates welded to the sides. Water is poured into the container until the needle reaches zero on the scale. Fry are then introduced into the scoop and the number on the scale to which the needle moves recorded. After counting the fry the scale points or divisions are calculated for 1,000 fish. This operation is repeated three times to determine the average value. Having recorded the values on the scale, the scoop is taken out and the fry released into the transporting container or pond. The quantity of water in the container should be checked from time to time with an empty scoop and, if necessary, the position of the needle reset at zero on the scale."

The quantity of fry is quickly estimated with the help of this equipment. A container 100 cm in diameter can hold 20,000 fry with an average weight of 10 mg. Solov'ev recommends that the apparatus be constructed of nonoxidizing durable material and painted with anti-corrosion paint, lacquer, or enamel. It should be washed thoroughly after use.

The volume and number of apparatus depend on the number of fry to be estimated. In the case of very large numbers an apparatus with a water capacity of 10 l is used. Such an apparatus holds 10,000 fry six to eight days old, and the estimation of each batch takes no more than five minutes.

Accuracy of estimation also depends on the amount of irrelevant material which enters the apparatus with the fry, i.e., tadpoles, insects, plant residue, and other admixtures. This extraneous material displaces some of the water in the measuring apparatus and therefore interferes with the estimation.

To control ectoparasites the fry transplanted from hatching ponds should be passed through a 0.2% ammonia bath. If the weather is warm, cool periods in the day are selected for this treatment. When the water temperature is in the range of 7 to 18°C the duration of the bath is 1.0 minute, but at a water temperature of 18 to 25°C, 0.5 minute. Since an ammonia solution is an unstable compound it cannot be used for more than five minutes. After two uses a fresh solution should be prepared and the old one poured out at a place where infection cannot spread to the ponds.[9] After the ammonia bath the fry are immediately taken to the transplantation (nursery) ponds or rearing ponds.

The catching of fry, passing them through debris-separating traps, ammonia baths, counting, and releasing into ponds should be done very carefully to avoid damage to the fish. Otherwise considerable losses will occur in the transplantation and rearing ponds.

In most commercial establishments the fry are transplanted directly from the hatching ponds to the rearing ponds. However, according to the system of Dubisch, widely practiced in a number of European countries (Czechoslovakia, German Democratic Republic, Poland, etc.) and in the western districts of the Ukraine, the fry are transplanted in two installments, first at the age of 3 to 5 days from hatching ponds to nurseries, and then at the age of 30 to 40 days (end of June—first half of July) to rearing ponds where they are kept until autumn. In the early spring vetch-oats or some other grass mixture containing vetch is sown in the depressions of the nurseries at a seed rate of half that commonly followed in feed conditions. When flooded with water vetch rots rapidly, creating favorable conditions for the development of bacteria and consequently, zooplankton. The fry find good conditions in the remaining sparse grass stand of oats or any other crop sown with vetch. With a stocking density of 50,000 per hectare they attain a

[9]Fingerlings, yearlings, replenishing stock, and spawners are also passed through these baths to prevent the diseases trichodiniasis, chilodonelliasis, dactylogyrosis, and gyrodactylosis. The concentration of the water solution of ammonia should be 0.1% for the replenishing stock and spawners.

weight of 5 to 7 or even 8 to 11 g. With denser stocking (300,000 to 500,000 per hectare) the weight of an individual fish reduces to 1.15 to 0.6 g. The fish productivity of grass-sown ponds is usually not less than 300 kg/ha and often reaches 480 kg/ha.

The double transplantation of fry, first to nurseries and then to rearing ponds, has the following advantages:

1. The rearing ponds are better prepared since they remain dry for a longer period of time, making it possible to carry out the necessary reclamation and anti-epizootic measures and ensures mineralization of organic residue. The food organisms required by the fry, especially in the first month of their life, are generally more abundant.

2. Early crops planted in the depressions of the rearing ponds provide additional agricultural produce and the postharvest residue increases the natural food base for the fish.

3. The number of fingerlings is notably higher in autumn.

After removing the fingerlings to the rearing ponds the nurseries are dried and planted with any agricultural crop or used for green manure which is ploughed back into the soil in autumn. Either procedure creates better conditions for the maintenance of these ponds.

The relationship between the weight and age of fry in the nurseries is as follows:

Number of rearing days	Weight, g
10	0.02–0.04
15	0.45–1.15
20	2.0 –4.0
30	6.0 –6.5
40	6.7 –8.0

Rearing of Fingerlings. Fry stocked in the rearing ponds remain there until autumn—in the central regions of the USSR up to the first or middle of October, and in the southern regions until November.

The primary purpose of the rearing pond is to obtain fingerlings of a definite size, standard weight, and requisite fattiness (condition) in the given zone, which ensures a better hibernation during the first winter and a weight of not less than 0.5 kg by the second summer. For the central belt of the European part of the USSR and western Siberia, the fingerlings of carp which undergo normal hibernation and growth rate (increase in weight) in the second year are expected to weigh 25 to 30 g. For fish reared in ponds constructed on river beds, where predatory fish from the river can also penetrate, the standard weight of the stocked carp material as demonstrated by experience in Ukrainian SSR should be increased to 60 g or more.

In the central belt of the European part of the USSR the coefficient of condition of fingerlings before hibernation with an average weight of 25 to

30 g or more and of standard length should be 2.7, for fingerlings weighing 15 to 25 g—2.9, and for fingerlings weighing less than 15 g—3.1. However in this zone and others with similar climatic conditions, fingerlings weighing less than 25 g are unacceptable. Otherwise, an average weight of 25 g or more is considered standard, a weight of less than 25 g down to 15 g sub-standard, and a weight of less than 15 g waste.[10]

In Yugoslavian pond fisheries (fish nursery Zdenitsa) under-yearlings weighing about 100 g are used. It is thought that fingerlings of this weight suffer less damage from infectious diseases such as dropsy.

L.E. Tsuladze recommends a weight of 70 to 80 g for fingerlings in the conditions of Georgian SSR where the period of vegetative growth is longer.

Thus, given the variety of soil-climatic conditions in different zones of the country, the norms fixed must differ in accordance with experimental studies and practical experience. At present the weight index for individual fish is almost identical throughout the country but this should be considered a temporary phenomenon. The following guidelines should constitute a basis for establishing a weight index for rearing young fish of economically valuable species in pond culture, especially carp: (1) fingerlings should acquire the commercial weight of carp in the second year of life; (2) losses in the wintering pond should be minimum; (3) fish quantity per unit area of rearing ponds should be maximum; and (4) the natural food resources of the pond should be utilized to the fullest extent.

To achieve standard and higher weights in fingerlings and the requisite condition index the following measures are important:

1. Earlier spawning and stocking of fry in nursery and rearing ponds, utilizing the warmer waters from hydroelectric power stations (wherever this is possible).

2. Under the conditions of a cold summer, considerably increase the water level in the ponds (10 to 15 cm higher than the planned level) and take measures to reduce water losses through water drainage outlets and percolation through the dam. Given a warm summer the fry should be fed two to three times a day, feed being supplied in accordance with consumption. Even a brief warming (up to three weeks in the season) in a cold summer will make it possible to raise fully fertile under-yearlings if these recommendations are followed.

3. Fresh water should be added regularly to the ponds, especially in the case of dense stocking, and the oxygen content maintained at the required level. A reduction in oxygen content to 2 mg/l reduces food intake by four times.

[10]According to the current standard of the Ministry of Fisheries of the USSR fingerlings weighing less than 10 g are considered waste

4. When the weather in autumn is good, the catching of fish from the rearing ponds should be done as late as possible in the season.

5. In the case of dense stocking of fry it is better to maintain the growing fish on natural foods, increasing their quantity through applications of manure. This is especially important in the first half of the vegetative period, when the fry are not able to utilize the feed given them in accordance with their size. Only in the second half of summer, when the fingerlings have become hardy and grown, is it appropriate in a number of cases to give feed and apply manure, thus maintaining a proper ratio. Feeding should be started when the growing fry weigh 3 to 4 g; earlier feeding is less effective.

Body weight is not the only index of quality of stocked material. Winter-hardiness depends not so much on this index as on the condition factor. This important qualitative index requires changes not only in the conditions of hibernation but also in the rearing of fry in rearing ponds. To prepare fingerlings for hibernation, especially in regions with a prolonged winter continuing for 6.0 to 6.5 months in a year, it is essential that conditions in the rearing ponds be organized carefully. It is known that carp is less active in winter; it does not feed and lives on reserve matter accumulated over the vegetative period.

Polyakov (1964) believes that fingerlings weighing more than 20 g and raised in the central regions of the European part of the USSR only on natural foods should contain not less than 1.5 to 2.0% fat. In fish raised under dense stocking with artificial feeds the fat content should constitute 3.0 to 4.0% of the live weight. The protein content should not be less than 12%. The accumulation of fat through a reduction in protein content is not beneficial. It has been proved in experiments that a fat content above 8.0 to 9.0% is excessive and a decrease in proteins to 5.0 to 6.0% a cause of mortality.

V.M. Il'in and other researchers believe that for successful hibernation the fat content of fingerlings during autumn should not be less than 3.5 to 5.0%. Brizinova (1958) is of the opinion that fingerlings with a fat content of 3.0 to 3.5% are fully prepared for hibernation. Shäperclaus (German Democratic Republic) found that fingerlings with a fat content below 3.0% are more susceptible to disease. According to Wolny (1963), for a successful hibernation the fat content toward autumn should be 4.0 to 6.0%. For feeding fingerlings R.I. Mukhina (1958) recommends a fat content of not less than 3.0 to 4.0%. These data reveal that the index of winter-hardiness differs according to zones and conditions of rearing.

The conditions of feeding, density of stocking, and other factors influence protein metabolism and, in turn, the winter-hardiness of stocked material.

The fat content changes during the course of the vegetative growth period of fingerlings. With a reduction in water temperature, seasonal changes in metabolism commence, the growth rate decreases, and reserve energy-giving substances are deposited intensively. Among such substances fat occupies first place. This accumulation can be considered adaptive in fry since it prepares the organism for winter starvation.

The difference mentioned above in the fat content of fingerlings raised on natural foods and in dense stocking with artificial feed (not less than 1.5 to 2.0% versus not less than 3.0 to 4.0%) is probably explained by a difference in the protein and fat quality of carp fingerlings mainly raised on feed extraneously supplied. More studies are needed on the influence of rearing conditions (natural foods versus supplementary feeds), on the quality composition of reserve fat in fingerlings, and on their growth, development, and winter-hardiness, taking into consideration zonal peculiarities. The glycogen content in carp liver is also subject to strong seasonal fluctuations: in summer it is 3.0 to 5.0% of the fresh weight of liver, toward autumn 10 to 15%, and toward spring reduces to 4.0 to 5.0%.

Studies conducted by the Department of Pond Fisheries of the Timiryazev Agricultural Academy have shown that the age of spawners also exerts an influence on the winter-hardiness of fingerlings. Different pair combinations were used in an experiment (eight-year-old female and five-year-old male and vice versa); the progeny from the eight-year-old female and five-year-old male proved more tolerant to winter temperatures. The progeny of this pair of spawners consumed nutrients more economically during the winter period and the survival rate in spring was 73.5 to 99.3%. In the reverse combination the survival rate was 66.0 to 75.6%.

The growth of fingerlings during the vegetative period is recorded by control catches every ten days. The catch is made with a seine, push net, or any other suitable gear at places of maximum fish concentration and, depending on the area of the pond, sampling done in one or several spots. To estimate growth rate over a ten-day period a minimum of 100 fish are taken, their weight in toto recorded, and the weight of a single fish calculated. The larger fish are then weighed and the weight of a single fish determined. Finally, the smaller specimens are weighed and the minimum weight of a single fish determined.

To calculate the average weight of fingerlings in autumn at the time of transplantation to wintering ponds, a minimum of 100 fish are taken from each pond, graded large, small, or medium, their percentage ratio determined, and the coefficient of condition of each carp calculated. Diversity in the growth of individuals reflects insufficient preparation of the reared stock for hibernation. At the time of stocking in the wintering ponds this diversity gives a distorted picture of the average weight expected toward spring, since many fingerlings of small weight die due to a deterioration in condition. All these indexes are recorded in a special register (Table 6).

The method for determining the indexes L, l (or d), and K have been explained earlier; H = maximum body height; C = head length (all measure-

Table 6. Register* for control catches for rearing pond No.——of fish farm of collective or state enterprise——region——district——month——day——year——

S. No.	Carp breeding (scale, mirror, single row scaled, and leather carp)	L, cm	l, cm	H, cm	C, cm	Q, gm	K	Remarks

*In the course of research work or breeding studies, these parameters are determined in each control catch and for each sample of fish.

ments done according to the method shown in Figure 20); and Q=weight of fish. In the column headed "Remarks" the condition of the fish is recorded, the gear used for catching, and the number of fish caught. The condition factor "K" under commercial conditions should be estimated once in the season, on August 1st, on the basis of the measurements and weight of 50 specimens from each rearing pond. To determine the growth rate in small ponds with a normal rate of fish stocking, up to 50 fish are usually taken. However, such a quantity does not suffice for a proper assessment of the course of fish rearing in large ponds or even in small ponds with a high density of stocking. In my experimental work with small ponds, fairly accurate results were obtained by counting, sizing, and weighing 10 to 12% of the total quantity of fish stocked.

Control catches up to August lst are essential for ascertaining whether the fingerlings are lagging behind in growth and condition, and if so what measures should be effected for the remaining vegetative growth period before the fish are transplanted to the wintering ponds (supplementary feeding, thinning the stocks, etc.), which will ensure their achieving standard weight and established norms of condition—important elements for winter-hardiness.

The fingerlings are wiped with a muslin cloth and then weighed on a chemical balance with a precision of 0.1 g. The fish should be handled very carefully to avoid damage to them. Their condition is assessed and the absence/presence of diseases checked. After recording their measurements and weight they are returned to the ponds.

The established average parameters of fish weight through control catches are compared with projected indexes. In the case of growth-lag compared to the prior decade, the reasons are explored (deterioration in the hydrochemical regime, insufficient feeding, disease, etc.) and measures taken for their elimination.

Together with the control check on the growth of fingerlings observations are also carried out on the hydrochemical regime, especially the gas regimes of the rearing ponds. Samples for hydrochemical analysis are taken on the days of control catches (and if required the condition of the pond observed) immediately before catching, i.e., before the water is churned. Hydrobiological samples are collected simultaneously.

The weights of fingerlings under favorable temperature, hydrochemical conditions, and satisfactory feeding in various zones and months are presented in Table 7.

The weight of a single fish (carp) as given in Table 7 is not always achieved in practice primarily because of poor farm management. Weight can be significantly increased not only through technically correct farm management, but also by increasing the growth period of the fry, employing the commercial method of egg fertilization, using warm industrial and

Table 7. Weight of carp fingerlings raised in rearing ponds in different months of the vegetative (growth) period under climatic zones of the USSR (g)

Months of growth period	Moscow district	Bashkirian ASSR	Belorussian SSR	Southeast European part of the USSR	Southern zones of western Siberia	Ukrainian SSR	Moldavian SSR	Uzbek SSR
May	—	—	—	5.0	—	—	—	—
June	2.7	2.0	3.0	10.0	1.3	7.0	1.2	6.0
July	12.0	14.5	9.6	20.0	7.0	16.0	3.7	20.0
August	24.0	23.0	21.0	25.0	19.0	23.0	17.0	25.0
September	29.7	25.0	27.0**	30.0 and more	20–25	28.0	27.0	27.0
October	30.0*	—	—	—	—	30.0 and more	28–30 and more	28–30 and more

*Up to October 16.

**Up to 30 g or more.

natural waters to obtain progeny in earlier periods (March and April), and by rearing fry in special ponds or tanks with warm water until stable high temperatures are established in the rearing ponds. The variety undertaken for breeding is also important. Under the conditions of experimental stocking in the rearing ponds of Ukrainian SSR the weight of a single fish (unit weight) toward autumn was much higher in carp from a single breed (Krapin, 1955). For example, the unit weight of the progeny of a single breed of scale or mirror carp was 23.0, 43.1, and 60.5 g, versus 15.1 and 20.5 g for the progeny of mixed breeds raised in identical conditions. Even additional feeding with mustard cake in the case of mixed stocking did not improve the situation; under single-breed stocking the unit weight of the fingerlings on natural foods was higher.

The experiments of Fedoseeva (1965) with two-year scale carp and mirror carp with scattered scales in the central nonchernozem zone (Moscow district) revealed that raised separately the fingerlings of scale carp attained a weight of 66.8 g by autumn, and those of mirror carp with scattered scales 49.1 g. Raised together the weight of the former was 44.1 g and the latter 32.3 g. Fish productivity also reduced significantly of fingerlings of scale carp from 565.5 kg/ha in isolation to 425 kg/ha in combination with mirror carp with scattered scales. Fish productivity of fingerlings of mirror carp with scattered scales also reduced from 382 kg/ha in isolation to 268 kg/ha in combination.

According to Korneeva et al. (1969) the production indexes were slightly higher under isolated rearing of different carp breeds in fish cages or enclosures compared to mixed rearing. Under Moscow conditions scale carp and mirror carp with scattered scales yielded better results in growth and survival in both isolated and mixed rearing.

Transplantation of Carp Fingerlings from Rearing Ponds to Wintering Ponds. Fingerlings are transferred from rearing ponds to wintering ponds not later than the second half of October (in the central belt of the USSR). But if the water temperature is still sufficiently high (higher than 6 to 8°C) and the necessary valuable foods still available in the rearing ponds, an early transplantation to the wintering ponds could lead to a prolonged starvation of the fingerlings and increased losses during winter. At the same time transplantation should never be delayed up to the onset of frost as negative temperatures inevitably lead to freezing of the gills and hence quantum losses during the hibernation period.

A correct maintenance of the wintering ponds during summer exerts a positive influence on the hibernation of fingerlings. The ponds should be dried and cleaned before filling them with water in autumn. It is better to dry and plough the wintering ponds after netting the yearlings; channels and water courses should be thoroughly cleansed to ensure total drying of the pond bottom. Plants, rubbish, and dirt which may have accumulated over the summer period should be removed. Measures should be taken to prevent the entry of thawed waters, which contain little oxygen, by diverting them to other canals. The hydroengineering structures should be repaired, especially the water supply system, to ensure an uninterrupted flow of water during winter in the required quantity and quality. Screens should be fixed for sanitary-prophylactic purposes on the water supply channels and troughs, and the bottom and slopes covered with unslaked lime in the summer and again in autumn before filling the pond with water.

The network of water discharge channels should also be coated with lime.

A correct maintenance of the wintering ponds during summer (complete drying, organization of management practices, etc.) is one of the most important conditions for ensuring a favorable hibernation of fingerlings.

Ten to fifteen days before stocking the fish the wintering ponds are filled with water. In regions with a prolonged winter the ponds should be filled much in advance after the necessary reclamation work has been completed at the end of August. Simultaneously, they are treated with superphosphate at the rate of 30 kg P_2O_5 per hectare, potash up to 15 kg K_2O per hectare, and if necessary chemical nitrogen fertilizers. In addition, green manures should also be distributed along the periphery. Dung and composts should not be applied since these manures pollute the water. Before the onset of frost all the water plants brought in as manure should be removed.

To transfer fingerlings from rearing ponds to wintering ponds the water is discharged from the former. The fingerlings concentrate in the water collection channels and fish pits and in small ponds are caught with dip nets. In large ponds the water is discharged and the fish simultaneously caught with small-mesh seines or drag nets; after the water has been drained from the connecting channels and fish pits the remaining fingerlings are collected.

Catching in rearing ponds is very effective with fishing gear installed behind the outlet. This concrete construction is one of the several fish traps for fingerlings designed by Z.F. Andryuschenko. The trap is 10 m long, 1.0 to 2.5 m wide, and 0.8 m deep. Two slots are made in the outward opening—one for placing a net (screen) and the other for a shutter panel. Clean unpolluted water is pumped into the trap to create a better gas regime and the required depth controlled by the shutter panel. The fish are cleansed of dirt after removal. With such a fish trap six people can catch and count 300,000 fish in a single day.

The fingerlings collected are kept in containers filled with water (buckets, etc.) and graded only when weight proves substandard for more than 10% of the total fish. Fingerlings with defects are graded separately—diseased, damaged, defective—and small "admixture fish" ("verkhovka", brook trout, etc.), which may have entered the pond through the water supply source, identified. For the central belt of the European part of the USSR three

Figure 35. Table for grading of carp fingerlings and yearlings.
1—fish larger than standard size; *2*—standard size fish; *3*—substandard; *4*—remnants.

Figure 36. Size grading of carp fingerlings in a trough (diagrammatic).

classifications of fingerlings are acceptable: (1) weighing up to 15 g; (2) from 15 to 25 g; and (3) 25 g and above. Fingerlings weighing less than 10 g, or above this weight but not normal in terms of condition, are not counted in the plan of the farm and are stocked in separate wintering ponds.

Fingerlings are graded on a specially constructed grading table (Figure 35) or with the help of a wooden or iron box with vertical net screens (Figure 36). The mesh size of the screens should be adjusted to the size of the fish being graded. The box is inundated with water during the process of grading. The flow is so regulated that the fish move toward the inflow with the largest being detained in the first section, the medium-sized in the second section, and the smallest fish in the third depending on the mesh size of the net. By selecting the required mesh size an even more detailed grading can be achieved.

The condition (fattiness) of the fingerlings determines their inclusion in one or the other group of winter-hardiness. An average sample of 150 fish is taken (50 fish at the beginning, 50 midway, and 50 at the end of catching from the rearing ponds), measured with a precision up to 0.1 cm, and weighed with a precision of 0.1 g. Before stocking in the wintering ponds the fingerlings must be passed through a prophylactic salt or ammonia bath (see above). The salt bath should not be given if the temperature is less than 4 to 5°C (see page 54). The fingerlings are subjected to a special ichthyopathological scrutiny. For this 15 specimens from each pond, either before or after the salt bath, are given a full parasitological examination. If an epizoon is detected, all of the material is passed on to the Veterinary Inspector who decides whether the fingerlings should be transferred to wintering ponds. If he decides favorably, the fingerlings are counted and the norm of productivity determined for each rearing pond. Counting is done with the usual conical bucket of 8 to 10 kg capacity by the volume-weight method. The fingerlings in each conical bucket are counted and weighed. Catching, grading, and calculations are done carefully to avoid damage to the fingerlings.

Studies on the damage caused to fingerlings during their transplantation from rearing ponds to wintering ponds have shown that out of 7,428 fish, 1,231 fish or 16% were injured. Of the damaged fish, 65% had injuries on the fins, 16.5% damage by abrasion, 11% shedding of scales, and 7.5% scratches or blood clots. Most of these injuries cause greater losses in a shorter period than starvation. In terms of consequences physical injuries to fish can be compared to any severe infection.

If the conditions allow, all the fingerlings from one rearing pond should be transplanted to an adjacent wintering pond. This facilitates the isolation of any disease which may have occurred in the various rearing ponds.

The norms for stocking of standard fingerlings of normal condition in wintering ponds are not the same throughout the country; they vary from 400,000 to 800,000 fish per hectare.

At the fish farm of the collective enterprise "Rossiya" (Bryansk district) a stocking density of an average of 800,000 fingerlings per hectare in wintering ponds has given an annual spring yield of yearlings of 92 to 97% over a period of nine years.

Hibernation of Fingerlings. The process of maintenance of fish in wintering ponds is complex and requires a thorough preparation of the ponds and conditioning of the fingerlings for their first hibernation, as well as continual observation of the fish and water supply.

The coefficient of condition or fattiness of fingerlings stocked for hibernation (using the value 1.0 as the basis) should be 2.7 to 2.9 and 3.1 for different groups. If toward spring the coefficient of condition is nearer to 1.5 to 2.2, this constitutes the lowermost limit of critical emaciation.

To obtain favorable results from hibernation the following conditions should be fulfilled:

1. Wintering ponds should be properly prepared before fish transplantation (see above).

2. Water exchange for the entire winter period should be adjusted at a uniform rate, taking into consideration that an increased flow leads to fish exhaustion, due to hastening the process of normal emaciation during starved metabolism, which is particularly undesirable for fingerlings of a low weight. A water exchange every 12 to 20 days is normal; with a drop in oxygen content, the water might have to be exchanged every 8 to 12 days. A more intensive replacement of water is not necessary.

If the water for the wintering ponds is taken from artesian wells, the ponds are filled two weeks before stocking fish to allow time for oxygen enrichment. Before using such water gas and salt analyses should be done.

3. An aerator should be installed near the inflow of water and, if the quality of water deteriorates in the head pond, the process of aeration intensified.

4. During frosts ice should be removed from the water outlets to avoid a disruption in the normal water exchange. Ice retards the flow of water and may even block the water outlet. Water outlets are protected from ice by installing straw or reed fences.

5. Troughs and water supply channels should be kept warm from the top and sides by laying dry branches, straw, or hay on them to avoid freezing, obstruction of the normal water supply, and an accumulation of dirt. If the source of water supply is spring water with a temperature of 5 to 7°C, the water should be cooled before its transmission to the ponds. Spring water is least diseased since it contains insufficient oxygen and large quantities of carbon dioxide.

6. Hydroengineering installations should be kept under systematic surveillance.

7. Water temperatures should be measured at a depth of 5 cm from the bottom three times each day: at 7:00 a.m., 1:00 p.m., and 7:00 p.m.

The most favorable water temperature at the bottom of large ponds is 0.5 to 2.0°C. With other conditions being favorable the carp can be kept for a long duration at 0.1 to 0.3°C. Given unsatisfactory conditions, especially poor feed, with a decrease in water temperature to below 0.3°C the fingerlings catch cold and die. Temperatures ranging from 0.5 to 1.0°C in the wintering ponds are considered beneficial. Sebentsove (1940) believes that under these temperatures carp stop feeding, making dense stocking of fish possible. He also believes that carp in the first year of hibernation do not enter a state of "winter dormancy", as proposed by other authors, but are characterized by a differently expressed activity.

It is important that the water flowing into the wintering ponds be uniform in temperature. A sudden temperature change adversely affects hibernating fish; they come out of hibernation and are forced to move under conditions of minimum metabolism. A high water temperature (above 4°C) activates the "dormant" fish into a search for food. An increase in temperature from a fraction of a degree up to 10°C, especially toward the end of hibernation (spring), exacerbates fish fatigue.

8. Gas analysis of the water (mainly for oxygen and carbon dioxide content) should be done once every ten days during the period of hibernation. If this condition is unsatisfactory, an analysis should be done every day until the regime is properly balanced. Samples are taken at the source of inflow and near the outlet of each wintering pond at a depth of 5 to 10 cm above the bottom, and from the head pond. If the oxygen content of the water falls below 3 ml/l, aeration must be done (see Chapter X). Aeration is increased if the difference in oxygen content near the source of water inflow and outlet is more than 20%; this indicates that easily oxidizable substances are present in the water, especially H_2S. If the gas regime of the water is satisfactory, but the fish approach the ice holes, such fish are examined for possible disease or poor physical condition. For this purpose 30 to 50 fingerlings suffice.

The oxygen regime of the wintering ponds may deteriorate during thaws and severe frosts. Either bottom waters or ground waters "pressed out" from the soil, both of which are poor in oxygen content, enter the source of water supply or the wintering pond itself. A poor flow in the wintering ponds or their location on soils rich in organic matter are other causes of a deterioration in the oxygen regime.

9. Salt analysis of the water should be done three times during the winter season (just before ice formation, in January–February, and before ice melting). An analysis for H_2S, Fe^{2+}, Fe^{3+}, and pH should be done four or five times in the winter season. Cases are known, however, of the iron content increasing to 16 to 24 mg/l (expected norm 1.0 to 1.5 mg/l) without harmful consequences for the fish. Apparently the harmful effect of iron arises not from its quantity alone, but from a given quantity in conjunction with other factors in the aquatic medium. If one has to use water with a high iron content in the ponds, it should be vigorously aerated beforehand so that the ferrous salts are changed into harmless oxides which are precipitated.

10. L-shaped ice holes are best with a length of 2.0 to 2.5 m and a width of 1.0 to 1.5 m. These should be cleaned daily of ice and covered with panels which permit observations of fish behavior during winter.

11. Control catches should be carried out every month, starting from January. These catches are especially important in March and April. Not only is the general condition of the fingerlings assessed through such catches but also their health and fattiness determined. If *Chilodonella* or *Ichthyophthirius* are detected (in a single field of vision under a microscope 10 to 15 specimens

of the first and 3 to 5 specimens of the latter), treatment of the fish is initiated and the conditions of hibernation improved.

12. Should progressive emaciation of the fingerlings be observed in spring, feeding is started before the ice has completely melted (see Chapter XI), and catching in the hibernation ponds done even with ice present.

13. Water exchange is stopped at the time of spring floods and the inflow of turbid thawed water into the wintering ponds prevented, since such water contains myriad suspended particles. Clean, transparent melted water is not only harmless but well saturated with oxygen.

14. To check the condition of the fish in each wintering pond wooden boxes 1.0 m × 0.5 m × 0.5 m covered with metallic net are installed. About 100 fish are collected in each box and the boxes lifted two to three times in the winter period, the fish examined and measured, their weight recorded, and the coefficient condition calculated.

Successful fingerling hibernation depends on the proper follow-up of the foregoing measures. Mortality of fish in wintering ponds is primarily due to unsatisfactory indexes of weight and condition of fingerlings transferred from rearing ponds. Underweight fish become highly emaciated during winter, while larger fish accumulate considerable fat toward autumn. Polyakov (1964) (using the value of 1.0 as a base) considers an average coefficient of condition of 1.8, average fat content of 0.35%, protein content of 6.8%, and moisture of 89% of live weight critical values, i.e., those which lead to mortality as a result of emaciation.

The norms of yield of carp yearlings in spring vary in different zones of the USSR according to the Ministry of Fisheries (expressed as % of total autumn stocking) from 60 to 80%. For hybrids of carp and Amur sazan the yield expected in regions north of 55°N is 85% and in regions south of 55°N about 90%. In the northwest, Ural, eastern and western Siberia, and other regions with severe climatic conditions and long periods of hibernation these norms drop to 5%. Improvements in the techniques of managing fish ponds can sharply increase the output from wintering ponds and decrease losses to 5 to 10% or even nil.

Losses in weight of fingerlings during winter are not similar and usually fluctuate from 5 to 10% throughout the season for different weight groups with various degrees of fattiness.

The influence of single-breed and mixed-breed stocking of carp fry in rearing ponds on the autumn weight of fingerlings has already been discussed. It also has an impact on the results of hibernation. Fingerlings raised from a single-breed and stocked in wintering ponds give a higher yield in spring (82%) compared to fingerlings of mixed-breed (up to 44%). One can assume that with mixed-breed stocking, probably conditions are created which influence the biology of carp. This situation requires further study.

Besides stocking in special wintering ponds, fingerlings can be stocked for hibernation in rearing and foraging ponds, outdoor fish tanks, and in hibernation houses with tanks (Figure 37).

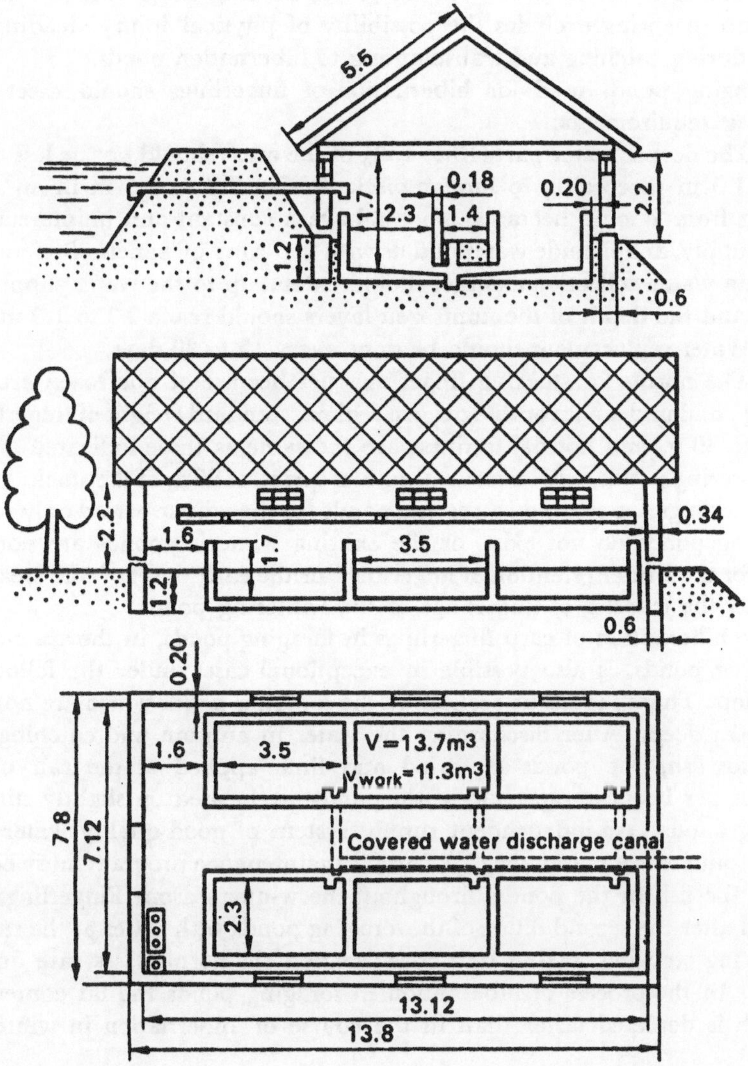

Figure 37. Diagram of hibernation house with cement tanks for storing live fish.

The hibernation of fingerlings in rearing and foraging ponds is permitted only under exceptional conditions, since round-the-year filling with water deteriorates the sanitary conditions and hydrochemical regimes of these ponds, reduces natural fish productivity, and eliminates the possibility of determining the output from rearing ponds over the vegetative growth period and hibernation period. Furthermore, not all rearing and foraging ponds are suitable for fish hibernation. However, keeping fingerlings in rearing ponds up to spring excludes the possibility of physical injury (leading to losses) during catching and transplanting to hibernation ponds.

Foraging ponds used for hibernation of fingerlings should meet the following requirements:

1. The deeper water part (10 to 15% of the area) should not be less than 1.3 to 1.5 m (according to foreign pisciculturists 2.0 to 2.5 m in an area ranging from 2 to 15 hectares), should have a constant and uninterrupted water supply, and include water outlets with two rows of panels. Preferably, the main water collection channel should extend up to the water supplying device and the depth of the nonfrozen layers should reach 0.7 to 1.0 m.

2. Water replacement should be done every 15 to 30 days.

3. The norms for stocking fingerlings per hectare of nonfrozen area of rearing pond under a favorable oxygen regime, standard weight of fingerlings at 25 to 30 g, and normal fattiness are the same as those indicated above for wintering ponds, with a survival rate in spring of 90% of the stock.

Fingerlings are stocked in rearing ponds for the winter period only when wintering ponds do not exist, or the existing wintering ponds are not yet ready for the winter retention of fingerlings, or the early onset of cold weather prevents the transfer of fingerlings to the wintering ponds.

The hibernation of carp fingerlings in foraging ponds, in the absence of wintering ponds, is also possible in exceptional cases under the following condition: There should be sections in the foraging ponds which are not less than 3 m deep. After discharging the water in autumn and catching the two-year fish, the ponds are dried and lime applied at the rate of 10 quintals per hectare, especially if the ponds are located on slightly silty or peaty grounds. An independent supply system of good quality water and proper outlets are mandatory. A careful maintenance program must be set up for the fish in the ponds throughout the winter season. Fingerlings are stocked after the second filling of the foraging ponds with water at the rate of the spring stocking, taking into consideration the normal loss rate during winter. In the process of hibernation in foraging ponds the fat content of the fish is depleted faster than in the course of hibernation in wintering ponds.

Experience with hibernation of carp in rearing ponds has shown that the bottom fauna of such ponds is poorer in summer than in ponds which are kept dry. At the time of water discharge during spring the eggs and food organisms occurring at different stages of development are washed out. Only toward June does the biomass of benthic organisms equalize in ponds used for hibernation and those kept dry in winter. Thus in ponds used for carp hibernation a sharp deficiency in natural foods is seen at the time of spring stocking, which is not observed in rearing ponds left dry throughout winter.

Hibernation of carp fingerlings in specially made live fish tanks located in a covered space (hibernation houses) has yielded very favorable results. For example, in Latvian SSR where carp fingerlings underwent hibernation in tanks in hibernation houses (see Figure 37) the yield was 88.4% (winter of 1957 to 1958) and 90% (winter of 1959 to 1960) versus 56 and 60% respectively in open wintering ponds. The recommended water temperature is 1 to 2°C and the area of each tank should not be less than 6 m^2 with a depth of 1 m. In tanks of lesser area water pollution is greater, which may cause parasitic diseases and consequent losses. Even in covered tanks successful hibernation depends on the physiological condition of the fingerlings; their condition should not be less than the index required for stocking in wintering ponds.

The permissible optimum density of stocking of standard carp fingerlings in the tanks is 2,500 to 3,000 fish per 1 m^3. The water supply should be continuous, the water of high quality, and replacement done twice in 24 hours. The water consumption in closed tanks is more economical. Water i supplied from a higher placed hibernation pond through a pipe laid over a damp incline and directed toward the bottom of the hibernation tank. The water from the pipe passes into the stone-concrete set-up in the pond where fish screens are installed, continues to the sedimentation filter tank of the hibernating house, and finally, from the sedimentation filter tank is distributed through open troughs to the tanks. A peculiarity of hibernation in closed tanks is the time of stocking—toward the beginning of snowfall. Until this time the fingerlings from rearing ponds are kept in specially prepared reserve ponds and fed food mixtures. An early transplantation, when the temperature is still high, leads to an active movement of the fingerlings, which injure themselves against the sides of the tanks.

Thus, the hibernation of fingerlings in electrically heated and lighted hibernation houses has a number of advantages, especially in those cases and in those zones where hibernation in the usual wintering ponds leads to high losses (daily checking of the fish, immediate adoption of measures against diseases, removing harmful effect of thawed waters to a great extent, and eliminating those processes involved in bottom accumulations).

Under the conditions of western Siberia good results have been obtained at the training farm of the Novosibirsk Agricultural Institute (N.N. Moiseev, 1970). The walls of the hibernation houses are made of cement and are 15 to 20 cm thick and 3 m high. The roof is made of ferro-concrete slabs covered on top with a layer of peat and soil. Lighting and ventilation occur through two hatches in the roof and the temperature is maintained at about 0°C throughout winter by regulating the size of the openings. Two rows of 16

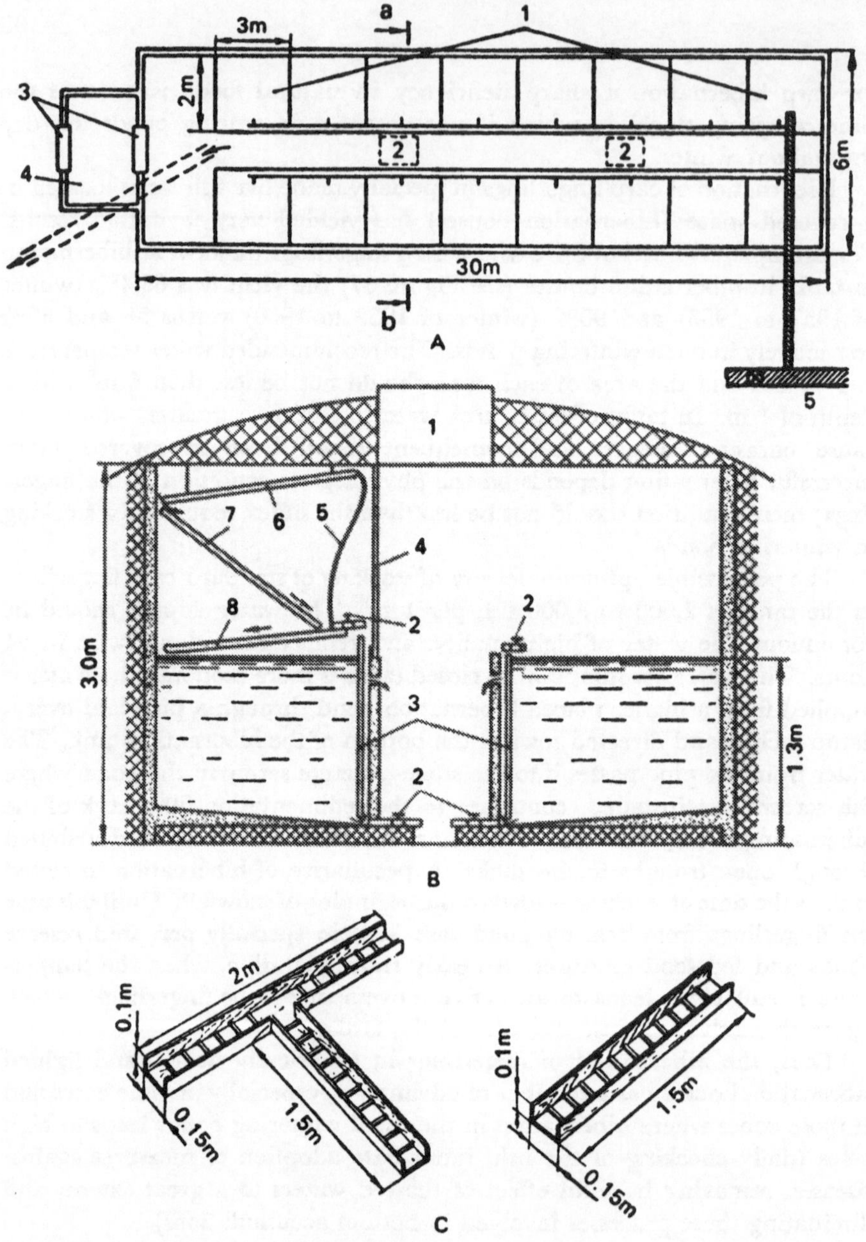

Figure 38. *A*—Hibernation house of training farm of Novosibirsk Agricultural Institute. *1*—tanks; *2*—hatches; *3*—gates; *4*—vestibule; *5*—bottom water outlet of head pond.

B—Cross section along *a* to *b*.

1—hatch; *2*—tap; *3*—water discharging mechanism; *4*—pillar; *5*—rubber pipe; *6*—first chute; *7*—second chute; *8*—T-shaped chute.

C—Aeration chutes.

Right—chutes I and II; *Left*—T-shaped chutes.

tanks are arranged under a single roof. Each tank is 1.3 m deep and the working volume 6 m^3 with a slope on one side.

The premises are not heated. Water is supplied to the tanks through pipes hooked up to the warmer lower layers of the head pond, which protects them from ice formation. The water is supplied through rubber or polyethylene tubes to the upper chute (the first inclined trough; see Figure 38 A, B), then the second inclined chute, the lower T-shaped chute, and finally reaches the tanks. Each chute is provided with a large number of planks (plates) for water aeration (Figure 38). The water after processing enters a special channel in the head pond through a pipe laid along the outlet channel.

The most effective density of stocking of fingerlings is 1,000 to 1,600 per m^2; a higher density leads to higher losses. All the fingerlings should weigh more than 15 g, their condition factor should not be less than 2.8 to 2.9, and their fat content not less than 6.0 to 8.0%. When rearing is done on natural foods, the fat content should not be less than 4.0 to 5.0%.

The hibernation of fingerlings in such places makes it possible to: (a) reduce losses among hibernating fingerlings; (b) significantly reduce the required hibernation area; (c) effectively control the process of hibernation which, according to N.N. Moiseev, makes it possible to realize not less than 90% of the total stocked fish after hibernation; and (d) transplant yearlings 15 to 20 days earlier to foraging ponds, compared to the situation in the usual wintering ponds, which augments productivity.

HIBERNATION OF CARP OF OLDER AGE GROUPS IN PONDS (two-year fish, replenishing stocks, and brood fish) is handled in special wintering ponds. The norms for stocking are 250 to 300 quintals per hectare and under favorable conditions 500 to 600 quintals per hectare. Under normal conditions the older age groups can undergo hibernation with almost no loss. The carp of these groups are distinguished by a higher fat content, giving rise to better fattiness and hence a more favorable progress during hibernation.

In scale and mirror carp two years old the adipose tissue is located peripherally, in direct contact with the outer environment, in between the covers of the most important systems of the body: the nervous system, bones, muscles, blood circulatory system, etc. (Korolev, 1954). It would appear that the disposition of the adipose layer is an important factor for withstanding low temperature levels. That is why older and fattier age groups of carp undergo hibernation with either insignificant or no loss whatsoever. In scale carp the fat is located under the scales in a continuous layer, while in mirror carp it is well developed only in those regions of the skin which are covered with scales. Many authors claim that scale carp, sazan, and their hybrids are more winter-hardy than carp with a partial scale cover.

The adipose tissue is better developed in larger-sized carp of the same age. This accounts for the greater winter-hardiness of larger specimens.

Table 8. Register for recording observations on progress in hibernation of fish in ponds——————of fish farm——————
(or progeny of fish farm)——————, collective farm——————region——————district——————

Parameters	Months and decades																	
	November			December			January			February			March			April		
	I	II	III	I	II	III	I	II	III	I	II	III	I	II	III	I	II	III
Water temperature (°C), fluctuations over decade																		
Thickness of ice layer (cm)																		
Actual depth (cm):																		
at inlet																		
at outlet																		
Water exchange:																		
at inlet per day (l)																		
at outlet per day (l)																		
Water exchange (days)																		
Oxygen content in water (ml/l):																		
in source of water supply																		
at inlet (at a depth of 0.5 m)																		
at outlet (near bottom)																		
Free carbon dioxide content in water:																		
in source of water supply																		
at inlet (at depth of 0.5 m)																		
at outlet (near bottom)																		
Presence of hydrogen sulfide (strong, poor)																		

Presence of fish at ice holes (yes or no), with what external symptoms and in what condition						
Size of fish at ice holes						
Coefficient of condition of fish at ice holes						
Diseases of fish (yes or no)						
Number of dead fish collected over decade (distribution of dead fish)						

Note: The proforma of this register is taken from "Compilation of Production Instructions for Pond Fisheries," 1949. Some changes have been incorporated by me (F.M.).

During the period of hibernation of fingerlings and fish of older age groups systematic observations are carried out in each hibernation pond and the results entered in a special register (Table 8).

The entire progress of hibernation is assessed in spring after removing the fish from the hibernation ponds.

Catching in Wintering Ponds. Transplantation of Yearlings from Wintering Ponds to Foraging Ponds. Depending on climatic conditions yearlings are caught from the wintering ponds from March to the first half of May. They are counted, the percentage of recovery estimated for different wintering ponds together with the actual percentage of yield recovery, weighed, and the average weight determined on the basis of average random samples together with the extent of weight loss over the winter period, and thereafter transferred to the foraging ponds. Transplantation is done as early in the spring as possible, sometimes before the ice has completely melted, which reduces the period of starvation of the fish and increases the period of foraging. The yearlings are caught with some kind of straining fishing gear (seines) and released in the ponds, after which the remaining fish in the collecting channels and fish pits are collected.

The preparation of the foraging ponds for stocking includes their drying and reclamation in autumn (see Chapter X). The ponds are filled with water at the time of floods up to the maximum level. If for filling the foraging ponds up to the required level floodwaters are not sufficient, they are additionally filled from the head pond. Before filling the foraging ponds the system of fish collection channels and water supply, dikes, weirs, coffer dam, water outlets, and the top sluices are put in proper working order. After completing these tasks and filling the pond with water the carp yearlings are transplanted. The preparation of foraging ponds in river beds and the preparatory reclamation and repair works which ensure a normal course of pond pisciculture (discharge capacity, eradication of predatory and trash fish, etc.) need special attention. Before stocking yearlings in foraging ponds they are passed through a prophylactic bath as described earlier. The rate of stocking is determined by the following formula:

$$A = \frac{G \times P \times 100}{(V - v) \times r},$$

where A = required level of stocking (no. of fish);

G = area of pond (ha);

P = natural fish productivity (kg/ha). This value is taken on the basis of actually obtained growth rate per kg toward autumn of the previous year;

V = ultimate unit weight of two-year fish toward autumn (see below);

v = initial unit weight of yearlings at time of stocking in spring (kg);

r = yield of two-year fish toward autumn (%).

Taking losses into account, for each quintal of marketable produce not more than 250 yearlings should be stocked. Exceeding this norm is an irrational use of the frequently deficient stocking material and increases the cost of production of marketable produce.

REARING OF TWO-YEAR FISH. In foraging ponds carp feed on benthos, mainly chironomids and lower crustaceans. The role of zooplankton in the feeding of carp in the second year of life is still important. The monthly growth rate is not the same under different climatic zones (Table 9).

The main growth of two-year carp takes place in July to August (65 to 95%). Fish growth is checked throughout the entire vegetative growth period at ten-day intervals by examining control catches. Prior to such catches (i.e., before the water is disturbed) water samples are taken for gas analysis and hydrobiochemical samples for investigating the status of natural foods. The first control catch is done 21 days after fish stocking. Not less than 1% of the stocked fish are caught with drag nets from different places in a small pond, their condition checked, the fish counted, and the average, maximum, and minimum weights recorded.[11] According to the findings of pisciculturists in the German Democratic Republic reliable data are obtained by collecting 5 to 10 fish from 5 to 10 points in the pond. The data obtained constitute the first production document and are recorded by the group leader and persons conducting the catching, weighing, and measuring of the fish. These data are recorded in a special register (see Table 6) and compared with planned indexes for a given period. If the average individual weight is less than that projected the causes are ascertained and measures taken accordingly (improvement of the hydrotechnical regime, manuring, feeding, etc.).

The carefully examined fish, in the absence of any external epizoon or symptoms of disease, are released back into the foraging pond. If disease or any other aspect in which the fish farm authorities might be interested is detected, the fish are sent to the respective laboratories for study after fixing them in 4% formalin (if, for example, the nature of its feeding is to be explored) or sent live (for investigating parasites, deposition of fat content, etc.).

Knowing the approximate monthly growth rate for one or the other zone, the total growth rate of the fish for each month of the vegetative growth period can be determined through control catches. For example, in the central belt of the European part of the USSR, if the final unit weight of two-year fish is

[11]For research and breeding work each fish is measured and the coefficient of condition determined.

Table 9. Increase in weight of two-year carp in foraging ponds in different geographic zones (%)

Months	Central belt of European part of the USSR	South western RSFSR	Krasnodar territory	Bashkirian ASSR	Ukrainian SSR (southern region)	Georgian SSR	Moldavian SSR	Tomsk district	Southern regions of western Siberia
April	—	—	1	—	4	—	1–5	—	—
May	10	12.5	4	3.7	9	8.4–15.0	5–8	—	1.5
June	25	25.0	10	25–30	16	15.5–18.5	15–20	15.6	23.0
July	30	25.0	30	35	25	26.4–31.9	25–35	22.7	40.5
August	30	25.0	40	35	32	28.0–33.4	25–40	50.0	31.0
September	5	12.5	15	2–3	14	7.2–18.2	10–12	11.7	4.0
October	—	—	—	—	—	0.2–1.7	1–8	—	—

500 g, its planned (projected) weight should be as follows (g): On June 1st—50; July 1st—275; August 1st—325; September 1st—475; and at the end of September—500.

A curve of actual growth rate (weight) of carp is drawn from the data obtained from the central catches, compared with the data of previous years, and assessed in terms of the projected growth rate (increase in weight) for the current year.

AUTUMN CATCHING FROM FORAGING PONDS. Depending on the climatic conditions of the area, some water is discharged from the ponds and the fish caught with seines mostly at the end of September—early October, when the growth rate of carp declines significantly. After this all of the water is discharged and the fish remaining in the collection channels and fish pits retrieved with dip nets.

Fish traps facilitate the catching of fish from foraging ponds (keeping the ratio between fish weight and water weight at 1:4 to 1:5). Such traps differ in construction. One, developed by Solov'ev, consists of a rectangular trap with openings on the sides for water discharge; it is placed behind the bottom water outlet on the outside of the dike. The catch from this trap should be combined with the catch taken near the inlet. At the time of water discharge from the pond the fish pass through this outlet with the water and are caught in the trap. They are collected from the latter in dip nets or even a small drag net when the trap is relatively large. The size and shape of the fish trap depend on the quantity of fish to be collected and on pond relief.

In recent years a method for partial catching (a number of times over the summer period) has been developed for catching carp in foraging ponds with a unit weight of 400 to 500 g. In the Moscow districts where the unit weight of yearlings is 70 g, marketable carp can be obtained toward the beginning of August (two months earlier than the usual period). In Moldavia, where foraging ponds are stocked with carp yearlings of different weights (from 25 to 100 g) and size, the larger specimens achieve a marketable weight earlier than the others. They can even be caught at the beginning of August. Fish which have not achieved marketable weight by the time of partial catches remain in the pond and due to the progressive reduction in numbers grow more rapidly and achieve standard weight by autumn.

Such "differential" catches make it possible to supply the people with fish in phased periods. In this connection one must re-examine the level of standard weight for fingerlings, which varies in different zones, and determine the weight of larger groups which achieve marketable weight toward the end of July—beginning of August. Their percentage in the total stock of fingerlings depends on the demand for marketable fish in the earlier periods. In Moldavia, for example, in one experiment 40% carp yearlings weighing 25 to 30 g were stocked for feeding with 40% fish weighing 50 to 60 g and 20% weighing 90 to 100 g.

One shortcoming of this method is that the opportunity for selecting the most valuable specimens for the replenishment stock is lost. Therefore, in the case of differential catches specimens should be left for a subsequent selection and replenishment on the basis of external appearance, weight, and rapid growth rate.

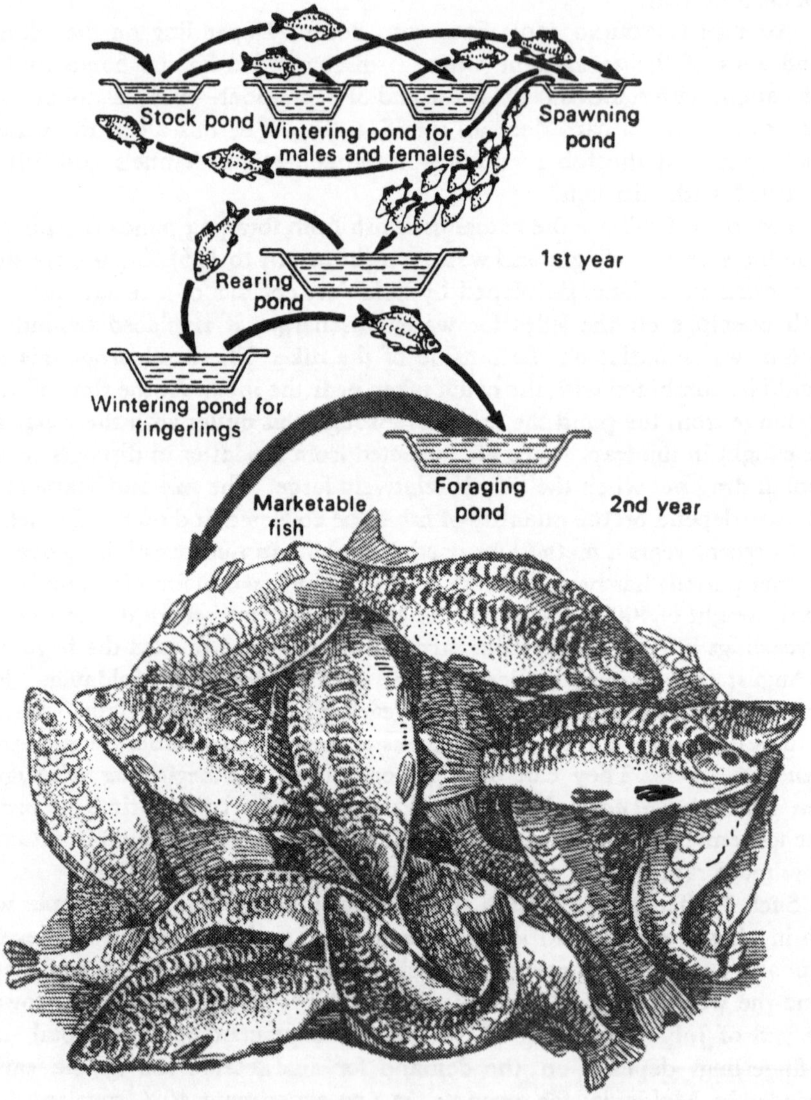

Figure 39. Scheme of production processes in a warm-water fish farm with two-year rotation.

The fish removed from the foraging ponds are weighed, their total weight increase over the vegetative growth period determined, the average unit weight established, and the actual fish productivity recorded. The output of fish over the vegetative growth period is determined as a percentage of the total number of yearlings stocked in spring. The underweights (fish not growing to marketable weight) are transplanted to the wintering ponds for use in the spring of the successive year (after returning them to the foraging ponds for further feeding). Diseased fish are discarded.

The production processes in a full-system carp farm with a two-year rotation terminate with the catching and sale of marketable fish from the forgaging ponds (Figure 39).

The output of two-year carp from completely drainable and diked foraging ponds is 90 to 95% of the total number of yearlings planted in spring under conditions of good fish management. In large fish farms it is difficult to market all the fish collected from the foraging ponds. Hence part of the catch is temporarily kept live in earthen, cement, or wooden enclosures or tanks located below the head pond, close to the foraging ponds. Earthen tanks are preferred since the fish injure themselves against the sides and bottom of wooden or cement enclosures or tanks. Their areas vary from 100 to 1,000 m^2. The enclosures are rectangular in shape (ratio between width and length 1:3 to 1:4) with an average depth of 2.0 m, depth of unfrozen layer of water 1.0 to 1.5 m, air opening 0.3 m, thickness of ice cover 0.2 m, and ratio of fish by weight and water 1:15 and 1:10. The recommended norms for stocking of carp for winter storage are 75 to 125 kg per 1 m^3 of water, depending on the capacity and quality of the water source. The quantity of water required for the tanks for each quintal of fish at a water temperature of 5°C is 0.05 l/sec. The permissible minimum oxygen content at the outlet should be 3 mg/l.

The maintenance of fish in fish enclosures or tanks includes observations on the hydrochemical conditions of the water, removal of weeds and waste material, and removal of dead fish.

References

Albrecht, M.L. 1966. Winterrhue und kohlehydratstoffwechsel des karpfens (*C. carpio*), *Dtsch. Fisch. Ztg.*, vol. 13, no. 4.

Avdeeva, A.T. and Yu.S. Dmitrienko. 1968. Osobennosti neresta karpa v Karelii (Peculiarities of spawning of carp in Karelia). *Materialy XIV Konf. po Izucheniyu Vnutr. Vodoemov Pribaltiki.* Riga.

Bigiev, L.L. and T.A. Yakovchuk. 1967. Povyshenie kormovoi bazy nerestovykh i lichinochnykh prudov putem vneseniya produktov kislotnogo gidroliza restitel'nykh tkanei (Improving the food base in the hatching and larval ponds by adding the products of acid hydrolysis of plant tissues). *Sbornik Nauchno-tekhn. Inform. VNIRO,* no. 9.

Bogatova, I.B. 1969. Pitanie segoletkov karpa v prudakh i pishchevoe znachenie kolovratok (Feeding of carp fingerlings in ponds and the food value of rotifers). *Trudy VNIIPRKh.* Moscow, vol. 16.

Bogoslovskii, A.S. 1955. Kolovratka mozhet byt prichinoi gibeli lichinok karpa i drugikh ryb (The rotifer as a likely cause of mortality for carp larvae and other fish). *Rybn. Khozyaistvo,* no. 2.

Brizinova, P.N. 1958. Izmenenie zhirnosti v ontogenze ryb (Changes in fattiness during ontogeny of fish). *Trudy Soveshch. Ikhtiol. Komissii AN SSSR,* no. 8.

Brizinova, P.N. and V.S. Kirpichnikov. 1952. K probleme povysheniya zimostoikosti segeletkov karpa, Amurskogo sazana i ikh gibridov (On the problem of increasing winter-hardiness of fingerlings of carp, Amur sazan, and their hybrids). *Zool. Zhurn.,* vol. 31, no. 6.

Chaplina, A.M. 1948. K voprosu o pitanii mal'kov sazana i zerkal'nogo karpa (On the problem of feeding young sazan and mirror carp). In *Vestnik. In-ta Gidrobiologii Dnepropetrovsk. Un-ta.*

Cherfas, B.I. and K.N. Budnikov. 1932. K voprosu o vesovom standarte posadochnogo materiala v karpovykh prudakh (On the problem of standard body weight of stocking fish material in carp ponds). *Rybn. Khoz-vo SSSR,* nos. 8-9.

Deeva, T.A. 1960. Vyrashchivanie segoletkov v kholodnoe leto (Raising fingerlings in a cold summer). *Rybovodstvo i Rybolovstvo,* no. 3.

Emel'yanov, S., P. Konovalov et al. 1969. Inkubatsii ikry v prikleennom sostoyanii (Incubation of adhesive fish eggs). *Rybovodstvo i Rybolovstvo,* no. 3.

Fedoseeva, E.N. 1965. Izuchenie porodnykh osobennostei cheshuichatykh i zerkal'nykh karpov pri razdel'nom i sovmestnom vyrashchivanii (Study of the breed characteristics of scale and mirror carps in isolated and mixed raising conditions). Avtoref. Kand. Diss.

Gamayun, E.P. 1968. Vozrast proizvoditelei i zimostoikost' segoletkov (Age of spawners and winter-hardiness of fingerlings). *Rybovodstvo i Rybolovstvo,* no. 1.

Georgiev, N. 1965. Ednoletno otglezhdane na pazaren, sharan, *Rybno Stopanstvo,* no. 1, no. 12.

Grimal'skii, V.L. 1961. K voprosu o ritmichnosti pitaniya molodi karpa (On the problem of the rhythmic nature of feeding of carp fry). *Trudy Kishinevsk S.-kh. In-ta im. M. V. Frunze,* vol. 25.

Grimal'skii, V.L. and S.N. Tyutyuhik. 1961. Rol'kislorodnogo faktora v protesesse vyrashchivaniya tovarykh segoletkov karpa (Role of the oxygen factor in the process of raising marketable under-yearlings). *Trudy Kishinevsk. S.-kh. In-ta im. M. V. Frunze,* vol. 25.

Il'in, V.M. 1970. Vyrashchivanie posadochnogo materiala. V pomoshch rybovodu (Raising of Stocking Fish. Guidelines for Fish Growers). VNIIPRKh.

Instruktsiya po polucheniyu lichinok karpa i sazana zavodskim metadom (Instructions on Obtaining Carp and Sazan Larvae by the Commercial Method). Ministry of Fisheries of the USSR, 1969.

Iozepson, U.P., T.P. Glagoleva and R.F. Platnietse. 1968. Vliyanie mikroe'lementov na fiziologicheskoe sostoyanie godovkikov karpa v period zimovki (Influence of trace elements on the physiological condition of carp yearlings during hibernation). *Sbornik Nauchn.-tekhn. Inform. VNIRO.* Moscow, no. 7.

Irikhimovich, A.I. and A.A. Gerbovskii. 1959. Proizvodstvennoe znachenie posadki rasnorazmernykh karpov-godovnikov v nagul'nye prudy (Commercial significance of stocking different sized carp yearlings in foraging ponds). *Rybn. Khoz-vo,* no. 12.

Irikhimovich, A.I. and S. N. Tyutyunik. 1967. Karpovy konveyer (The carp conveyor). *Rybovodstvo i Rybolovstvo,* no. 4.

Irikhimovich, A.I. and N. Navrodskii. 1969. Bunker-sortirovshchik (Grading ponds). *Rybovodstvo i Rybolovstvo,* no. 6.

Jarašek, I. 1966. Specialni komorove rybniku se osvědčily, *Cs. Rybarstvi,* no. 9, p. 132.

Kanaev, A.I. 1962. Mechenie tatuirovkoi (Tagging by singeing). *Rybovodstvo i Rybolovstvo,* no. 2.

Kirpichnikov, V.S. 1963. Znachenie estestvennogo i iskusstvennogo korma pri vyrashchivanii karpov-segoletkov (Significance of natural foods and artificial feed during raising of carp fingerlings). *Sbornik Gidrobiol. i Ikhtiol. Vnutr. Vodoemov Pribaltiki.* Riga.

Kirpichnikov, V.S., N.A. Mosevich, V.I. Protasova and O.N. Bauer. 1960. Optimalnyi rezhim zimoval'nykh prudov (Optimum conditions of wintering ponds). *Trudy Vsesoyuzn. Gidrobiol. Ob-va,* vol. 10.

Konradt, A.G. and A.M. Sakharov. 1966. Zavodskyi sposob polucheniya bichinok promyslovykh ryb (The commercial method of obtaining larvae of economically valuable fish). *Izv. VNIORKh.* vol. 61.

Konstantinov, A.S. 1954. Opyt massovogo razvedeniya *Chironomus dorsalis* i dal'neishee issledovanie ego biologii (Experience in mass-scale rearing of *Chironomus dorsalis* and further studies on its biology). *Trudy Saratov. Otd. Kasp. Filiala VNIRO,* vol. 3.

Korneeva, L.A. et al. 1969. Rost segoletkov karpa razlichnykh genotipov pri vyrashchivanii v sadkakh na teplykh vodakh (Growth of carp fingerlings of different genotypes during raising in enclosures in warm waters). Sbornik *Rybovodstvo v Teplykh Vodakh SSSR i za Rubezhom.* Moscow.

Korolev, N.F. 1954. Topografiya otlozhenii zhirovoi tkani y prudovykh karpov i nekotorye voprosy ikh zimoustoichivosti (Deposition contours of the adipose tissue in pond carp and some problems of winter-hardiness). *Uch. Zap. Vitebsk. Vet. In-ta,* vol. 13.

Korovin, V. 1968. Prisposoblenie dlya tavrirovaniya ryb (An instrument for marking fish). *Rybovodstvo i Rybolovstvo,* no. 5.

Korovina, V.M. 1960. Vliyanie termicheskogo rezhima v period sozerevaniya ryb na stoikost ikry k povyshennoi temperature (Effect of thermal regime on the period of fish maturation on the resistance of eggs to high temperatures). *Nauchn-tekhn. Bull. GOSNIORKh,* no. 10.

Krapin, V.M. 1955. Ekologicheskie osobennosti rosta i vospitaniya molodi karpa (Ecological peculiarities of the growth and maintenance of carp fry). *Sbornik Nauchn. Rabot. Ukr. Stantsii Rybovodstvo,* no. 1.

Krayukhin, B.V. 1955. Vliyanie nizkikh temperatur pishchevuyu reaktsiyu segoletkov karpa (Effect of low temperatures on the food intake of carp fingerlings). *Vopr. Ikhtiologii,* no. 5.

Lillelund, K. 1967. Experimentalle untersuchungen über den einfluss cornivorer cyclopiden auf die sterblichkeit der fischbrut, *Z. Fisch.,* vol. 15, nos. 1-2.

Mamontova, L.N. 1960. Artezianskaya voda dlya zimoval'nykh prudov (Artesian water for wintering ponds). *Rybovodstvo i Rybolovstvo,* no. 6.

Merla, G. 1964. Beodachtungen über die vorher zur K_2-oberwinterungdienten, *Dtsch. Fisch. Ztg.,* vol. 11, no. 2.

Mikheev, P.V. 1949. Zinnovanie karpa v plavuchikh sadkakh (Hibernation of carp in floating enclosures). *Rybn. Khoz-vo,* no. 11.

Nikitin, P.P. 1961. Zimovanie segoletok karpa v basseinakh zimoval'nykh domikov (Hibernation of carp fingerlings in tanks of hibernation houses). *Sbornik Rybn. Khoz-vo Vnutr. Vodoemov Latv. SSR.* Izd. Akad. Nauk Latv. SSR, vol. 6.

Nikitin, P.P. and A.P. Volkova. 1961. O srokakh neresta karpa v prudovykh khozyaistvakh Latv. SSR (Spawning periods of carp in fish farms of Latvian SSR). Sbornik *Rybn. Khoz-vo Vnutr. Vodoemov LatvSSR.* Izd. Akad. Nauk LatvSSR, vol. 6.

Pasovskii, A.I. 1953. Ammiak kak sredstvo bor'by s daktilogirozom karpa (Ammonia as a means to control dactylogyrosis in carp). *Trudy Ukr. NII Ozern-Rechn. Rybn. Khoz-vo,* no. 9.

Persov, G.M. 1953. Dozirovania spermy i izberatel'nost v protsesse oplodotvoreniya u ryb (Spermatozoa quantity and selectivity in the process of fish fertilization). *Rybn. Khoz-vo,* no. 12.

Polyakov, G.D. 1964. Biologicheskie osnovy zimovki karpa (Biological basis of hibernation in carp). Sbornik *Dokl. na Seminare Prepodavatelei S.-kh. Vuzov po Prud. Rybovodstvu.* Timiryazev. Agric. Acad.

Polyakov, Yu. and K. Sadlaev. 1960. Raschet zagruzki zhivorybnykh sadkov (An estimation of loading of live fish cribs). *Rybovodstvo i Rybolovstvo,* no. 4.

Popova, G.A. 1956. Ammiachnye vanny kak metod bor'by s ektoparazitami mal'kov karpa (Ammonia bath as method of controlling ectoparasites of carp fry). *Rybnaya Promyshlennost,* pt. 2.

Putina, E.P. et al. 1970. Vliyanie drobnykh in "ektsii suspenzii gipofiza na sozrevanie polovykh produktov sudaka (Effect of portioned injections of pituitary extract on the maturation of sex products of sander). *Vopr. Ikhtiol.,* 10, 6, 65.

Samoilov, M.D. 1968. Uluchshenie rybovodnykh kachestv nerestovykh prudov putem regulirovaniya vysshei rastitel'nosti (Improvement in the fish-rearing qualities of carp hatching ponds through regulation of higher vegetation). Avtoref. Kand. Diss., Kiev.

Satsunkevich, I.L. 1965. Peredovoi opyt prudovykh khozyaistv Belorussii (Advanced experience in fish farming in Belorussia). Sbornik *Rybn. Khoz-vo.* Kiev, no. 1.

Sebentsov, B.M. and T.I. Imerligishvili. 1940. Normy posadki karpa v zimoval'niki v zavisimosti ot rezhima vody (Stocking norms for carp in the hibernation house depending on the water regime). *Trudy Voronezh. Otd. VNIIPRKh.* Moscow, vol. 3, no. 4.

Shikidze, A.L. 1969. Osobennosti razmnozheniya i sozrevaniya karpa v usloviyakh Uzbekistana (Peculiarities of reproduction and maturation of carp under the conditions of Uzbekistan). Sbornik *Ekologiya i Biologiya Zhivotnykh Uzbekistana.* Tashkent.

Shiryaev, A.V. 1966. Osobennosti abioticheskikh uslovii sredy, skladyvayushchiesya v prudakh razlichnykh klimaticheskikh zon, v svyazi s odnoletnei kul'tury karp (Peculiarities of the abiotic conditions developing in ponds in different climatic zones, in relation to one-year carp culture). *Trudy VNIIPRKh.* Moscow, vol. 14.

Sigov, V.A. 1940. K materialam po ekologii zimovaniya karpovsegoletkhov (Data relating to the ecology of hibernation of the current year's carp brood). *Trudy Voronezh. Otd. VNIIPRKh.* Moscow, vol. 3, no. 1.

Sigov, V.A. 1946. Biometricheskie kriterii zhiznesposobnost' pervozimuyushchikh karpov (Biometric criteria of viability of carp hibernating for the first time). *Trudy VNIIPRKh.* Moscow, vol. 4.

Solov'ev, T.T. 1961. Pribor dlya acheta mal'kov (An apparatus for counting fry). *Rybovodstvo i Rybolovstvo,* no. 3.

Spanovskaya, V.D. and V.A. Grigorash. 1961. Sutochny ritm pitaniya nekotorykh karpovykh ryb (Daily rhythm of feeding of certain types of carp). *Vopr. Ikhtiol.,* 1, 2, 19.

Spravochnik po prudovomu rybovodstvu (A Handbook of pond Fisheries). Pishchepromizidat, 1959.

Surnova, B.D. 1968. Dinamika nakopleniya zhira u molodi karpa (Dynamics of fat deposition in carp fry). Sbornik *Puti Uvelicheniya Produktivnosti Zhivotnovodstva v Zap. Sibiri.* Novosibirsk.

Tatarko, K.I. 1965. Vliyanie temperatury na embrional'nov razvitie prudovogo karpa (Influence of temperature on the embryonic development of pond carp). *Gidrobiol. Zhurn.*, vol. 1, no. 1.

Tatarko, K.I. 1966. Vliyanie temperatury na rannye e'tapy poste'mbrional'nogo razvitiya prudovoga karpa (Effect of temperature on the early stages of postembryonic development of pond carp). *Gidrobiol. Zhurn.*, vol. 2, no. 3.

Tolmachev, T.T., V.A. Musselius and V.I. Laptev. 1969. Profilakticheskaya obrabotka ryby v zimoval'nykh prudakh rybokambinata "Para" (Prophylactic treatment of fish in the wintering ponds of the fish farm "Para"). *Rybn. Khoz-vo,* no. 8.

Verigin, B. 1966. Iz opyta yaponskikh rybovodov (The experience of Japanese pisciculturists). *Rybovodstvo i Rybolostvo,* no. 2.

Vinogradov, V.K. and L.V. Erokhina. 1966. Vliyanie kobal'ta na zimostoikost' segoletkov karpa (Effect of cobalt on the winter-hardiness of carp fingerlings). *Trudy VNIIPRKh.* Moscow, vol. 14.

Vladimirov, V.I. and K.I. Semenov. 1959. Kriticheskii period v razvitii lichinok ryb (Critical period in the development of fish larvae). *Doklady Akad. Nauk SSSR,* vol. 126, no. 3.

Woynarovich, E. 1962. Inkubatsiya ikry karpa i podrashchivanie desyatidnevnykh lichinok (Incubation of carp eggs and rearing of ten-day-old larvae). *Rybn. Khozyaistvo,* no. 9.

Wolny, P. 1963. Czynniki warunkujace Prawidlowe zimowanie karpi, *Gosp. Rubna,* no. 9.

Zagoskii, N. 1965. SZhK—e'fekten stimulyator v razv "zhdaneto. Edno zavoevanie na naukata p"t, *Ribno Stopanstvo,* vol. 12, no. 3.

Zlokazov, V.N. 1967. Razmnozhenie i zimovka karpa v usloviyakh Novosibirskoi i Kemerovskoi obl. (Reproduction and hibernation of carp under the conditions of Novosibirsk and Kemerovsk districts). Sbornik *Vopr. S.-kh. Rybovodstva i Gidrobiologii Zapadnoi Sibiri.* Barnaul.

Chapter V

PARTIAL SYSTEM: ONE-YEAR PRODUCTION FISH FARMS. USE OF LAGOONS (ESTUARIES) AND BAYS FOR ONE-YEAR PRODUCTION

Partial System: One-Year Production Fish Farms. In addition to full-system fish farms which produce not only marketable fish, but also seed fish or stocking material, as well as their own replenishment stocks and brood fish, one-year fish farms are also quite common. Such farms have only one category of ponds—foraging (production)—and receive stocking material each year from fish hatcheries or full-system farms to raise marketable fish in the course of just one season. The fish are sold in autumn, the necessary reclamation operations done, and the foraging ponds prepared for a similar use in the spring of the following year. The production processes in this case are similar to those described earlier for foraging ponds in fish farms with a two-year rotation scheme (Chapter IV). For one-year feeding it is better to use fully drainable ponds provided with the necessary outlets. If nondrainable ponds are used, they must be properly prepared before stocking fish in them. The release of fish in ponds which are unprepared, neglected, muddy, overgrown with emergent aquatic plants, or filled with large quantities of trash and predatory fish, especially the nondrainable type, is economically foolhardy. Fish cultured in such ponds cannot attain normal conditions of growth and are victims of predators; the survival rate in autumn will rarely exceed 25% of the fish stocked.

Preparations of nondrainable ponds for fish stocking include: (a) checking the water quality by chemical analysis for suitability for raising the particular fish species; (b) eradication of trash and predatory fish since the former compete for food and the latter prey on the fish being cultured; (c) clearing the ponds of stumps, roots, and various other kinds of plant residue; (d) removal of coarse emergent vegetation and excessive submerged soft vegetation; and (e) cleaning the pond to remove excessive silt deposits.

If these conditions are observed it is possible to obtain marketable fish under favorable autumn conditions at the rate of 75 to 80% of the stock.

For the development and normal operation of partial one-year production farms a widespread network of fish hatcheries is necessary, the capacity of which must correspond to the requirements for stocking material with the accepted degree of intensification. Fish hatcheries should be created in each region, territory, and autonomous Republic; they should be geographically arranged in such a way as to be convenient to farms culturing fish of one summer. Small and floodland lakes can be used for one-year foraging of pond fish.[1]

USE OF ESTUARIES AND BAYS FOR ONE-YEAR PRODUCTION. Lagoons, or limans as they are called around the Black Sea, or ilmeni (Caspian basin) are shallow water bodies connected with a river (or sea) through tides. These water resources have served for culturing valuable species of fish since ancient times. Lagoon (liman or ilmeni) fisheries are possible in creeks and are numerous in the USSR, especially in the regions of the southern seas. Their area, according to the incomplete statistics available, covers about 300,000 hectares. In Krasnodar lagoon fish farms already comprise an area of about 5,000 hectares, which could reasonably be increased to 60,000 hectares. Individual lagoons in Krasnodar range from 46 to 740 hectares, with larger ones reaching more than 6,000 hectares. The biomass of phytoplankton and zooplankton in them is, on the average, 0.5 g/m^3, and that of benthos 0.1 to 8 to 10 g/m^2; the quantity of both increases five to ten times with manuring. Experiments conducted by KRASNIRKh (Krasnodar Fisheries Research Institute) have shown that better results are obtained by the addition of 5 to 7 q/ha of phosphate and nitrogen fertilizers during the vegetative growth period. Common carp and phytophagous fishes (silver carp, bighead, and grass carp) have proved the most profitable in the lagoon fisheries of Krasnodar. Experience in Ukrainian SSR has shown that mullet raised in estuaries with stocking norms of about 500 fish per hectare yield a produce of 2 q/ha. Other species of fish can also be raised in estuaries.

The most economic rate of stocking of common carp is 1,500 to 2,000 fish per hectare. In this case the unit weight of the fish should be 450 to 500 g. A better norm of stocking of silver carp is 1,000 to 1,200 fish per hectare, for bighead 500 to 600 fish per hectare (with an average survival of 80% and total fish productivity 4.9 q/ha), and for grass carp 200 to 300 fish per hectare. The rearing of grass carp ensures not only a harvest of one quintal of produce, but also a more complete suppression of submerged aquatic vegetation, which favorably affects the feeding of common carp and silver carp. The total fish productivity from the lagoons of the Krasnodar territory could be increased to 10 to 15 q/ha on the average, making it possible to increase the quantity of fish culled currently from the Azov Sea. A further development of lagoon fishery, based on the principles of pond fisheries, would make it possible to understand a number of problems associated with a more rational management of farms. To be specific, methods ought to be developed for controlling small trash fish which enter the lagoons in large quantities (40,000 to 50,000

[1]This question is treated in the university course entitled "Fisheries in Lakes and Reservoirs."

per hectare). To prevent such entries before filling with water, protective screens should be placed slightly above the lagoon at a distance of 10 to 15 m from each other (Figure 40).

Shallow water bays of reservoirs are increasingly being used for the production of culturable fish. Along the Kuibyshev reservoir alone the

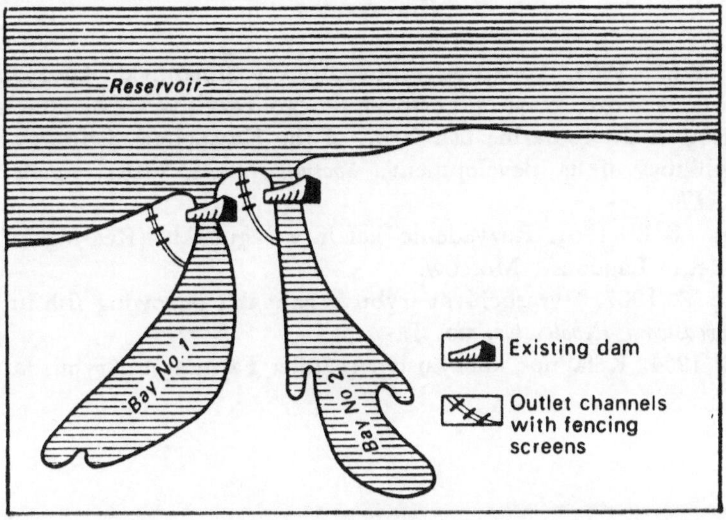

Figure 40. Bays of shallow waters along the Kuibyshev reservoir (from V.I. Belyaev).

number of such shallow water bays constitutes more than 30,000 hectares. These shallow waters are quite beneficial as foraging areas for fry recruited from the reservoir itself. In selecting each individual shallow water bay for foraging of valuable fish, problems related to its utility should be examined. In the first place only bays in which the depth does not exceed 3 m and with gently sloping banks should be selected. The area of such a bay should preferably vary from 30 to 100 hectares. In May-June, when the water filled in the reservoirs reaches the projected mark, the bays are also filled. In early spring, when the bays are still dry, a dam is built at the narrowest place at a distance up to 50 m from the stream part and an outlet canal constructed. Thus, an artificial water body is created which can be used for fish foraging. The methods of catching fish in such bays are similar to those used in foraging ponds at fish farms.

Water is pumped out and the fish collected. To ensure proper conditions for catching, the bottom is priorly cleared of trees, bushes, stumps, roots, and weeds, planed somewhat, the deeper parts filled with earth, and so on. It is better to develop fish culture in the zone of shallow waters in the reservoir as a full-system farm.

A large commercial fish farm has been developed at Suskarnskii Bay, along the Kuibyshev reservoir, with an area of 7,310 hectares (foraging area 6,060 hectares). The capacity of this farm for producing marketable fish is 88,000 quintals and 25.5 million fingerlings of common carp and phytophagous fish [*sic*].

References

Abaev, Yu.I. 1969. Tovarnoe rybovodstvo v limannykh khozyaistvakh Krasnodarskogo kraya i perspektivy ego razvitiya (Commercial fish rearing in the estuarine fish farms of the Krasnodar territory and the possibilities of its development). *Sbornik Nauchn.-Tekhn. Inform. KrasNIIRKh.*

Babayan, K.E. 1961. Razvedenie kefali v lagunakh (Rearing of Gray Mullet in Lagoons). Moscow.

Belyaev, V. 1967. Vyrashchivat' rybu v zalivakh (Growing fish in bays). *Rybovodstvo i Rybolovstvo,* no. 1.

Il'in, B. 1954. Kefal'noe khozyaistvo (Mullet Farming). Krymizdat.

Chapter VI

PRODUCTION PROCESSES IN FULL-SYSTEM AND PARTIAL-SYSTEM CARP FARMS WITH THREE-YEAR ROTATION

In the past fish farms were run on a five- to six-year rotation, which made the cost of production extremely high. Advancements in the biological sciences during the nineteenth century made it possible to accelerate rotation in carp farms to three years initially, and later to two years. In the USSR two-year rotation of the carp farm is followed, but in several Western European countries three-year rotation is preferred. It is often advantageous in a predominantly two-year rotation scheme to follow a three-year rotation in carp raising as well. This is especially true in the north (regions of 58° N and slightly above) where the vegetative growth period is much shorter. Secondly, in the process of raising two-year fish some do not attain the standard weight of 450 to 500 g. Such fish may be marketed at the body weight of 200 to 300 g, but it is better if they can be left in the ponds and in the spring of the following year stocked for further growth in the foraging ponds. The ultimate objective of three-year rotation is raising larger marketable fish with a weight of not less than 1 kg, but preferably 1.2 to 1.5 kg or even more than 2 kg. At the same time carp weighing 200 to 300 g serve as good stock for nondrainable ponds and lakes, where pike are cultivated to check the multiplication of trash fish. Carp yearlings should not be released in such ponds since they are destroyed by pike. Such ponds are better planted with underdeveloped two-year carp, which have not attained standard weight and are discarded at the time of catches from production ponds. Pike find it difficult to deal with carp in such ponds. Toward the end of the third year of raising, the three-year carp has attained a higher marketable weight. Thus, three-year rotation is useful in those cases where, due to climatic and other conditions, two-year rotation is not economical. Fish larger than two-year carp are readily marketable.

Compared to two-year rotation, three-year rotation has certain advantages and disadvantages. The advantages of three-year rotation are that in the third year of life carp yield a greater growth rate, become larger and more tasty, and only half the quantity of stock material (yearlings) is required. The disadvantages of three-year rotation are: (a) the period of rotation is longer and the production process more complicated; (b) the fish undergo hibernation twice, as a result of which losses increase; (c) the total harvest is slightly less; (d) the capital investment per unit of produce is more; and (e) three-year carp are more susceptible to the disease rubella (German measles) than under-yearlings and two-year fish.

On the other hand, the food and marketable properties of larger fish are better; their meat contains more fat and the moisture content is slightly less. Small fish contain relatively more inedible parts. The relationship between fat and water content in carp, depending on the age and weight of the fish, is presented in Table 10. The characteristics of the edible and inedible parts of carp of different weights are given in Table 11.

Table 10. Fat and moisture content in body of carp depending on age and weight

Age	Average unit weight, g	Percentage of Fat	Percentage of Moisture	Source
2 years	350	4.95	75.93	A.N. Eleonskii and N.F. Kuznetsov
2 years	740	9.30	72.30	
3 years	1,420	10.60	71.00	
3 years	1,850	11.40	70.50	E. Nauman
4 years	2,025	12.00	68.90	
4 years	2,175	13.30	67.20	

Table 11. Characteristics of edible and inedible parts of carp of different ages

Age/type of fish	Weight carp, g	Meat content, %	Viscera, %	Head, %	Fins, %	Bones, %	Scales, %
2-year scale carp	208.0	42.20	12.50	27.10	—	—	—
Same	320.0	44.60	17.50	18.80	4.90	9.0	5.20
Same	368.0	46.00	17.50	17.50	4.90	11.40	3.90
Same	442.0	50.20	18.90	18.50	4.50	2.70	5.20
Same	500.0	55.74	10.80	18.99	4.80	3.94	5.73
3-year scale carp*	1,218.0	53.67	15.17	17.54	4.11	4.51	5.00
Mirror carp with scattered scales	1,403.0	54.00	17.54	17.89	3.99	4.79	1.79
Leather carp	1,270.0	57.00	15.20	18.82	4.27	4.57	0.14

*Data for three-year carp raised at the fish farm "Bubyai" kindly supplied by A.P. Pechyukenas (Lithuanian SSR).

The three-year rotation scheme can be implemented in a complete carp farm with a three-year rotation plan and in a partial carp farm with a two-year rotation scheme.

The organization of a full-system fish farm with a three-year rotation differs from that of a fish farm with a two-year rotation by a larger number of individual categories of fish and a different percentage ratio of ponds. The additional categories of ponds are rearing ponds of the second order, which do not differ from the foraging ponds of fish farms with a two-year rotation. In the three-year rotation two-year carp are still not marketable and serve only as stock to be released into foraging ponds. These ponds are deeper since carp in the third year of life change over to feeding mostly on benthic fauna in deeper sections of the ponds. Furthermore, because two-year fish must hibernate in winter, farms with three-year rotation require wintering ponds with a larger area.

The ratio of ponds of different categories changes in farms of three-year rotation compared to two-year rotation in the following manner (%):

Spawning/hatching ponds	0.25–0.50
Transplantation (nursery) ponds	2.75
Rearing ponds of first order	10
Rearing ponds of second order	20–25
Foraging ponds	60–65
Wintering ponds	3
Fish tanks	1

With such a ratio of ponds of different categories the area actually producing marketable fish accounts for 60 to 65% of the total area of the farm versus 93.8% in the case of two-year rotation.

In the German Democratic Republic the following percentage ratio of ponds of different categories is currently being followed with three-year rotation:

Hatching ponds and ponds for fry (transplantation ponds)	2
Rearing ponds of first order	17
Rearing ponds of second order	25
Wintering ponds	5
Foraging ponds	51

Obviously, this percentage ratio, like that of the two-year rotation, is approximate and depends on the objective of the farm.

A partial fish farm, raising only three-year carp, comprises the following categories of ponds: rearing ponds of the second order, wintering ponds, and foraging ponds. Such farms do not maintain a stock of brood fish since spawning, nursery, and rearing ponds of the first order are absent.

Finally, there is one more type of partial farm raising three-year carp—the partial one-year production farm with just one category of pond, namely, foraging. The weight standards of different groups also change (Table 12).

Compared to two-year rotation, the weight indexes of fingerlings and

Table 12. Weights of different age groups under three-year rotation in fish farming

Young fry before stocking in nursery ponds		Fingerlings (in autumn)		Weight of 2-year fish, g	Weight of 3-year fish, g
Weight, g	Length, cm	Weight, g	Length, cm		
0.05–0.08	0.3–0.4	15	6–9	250	1,000
0.10–0.12	0.5–0.7	30	9–12	375	1,375
0.12–0.15	0.9–1.2	45	12–15	500	1,750

two-year fish significantly change in three-year rotation. In all other respects the methods for raising marketable fish and the production processes are similar in the two schemes. The yield of marketable fish from foraging ponds with three-year rotation is generally 100% or, in extreme cases, 97% of the stocked material.

In the German Democratic Republic marketable fish are raised in three-year rotation by using different weight groups of stock material (two-year fish) for this purpose. This ensures a constant supply of fish to the people through the summer period, starting from June to the end of September.

The required feeding and temperature conditions for this type of fish farming are given in Table 13.

Table 13. Changes in weight of two-year fish over vegetative growth period

Unit weight at time of stocking, g	Unit weight at end of the month, g			
	June	July	August	September
2-year fish with weight of:				
150	300	600	900–1,100	1,100–1,200
300	600–700	900–1,000	1,200	—
400	800–900	1,200	—	—
600	1,200	—	—	—

PART II

MIXED STOCKING, PLANTING OF ADDITIONAL FISHES, AND POLYCULTURE AT CARP FARMS

Fish belonging to different species and age groups feed on different kinds of natural foods. In spite of the variability of plant and animal organisms, what is produced in a pond cannot be utilized by any one particular fish species under culture. Hence, to increase the utilization of cheap natural foods, mixed stocking and planting of additional fishes are practiced in pond fisheries.

Chapter VII

MIXED STOCKING

Mixed stocking means the release of fish of one particular species or breed but different ages in the same pond. This type of stocking is meaningful since fish of the same species but different ages feed on different natural foods and, *in toto,* utilize more fully the natural food resources of the pond. For carp fry lower crustaceans, plankton, and organisms in vegetation thickets play a major role, and only in the second half and toward the end of the vegetative growth period are small larvae of Chironomidae, Ephemeridae, etc. consumed. In the second year of life the larvae of insects and larger fauna in general acquire a greater significance as food for carp, but lower crustaceans continue to be eaten. In the two-year rotation carp farms mixed stocking is done only in foraging ponds with large areas under shallow water, where juvenile fish find better conditions for feeding. In foraging ponds made unsuitable because of disease, mixed stocking is not permissible. In the case of mixed stocking under two-year rotation, which is widely practiced in the USSR, the main material for stocking comprises yearlings; juvenile fish are added in the foraging ponds. In the central belt of the European part of the USSR yearlings are stocked early in the spring (at the end of April—beginning of May) and juvenile fish added in June. According to the experiments of G.I. Shpet it is better to select young fish with a unit weight not less than 0.5 g for mixed stocking, since at the time of plantation of young fish the unit weight of yearlings will increase to 50 to 100 g.

It is apparent from the foregoing that great attention should be paid to the weight of juvenile and two-year fish at the time of mixed stocking.

Estimates of mixed stocking of carp of different age groups depend on the biological peculiarities of the foraging ponds and on the nature of feeding of the given age groups in different periods of time of the vegetative growth period. Therefore, in the estimates of mixed stocking it must be kept in mind that toward the end and even toward the second half of the vegetative period fingerlings of carp gradually change over to feeding on benthos, i.e., to a certain extent they compete with higher age groups which mainly feed on benthic fauna.

Mixed stocking is done only in those foraging ponds which contain very dense populations of natural foods. In experiments conducted on small fish farms at the Timiryazev Agricultural Academy, with ten repeated stockings of carp yearlings the average unit weight of the fingerlings exceeded 40 g toward autumn. The rate of stocking of fry was 1:10 to 14 (for one yearling 10 to 14 fry were stocked) compared to the normal stocking of yearlings in the foraging ponds. Practical experience has shown that with a high degree of intensity of stocking of carp yearlings the stocking of juvenile fish, if they are not fed additionally, should be calculated on the basis of normal stocking of yearlings in the same foraging pond. Smolian has recommended norms for mixed stocking by dividing the natural fish productivity with the larger (3/4) and smaller (1/4) fractions. The larger fraction refers to the main stocking material (yearlings) and the smaller fraction the additionally stocked juvenile fish. Part of the natural fish productivity, on the basis of which stocking of young age groups is determined, is increased by 10% since younger fish utilize food better. The stocking of each age group is estimated in accordance with the stocking formula for the rearing or foraging ponds.

Mixed stocking can be practiced for increasing the quantity of stocking material (carp yearlings), which is often in short supply. Such stocking is possible if there are foraging, nursery or hatching, or even foraging ponds in the collective or state fish farm. In the first case the foraging ponds are used simultaneously as rearing ponds. Yearlings of carp are raised in them in mixed stocking up to marketable weight and fry until they attain the standard weight of fingerlings. In the second case fry which are purchased every year from the state or collective fish hatcheries are also raised in the foraging ponds together with the yearlings. In both cases provisions should be made for winter maintenance of the fingerlings.

In autumn, at the time of catching, two-year fish are selected and marketed and fingerlings stocked in the wintering ponds. V.P. Shorkov has designed a unique fish grading trap (Figure 41), which eliminates injury to the fish. The construction is based on the tendency of the fish to move against the current. Fish of different age groups (fingerlings and two-year carp) move together with the water from the pond into Section I of the fish trap through the bottom outlet. The water passes into the discharge channel from this section through the water-discharging equipment and the fish remain behind the screen. After sufficient concentration of fish the flow of water from the head part of the fish trap is increased and under the impact of the current the fish from Section I start moving upward against the current toward the directing (grading) screen. Fingerlings easily pass through the mesh of the screen and enter Section II, while two-year fish enter the fish collection chute, in which only two-year fish can overcome the current. After passing through the fish discharge chute, the two-year fish enter Section III. Thus fingerlings and two-year fish are automatically graded in different sections of the fish trap without damage to them and are readily collected.

Early in spring, as soon as the ice breaks up, yearlings are transferred

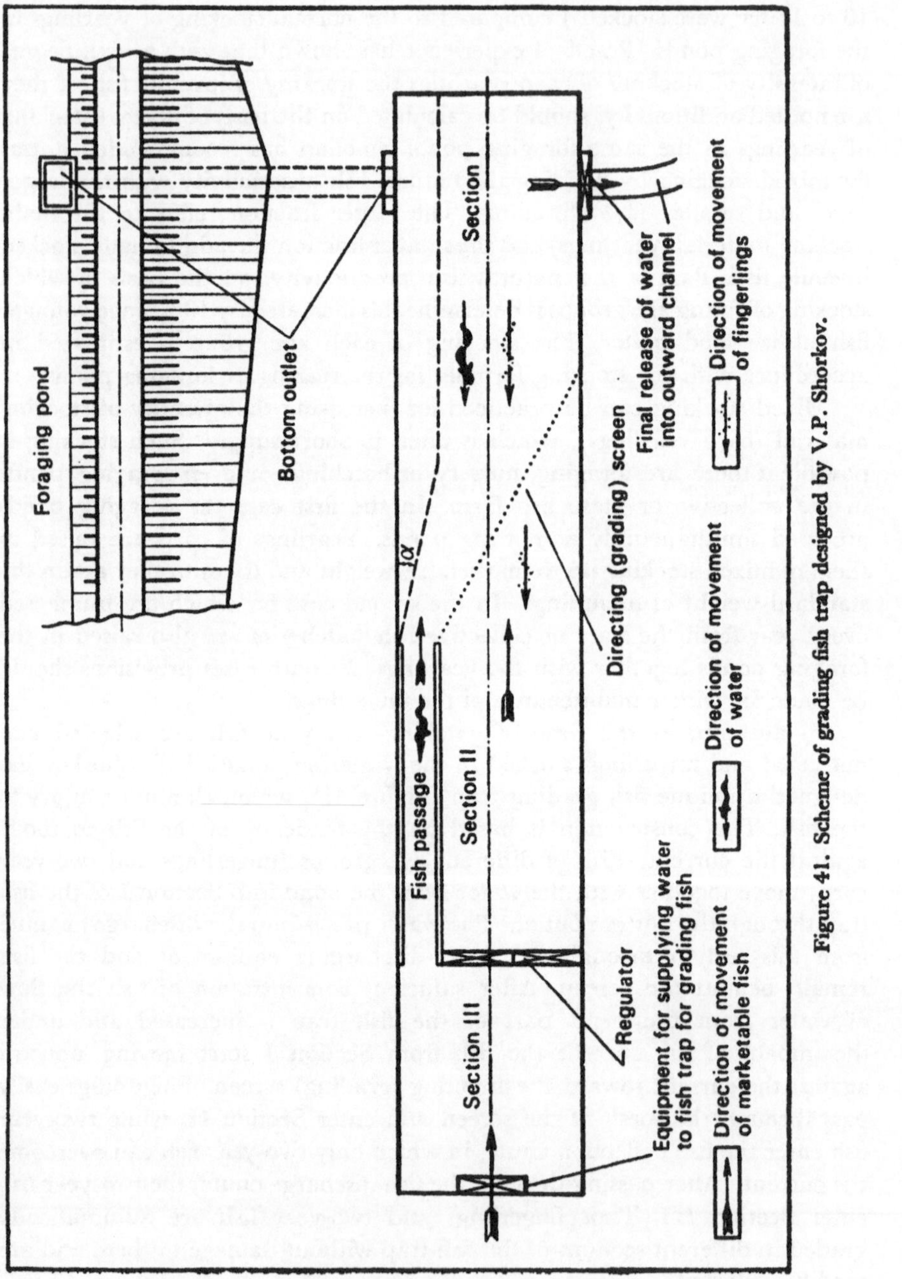

Figure 41. Scheme of grading fish trap designed by V.P. Shorkov.

from wintering ponds to foraging ponds in accordance with fixed norms, and later (toward the end of May–early June) fry are again procured and stocked additionally. The process is repeated in autumn. In this way, even without all the categories of ponds required in a two-year rotation of fish farming, it is possible to raise stocking material and marketable fish through mixed stocking.

Fingerlings raised in foraging ponds in mixed stocking serve two purposes: (1) to accomplish the target for growing fingerlings, and (2) to estimate the productivity of such ponds.

Chapter VIII

ADDITIONAL FISHES AND POLYCULTURE

For a more complete utilization of the natural food resources, not only are fish of the same species but different age groups (mixed stocking) placed in a single pond, but also fishes of different species (stocking of additional fishes), which differ in feeding habits. Additional stocking is done in such a way that toward the end of the vegetative period marketable fish are obtained not only from the main object of fish culture (common carp), but from the additional fishes (representing different species, genus, or family). A broader assortment of cultural products is thereby achieved.

The concept of "additional fishes" should not be confused with "polyculture". Pond fisheries until rather recently cultured exclusively only one species of fish. The same system prevails today but some farms are including additional fishes, which do not compete with carp for food. Common carp, the main object of fish culture in ponds, does not make use of "coarse" fauna (water beetles, bugs, larvae of dragonfly, tadpoles, small trash fish), readily consumed by sander, pike, trout, and other fish species, nor does it feed on higher aquatic plants and phytoplankton.

Pond fisheries, especially in the southern regions of the country, have begun to culture several fish species with carp to achieve a fuller utilization of the natural food resources in the ponds. The produce from such ponds is not less than the produce of the main object under culture—carp—and sometimes even greater. This is known as polyculture.

Among the fish species which can be reared with carp (additional fishes, polyculture), the preferred are grass carp, silver carp, and bighead,[1] followed by tench, Crucian carp, rainbow trout, pike, sterlet, trout perch, Vendace, cisco, peled, Lake Chudskoe whitefish, and a few other species. It is obvious from this list that not only compatible but also predatory fishes are used in additional culture.

In selecting fishes for additional stocking or polyculture one should keep in mind that fishes from the carp family feed in general on a similar type of food. However, the capacity for searching out or foraging for food, as well as the feeding niches are not identical. Depending on the fish selected the techniques of fish rearing and farm management need to be adjusted. In a

[1]See Chapter IX.

two-year rotation carp farm the age of the fishes and the norms of additional stocking have to be reviewed carefully.

In the USSR and a number of Eastern European countries (Czechoslovakia, France, German Democratic Republic, Yugoslavia, etc.) tench are additionally stocked with carp.

Tench (*Tinca tinca,* Figure 42) prefers quite grassy and moderately silty water resources with soft submerged vegetation in which the fish can shelter

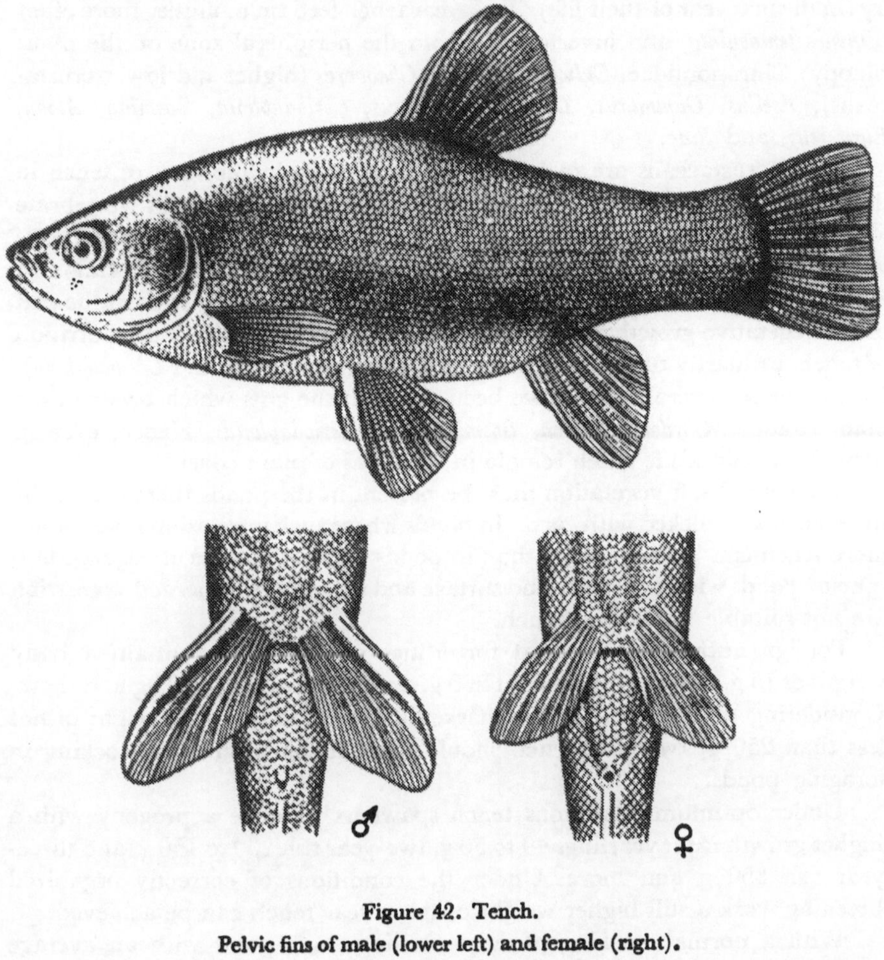

Figure 42. Tench.
Pelvic fins of male (lower left) and female (right).

during the day. Tench finds food in the thickets of vegetation and consumes food residue falling to the bottom, thereby preventing its decomposition and thus improving the sanitary conditions of the ponds. Carp generally remains in the open part of the pond, using the natural food available there, occupying the deeper layers only when searching for food. Thus a fuller exploitation of the natural food resources is realized with very little competition between the main and the additional fishes. In the first year of life tench feeds on zooplankton (*Daphnia, Ceriodaphnia, Bosmina, Chydorus, Cyclops, Diaptomus,* etc.). After attaining a length of 16 to 17 or even 11 to 12 mm it commences feeding on chironomid larvae. These larvae comprise, on the average, 81.8% of the total invertebrate food of young fish. In addition to chironomids, *Chaoborus* and other organisms have been found in the intestines of tench fry (in the first year of their life). Two-year tench feed on mollusks, more often *Bithynia tentaculata,* and invertebrates from the peripheral zone of the plant canopy: Chironomidae, *Culicoides, Corixa, Chaoborus* (higher and lower crustaceans), *Asellus, Gammarus, Daphnia, Chydorus, Ceriodaphnia, Bosmina, Alona, Euricereus,* and *Sida.*

Lower crustaceans are of secondary importance in the food of tench in their second year of life. The presence of large quantities of invertebrate forms (*Chydorus, Alona, Bosmina*) from the thickets and peripheral areas of vegetation in the intestines of tench is proof that they primarily remain in areas with a plant cover. In their third year of life, especially toward the end of the vegetative growth period, zooplankton play some role in the nutrition of tench, primarily the larger forms of Cladocera (*Daphnia* and *Ceriodaphnia*). In this period several forms have been found in the guts which occur in the plant canopy (*Chydorus, Alona, Bosmina,* and *Ceriodaphnia*). Hence, even in their third year of life tench remain in the areas of plant cover.

Submerged soft vegetation must be present in the ponds therefore at the time tench are stocked with carp. In ponds where such vegetation is abundant more tench can be introduced than in ponds where this type of vegetation is sparse. Ponds with a hard ground surface and devoid of submerged vegetation are not suitable for raising tench.

Foreign authors report that fingerlings of pond tench attain a body weight of 15 g, two-year fish about 200 g, and three-year fish 325 g and more. Considering that tench becomes marketable with an average weight of not less than 250 g, two-year tench should be used for additional stocking in foraging ponds.

Under optimum conditions tench spawners produce a progeny with a higher growth rate: yearlings 40 to 50 g, two-year fish 200 to 150 g, and three-year fish 350 g and more. Under the conditions of correctly organized breeding work a still higher weight of three-year tench can be achieved.

With a normal stocking of carp yearlings in a pond with an average growth of underwater soft vegetation and a similar quantity of two-year tench added to the pond, fish productivity increases by 20% without causing a reduction in growth of carp. In the case of additional fish culture in ponds with poor or rich soft vegetation, these norms would change accordingly. On the average, fish productivity of carp ponds with additional stocking of tench increases by 15%. As a result the additional stocking of foraging ponds is calculated in accordance with the formula used for stocking carp, with corresponding changes in the index of natural fish productivity (15% of productivity of carp), initial and final weight, and percentage of output—85 to 90% for two-year fish and 95% for three-year fish. If the farm produces its own material for additional stocking in carp ponds, the output of rearing ponds of tench fingerlings is taken as 70 to 75% and that of wintering ponds 90 to 94%. In this process, in accordance with the recommendations of the Ministry of Fisheries of the USSR, juvenile fish are stocked at the rate of 1,000 fish with a natural productivity of up to 150 kg/ha, 1,400 fish—150 to 200 kg/ha, and 1,600 fish—250 kg/ha. For dense stocking the quantity of tench should not exceed more than 15% of the carp stock.

In the German Democratic Republic tench is successfully raised in carp ponds by stocking yearlings as well as two-year fish. The density of stocking is regulated in such a way that the fingerlings attain a weight of 6 to 10 g and the two-year fish 70 g. Three-year fish reach 250 g. In a three-year rotation scheme tench is used in the first year spawning, in which case each pond should be 0.5 m deep and contain soft vegetation. From one hectare of pond area 10,000 to 60,000 young fish can be obtained. During the winter they are maintained in wintering ponds at the rate of 50 to 150 fish per 1 m^2. For feeding in the second year 20 to 30 tench are stocked per 100 carp (but sometimes only 5 to 10% of the number of carp) and kept together with the carp for hibernation in the wintering ponds. In the third year fish raising in the foraging ponds should be done at the ratio of 1:5 carp to tench or 150 to 200 two-year tench per hectare of pond area.

In Czechoslovakia the density of tench stocking varies from 10 to 30% of the number of carp stocked.

The fact that tench tends to scatter all over the pond and to dig into the bottom should be kept in mind. Its skin (especially that of fry) is exceptionally sensitive to mechanical injury, which complicates production procedures. In the course of transplantation of tench from one category of pond to another the water should be released slowly, without sound, and preferably during the night. Tench, particularly the young, are frightened by sounds and hence scatter or dig into the silt. With gradual water discharge from the ponds, they concentrate in the collection channels and fish pits, which makes collection easier.

GOLDFISH (*Carassius auratus gibelio*, Figure 43) is mainly used as an additional fish in carp farms of two-year rotation. The high food value of the meat of this fish is due to its high caloric value—969 kcal (versus 943 in carp) and edible parts comprising 69.7% of the total body weight (versus 59.1% in carp). Its simple habits and high survival rate make this fish a convenient object for pond culture. In the Veselovskii reservoir the quantity of fat per kg of edible meat is 5.52% versus 3.0% in sazan. This fish is immune to the red-spot disease of carp, branchiomycosis (gill rot), and other diseases. It should be noted, too, that the disease rate among carp reduces when they are reared with goldfish.

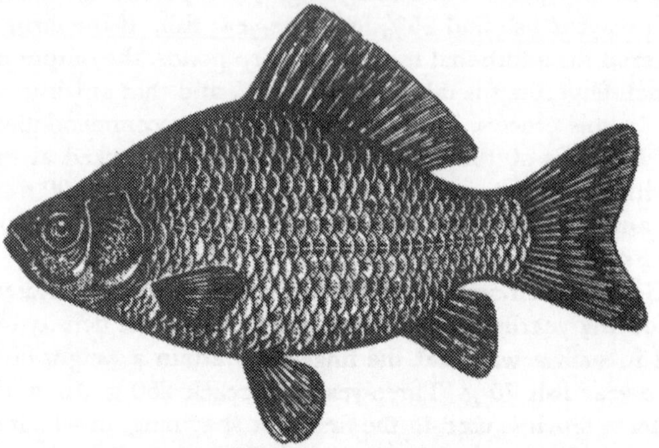

Figure 43. Goldfish.

The additional stocking of goldfish is possible only in completely drainable ponds, since in nondrainable ponds they multiply rapidly. A disproportionate number of goldfish to carp has an adverse effect on the latter. The young, in the early stages of development, compete for food with carp and hence their combined rearing with carp fingerlings is disadvantageous. Kharitonova (1968), working in the ponds of the Kiev and Lvov districts of Ukrainian SSR, found that the highest level of competition between carp fingerlings and goldfish occurred in June. For example, chironomids constitute 84.99% (by weight) of the food of young goldfish and 97.3% in the case of carp. Later on their total weight decreases notably and in July chironomids constitute 17.8%, in August 5.43%, and in September 2.95%. During these months the total weight of algae, number of cladocerans and copepods, and detritus content increase in the food composition. In two-year goldfish chironomids were found in the gut contents as follows (percentage by weight): April—2.62, May—nil, June—8.4, July—6.37, August—0.64, and September—0.80. Copepods, cladocerans, algae, and detritus are much more significant in the food of two-year fish.

Data from other authors indicate that in the second half of December, even with zooplankton available in the pond, two-year fish mainly feed on benthos—an observation confirmed by experiments conducted in Moscow. Only chironomids were found in their intestines. Anisimova (1963) has established that these fish feed on zooplankton and benthos, but their most common food is the larvae of chironomids. The spectrum of feeding of carp and goldfish under simultaneous raising in the ponds of Belorussian SSR was found to be similar by Dombrovskii (1964), but he states that supplementary feed additionally constitutes a large part of the ration. If the total weight of the extraneous feed introduced into an experimental pond in June was 59.1% in the case of carp, for goldfish it was 33.2%. These values for July were respectively 61.9 and 54.0%, August 93.9 and 78.3%, September 78.7 and 63.9%, and the average 75.9 and 61.4%. Hence the recommended additional stocking of goldfish on the basis of enhanced natural fish productivity should not exceed 20%. Kharitonova (1968), on the basis of corroborative data, recommends an increase of 15% over the estimated natural fish productivity for carp, yielding 25 to 27 kg goldfish per hectare.

Goldfish yearlings and two-year fish are stocked with carp yearlings if in the given zone they are not sexually mature (leading to spawning and overpopulation of the foraging ponds).

Additional stocking in foraging ponds is determined on the basis of the formula accepted for stocking carp with necessary changes in the parameters of natural fish productivity (20% of carp productivity), initial weight, and projected final weight (for two-year fish 0.15 to 0.20 kg and three-year 0.35 to 0.40 kg), as well as the percentages of output expected for two-year (90%) and three-year fish (95%).

Populations of goldfish in the European parts of the USSR are unique in that they almost exclusively consist of females. Only in the Far East is the ratio of males and females almost equal. The farther one proceeds West, the greater the deficit in males. In most of the regions of the European part of the country, given the condition of almost unisex populations, breeding occurs through the participation of males of other species—carp, Crucian carp, and tench—which are similar in size to the females. The resultant progeny consists only of females, and only goldfish. Furthermore, the progeny bred by the males of the foregoing species do not develop evenly. Anisimova (1963) found that in terms of growth rate better results are obtained by crossing female goldfish with male carp. The higher output during the incubation period observed in this process, compared to crosses with male goldfish, is compensated by the increased number of redds or pairs of spawners. The fingerlings of a goldfish brood weigh 15 to 20 g, two-year fish 150 to 170 g, and three-year fish 300 to 350 g. Still higher indexes of weight have been recorded. According to I. Anishchenko the yearlings of goldfish from Lake Sengileev (Ordzhonikidze district) have an average weight of 60 g and two-year fish of 280 g.

When the fish farm rears its own goldfish material for additional stocking in carp ponds, the output of fingerlings from rearing ponds is taken as 70 to 75% and the output of yearlings from hibernating ponds 85 to 90%. In this process the hibernation of carp stocked with fingerlings and two-year goldfish is combined. The norm of stocking for each age group or species should be reduced to half.

According to the recommendations of the Ministry of Fisheries of the USSR goldfish fry should be stocked at the following density, depending on the natural fish productivity:

150 kg/ha......1,000 fish
150 to 200 kg/ha......1,400 fish
200 to 250 kg/ha......1,600 fish

Under the conditions of dense stocking, the permissible addition of goldfish is not more than 15% of carp.

In autumn the goldfish are separated from the carp in live fish tanks. The loss of goldfish during maintenance is higher than that of carp. The density of stocking of goldfish in tanks should be 1 kg of fish to 10 liters of water under the conditions of a continuous water flow in the fish tanks.

SANDER (*Lucioperca lucioperca*) is a predatory fish (Figure 44). This species can also be used for additional stocking in carp culture. For rearing sander the carp ponds should be filled with clean water, preferably have a hard bottom, a good gas regime, some quantity of small trash or forage fish (such as ruff, "verkhovka," gudgeon, bleak, bitterling, roach, minnow, rudd, white bream, etc.), and a moderate growth of higher aquatic plants. Sander primarily feeds on small fish about 2 to 10 cm long. According to Mikheev and Meisner (1966), forage fish constitute 93% of their food. Sander feeds only on live and moving organisms. Supplementary feed in the form of various mixtures or monotypic food, larvae of insects, chopped fish, or intact dead fish are not eaten by sander. Sander fry grow best on insect larvae; growth is poor on zooplankton. Juvenile sander change over to a predatory mode of life at the age of one month. In this process they consume the most widely occurring fish species in the water resource; the body height of these forage fish should not exceed 10% of their linear measurement. The growth of young sander is most intensive when they feed on fish. In the central belt of the USSR they reach 30 to 60 g and even 120 g toward autumn.

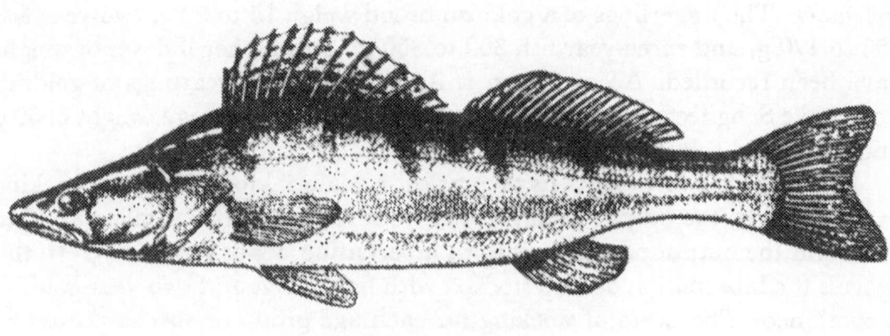

Figure 44. Sander.

Thus, on the basis of feeding behavior sander is a scavenger, clearing trash fish from the ponds and thereby improving the conditions of growth for the main object of culture—carp. If there are no trash fish or their quantity is low, sander and other predatory fishes used for additional stocking may be fed on the carp fry obtained in large quantities from reserve redds of carp spawners. The daily requirement of forage fishes, based on the food requirement during the summer period, is 1.5 to 2.0% of the body weight of sander.

Nikanorova (1965 and 1968) has defined different ecological groups of sander in the Seliger and Kaftine lakes (Kalinin district), located at a distance of 100 km from each other. The two groups differ with reference to period of spawning, spawning temperatures, duration of spawning, and fecundity. If spawning in Lake Seliger occurs at the end of May—first half of June, in Kaftine it takes place at the end of April—beginning of May. The peak spawning temperature range in Seliger is 13 to 18°C and in Kaftine 9 to 11°C. If spawning continues in Seliger for 20 to 25 days, in Kaftine it is completed in just 7 to 10 days. Considerable differences between these two groups have also been noted with regard to fecundity, which are summarized in Table 14.

Table 14. Fecundity of female sander

Age of female fish	Number of eggs laid by female in Seliger (in thousands)	Number of eggs laid by female in Kaftine (in thousands)
3+	84.4	268.7
4+	126.0	329.6
5+	198.0	452.1
6+	268.0	—

According to reports from the Puiga fish farm similarity in feeding of fingerlings of carp and sander is detectable only at the beginning of their rearing; however, different places of feeding were also noted for the two groups. Data show that for pond rearing in the central belt of the European part of the USSR better results are obtained by using early spawning sander from Lake Kaftine. Studies carried out with Don sander have shown that if the water temperature during incubation remains in the range of 10 to 16°C, the extent of losses in fish eggs is insignificant. Under different daily fluctuations and water cooling to 5 to 4°C sander eggs die. Until recently it was thought that in ponds without a water flow only sander from captive water sources (ponds and lakes) could be reared, that river sander does not breed under conditions of captivity. However, studies have now shown that not only sander from captive water sources, but also from fluvial sources (streams, rivers) and semimigratory sander can be successfully used in pond culture.

However, better results are obtained by using spawners and progeny raised under the conditions of pond culture. In the case of selection from natural water sources, large-sized or older spawners should not be taken since their adaptation to pond conditions is poor. It is better to select younger fish, age three to four years, weighing 1.0 to 1.5 kg, keeping the ratio of females and males 1:1. There should be no injury to their body, fins, or scales. Their eyes must be clear, with no symptoms of sickness due to parasitic cataract (dullness of eyes, which subsequently develops into eye sores and eventual destruction). A parasitic eye condition is often found in ponds frequented by sea gulls and those which contain pond mollusks. A blind sander is totally unable to hunt for food.

In the case of selection of already sexually mature sander from natural water sources one should keep in mind that such fish respond poorly to transportation (see Chapter XVII). The sex of sander can be identified only in spring before spawning on the basis of the fullness of the abdomen in females due to egg development, and a milt discharge with mild abdominal pressure in the case of males.

Under the conditions of the central belt of the European part of the USSR fingerlings can reach 120 g (usually 25 to 30 g), and in the southern regions 150 and even 300 g (usually 35 to 60 g and more). In the case of simultaneous raising with carp in foraging ponds sander yearlings are stocked additionally with yearlings of carp. In fish farms of three-year rotation, two-year sander are added to ponds containing two-year carp.

The weight of two-year sander in the central belt of the European part of the USSR, under favorable conditions (large amount of stocking material, sufficient supply of food, low density of stocking), reaches a maximum of 300 to 500 g, and under conditions of average food supply, 240 g.

In the southern regions of Ukrainian SSR simultaneous raising of Dnieper sander with a large quantity of trash fish increases the output of commercial fish by 35 to 43% and two-year fish reach an average weight of 400 to 600 g (maximum 900 g). The number of sander yearlings in relation to carp yearlings varies from 10 to 20% of the total number of carp under normal stocking. For the conditions of Moldavian SSR stocking of sander is recommended for foraging ponds containing carp, using 3- to 5-day-old larvae or 15- to 20-day-old fry 2.5 to 3.0 cm long and weighing 0.1 to 0.2 g before they change over to a predatory mode of feeding.

Under these conditions the survival of the larvae toward the vegetative period is about 30 to 40%, and the survival of fry 40 to 50%. The recommended density of stocking (at a density of 4,000 to 5,000 trash fish per hectare) is as follows: larvae not more than 300 per hectare and fry 200 per hectare. With a higher or lower quantity of trash fish the stocking of sander is also correspondingly increased or decreased.

According to the recommendations of the Ministry of Fisheries of the

USSR, the density of stocking of sander fry is calculated according to the number of trash fish in the pond:

Quantity of trash or forage fish, kg/ha	Density of stocking, number of fish/ha
up to 50	900
50 to 90	1,540
100 to 140	2,240
150 to 200	3,200
more than 200	4,000

The rate of stocking is increased by 30% when larvae are used as stock material.

The total fish productivity of foraging ponds under conditions of simultaneous raising of carp and sander is increased by 50 to 100 kg per hectare because of the creation of conditions for better carp growth, and an increase in the quantity of sander by 10 to 20 kg/ha. Catching sander, taking into consideration their sensitivity to the oxygen regime, turbidity of water, and injuries, should be done very carefully.

Pike (*Esox lucius*, Figure 45) is another predatory fish used as additional stock material in foraging carp ponds. This species is desirable for its valuable meat, which has a low fat content (proteins 18.7 to 19.0%, fat 0.5 to 1.2%) and the fact that it is a good scavenger in both natural and artificial water resources. As a predator it consumes: larvae of dragonfly, leeches, frogs, tadpoles, and fishes—perch, ruff, gudgeon, roach, Crucian carp, "verkhovka" (*Leucaspius* sp.), bleak, white bream, rudd, tench, pike, etc. Pike capacity for consuming different fishes depends on the environmental conditions as well as the occurrence of one or the other species in the water resources.

Figure 45. Pike.

I.F. Demchenko (1970) found that in Ukrainian SSR in August right up to autumn the main item of food of this fish is "verkhovka," accounting for 65 to 70% of the total food. Given a deficiency of forage fish in the pond pike feed on larvae of dragonflies, chironomids, leeches, etc. right up to autumn. Under aquarian conditions I have observed cannibalism in pike fry even in the earlier period of development, on the 24th day after hatching. Pike about 29 mm long change over to a predatory mode of life, and can even ingest another pike as much as 24 mm long. The latter was observed when juvenile fishes of other species as well as cyclopoid copepods and chironomids were placed in the aquarium together with smaller specimens of pike. Pike weighing more than 800 g are considered harmful predators; they are useful for controlling trash fish up to a unit weight of 500 g and maximum 800 g. Studies have shown (Sholz, 1932) that pike at the age of one to two years can ingest prey 10 to 15% of its own body weight, but not more than 30.7%. The intensity of pike feeding varies in different months and throughout the course of the day. It feeds more intensively in April–May and July, and in October–November.[2] The maximum feeding activity during the day occurs from 9:00 a.m. to 2:00 p.m. In winter the females feed more actively than the males.

For additional stocking with carp yearlings in foraging ponds pike fry 2 to 3 cm long are planted not earlier than on the 9th to the 10th day from the moment of their changeover to active feeding, and not later than the 18th to the 19th day after the emergence of larvae (termination of dormancy period). Additional plantation of pike fry is done in drainable ponds so that no specimens are left uncaught. In ponds constructed in river beds or totally nondrainable ponds fish catching is difficult. Pike stocking is not advisable therefore since the pike which are not caught become too large and pose a definite threat to the carp yearlings. On the other hand, many ponds in which fish catching is more systematically done are often populated with large numbers of small trash fish which feed on plankton and benthos and offer competition to the carp. Furthermore, trash fish (gudgeon, "verkhovka," roach, etc.) are also carriers of parasites for carp (argulus,[3] fish leeches, etc.). Thus stocking of pike fry not only makes it possible to increase produce, but also facilitates the eradication of trash fish from foraging ponds, making them more suitable habitats.

Pike grow fairly rapidly. According to my observations in the Moscow district the fingerlings of pike under favorable conditions grow to a unit weight of 350 to 500 g (average 400 g), and a unit weight of 156 g (maximum 180 g) in peat quarries of the lowland marsh type. The average unit weight under the conditions of Ukrainian SSR toward autumn is 500 to 800 g. Pike from the Ob-Irtysh basin toward the end of the first year attain an average length of 22 to 24 cm and weight of 130 to 150 g; two-year fish 35

[2]Intensity of feeding of pike differs in different zones in different months.

[3]Argulus is a branchial parasite, "fish louse," and one of the most dangerous pests for fish in pond culture; it mainly feeds on the blood of the host animal.

cm and weight of 500 g; three-year fish 47 cm and weight of 1,000 g; and four-year fish 53 cm and weight of 1,500 g.

The rate of additional stocking of juvenile pike is usually taken as 70 to 100 fish per hectare (without supplying forage fish) or 200 to 250 fish per hectare (with special rearing of forage fish). In the latter case sexually mature carp of Crucian carp are stocked which produce progeny in the foraging ponds. The rearing of forage fish ensures the normal growth of pike and gives a survival rate of 50 to 55% of the total stocked material, with an average weight of marketable fish of 250 to 300 g or more. In Czechoslovakia a mixture of filtered plankton and crushed spleen is used as feed for juvenile pike.

Experience in Ukrainian SSR has shown that as a result of additional stocking of pike and its ameliorative effect on the pond, fish productivity of carp increased 60 to 120 kg per hectare.

According to the recommendations of the Ministry of Fisheries of the USSR, the density of stocking of pike fry is calculated according to the quantity of forage fish in the pond:[4]

Quantity of forage fish, kg/ha	Density of stocking, number of fish/ha
up to 50	150
50 to 90	260
100 to 140	400
150 to 200	600
above 200	700

When larvae are used for stocking the corresponding norms of plantation are increased by 30%.

CATFISH (*Silurus glanis,* Figure 46) has recently been cultured with carp in foraging ponds. This fish has a rapid growth rate, tasty meat, fewer bones, and good resistance to disease. Catfish stocks in inland water resources are gradually being depleted. This fact and the food value of its meat (3 to 5% fat, 15 to 18% protein) have stimulated catfish cultivation in ponds.

Compared to other predatory fishes such as sander and pike, catfish has certain advantages. Under a satisfactory hydrochemical regime it can feed in small ponds and even in pits, channels, and quarries; it does not feed during winter even if the temperature is increased artificially. It is a very hardy fish and can withstand transplantation well from one type of pond to another. It is a heat-loving species, feeds on trash fish, waste from fish dressing, frogs, tadpoles, leeches, aquatic insects, even swimming birds, and

[4]I.F. Demchenko has given a slightly different calculation for stocking of juvenile pike on the basis of his experience in Ukrainian SSR (see Vopr. rybm. khoz-va Belorussii, vol. 7, 1970).

other aquatic animals. Even in the first year of its life this fish changes over to feeding on fish to some extent. Adult fish need almost exclusively on fish,

Figure 46. Catfish raised at the fish farm "Sarvash" (Hungary).

and in large quantities, but also consume frogs and crayfish. In the lower parts of the Danube catfish starts feeding on fish at a body length of 3 to 5 cm.

Balan (1968) has studied the conditions for breeding, growth, and feeding of catfish and, based on his findings, has given recommendations and norms for the production processes related to culturing in ponds. In view of the fact that special nurseries for rearing stock material of catfish are not required, the methods for selecting spawners and producing stock material are of special interest. One year before undertaking their breeding, brood fish between five to nine years of age are caught from natural resources, preferably from the deltas of the Volga, Kura, Danube, and Dnieper Rivers where catfish abound. Before spawning, the brood fish are fed small trash fish at the rate of not less than 30% of their weight and, to avoid injury, females and males transplanted. Spawning is carried out in small carp wintering ponds (500 to 700 m^2), filled the day before to a depth of 0.8 to 1.0 m. The usual spawning ponds are less suitable since the eggs in such ponds are scattered over the grassy substratum, their fertilization poor, and they are quickly covered with water. Catching larvae and grown fry from such ponds is also difficult. It is better to use nests made from the aerial roots of willow as a substratum, positioning them at a distance of 3 m from the periphery and close to the water supply troughs. Spawning is done in pairs by selecting males and females of approximately the same weight. The relative fecundity is 9,000 to 18,000 eggs per 1 kg weight of female fish.

The eggs can be incubated in the ponds, in special apparatus (Chalikov or Weiss), or in tanks. During incubation the eggs are protected from direct sunlight. Better results are obtained from incubation of eggs in ponds if they are kept at a depth of 15 to 30 cm. The duration of incubation is 80 to 82 hours with an output of 80 to 83% larvae. Catfish larvae are yellow after hatching and not transparent, contain a yolk sac, reach 7 mm in length, and even at this stage incipient barbels are visible. A characteristic peculiarity of these larvae is that they are afraid of light and hence concentrate in places covered with grass, roots, and so on. The transition to active feeding commences when the larvae are 7 to 10-days-old. From a single female weighing 8.5 kg the output of 10-day-old larvae is 50,000 to 70,000. The maximum loss of catfish fry, 20 to 24%, occurs at the age of one month.

The studies of A.I. Balan have also shown that for the maximum survival of catfish fry it is essential that they be reared for 30 days up to a weight of 2 to 5 g. This is better done in nursery ponds with an area of 500 to 1,000 m^2 and a stocking norm of 300,000 fish per hectare. The output in this case should reach 76 to 80%. If the fish are allowed to grow to the requisite stage, losses considerably decrease when they are reared together with carp fry. Ponds for rearing fry should have regions covered with soft underwater vegetation, serving as a substratum for the development of food liked by catfish fry, namely, phytophilic chironomids. Subsequently the fingerlings of both fish are reared together until they áttain a weight of 25 to 30 g with a stocking norm of 3,000 to 5,000 catfish fry per hectare and an output of 70% of the material stocked. Stocking in foraging ponds with yearlings of carp ranges from 100 to 150 yearlings of catfish weighing 25 to 30 g. The output of these fishes toward autumn is 95 to 100% with a unit weight of 900 to 1,100 g and an increase in fish productivity of 80 to 90 kg/ha.

According to the recommendations of the Ministry of Fisheries of the USSR the density of stocking of juvenile catfish depends (as in the case of pike and sander) on the quantity of forage fish in the pond:

Quantity of forage fish, kg/ha	Density of plantation, number of fish/ha
up to 50	200
50 to 90	360
100 to 140	560
150 to 200	800
above 200	1,000

The stocking norms are increased by 30% when catfish larvae are used for additional stocking.

The change in primary food items is very distinctly expressed in two-year catfish raised in ponds. Until the middle of June these fish feed on other fishes, from mid-June to mid-August on tadpoles and frogs, and thereafter up to the end of the season, fishes and large invertebrates, which account for

4.0 to 36.3% of the total weight of the gut contents at different times of the vegetative period.

The hibernation of catfish can be done with carp and in general with other nonpredatory fishes in the wintering carp ponds as well as in nondrainable ponds. A prolonged reduction in temperature (from 0.1 to 0.2°C) exerts no influence on the process of hibernation of catfish.

The norm of stocking of under-yearlings and two-year fish in wintering ponds is 2,000 to 3,000 kg per hectare with an output of 95 to 100%.

The rearing of catfish is becoming popular in a number of countries in Eastern Europe (Bulgaria, Czechoslovakia, German Democratic Republic, Hungary, Poland, etc.). In Bulgaria two-year catfish attain 400 to 1,200 g in weight with simultaneous raising with carp under favorable conditions of food, and reach up to 3.5 kg in the third year. The yearlings of catfish are stocked with carp yearlings weighing 50 to 80 g. Incubation of fertilized eggs is carried out in Weiss apparatus keeping the water flow at 0.7 to 3.0 l/min. The survival of larvae from the apparatus is 90%.

Spawning of catfish in Hungary is carried out in hatching ponds with an area of 100 m^2 and a depth of 1 m, in which reed nests are placed. Spawning takes place at a uniform rate at 20°C. The resultant eggs are transferred to incubators where they are kept for 2.5 to 3.0 days in tanks with flowing water at a temperature of 23 to 24°C. The tanks are protected from light. On the 4th day, after the absorption of two-thirds of the yolk sac, the larvae are fed natural foods and egg emulsion. On the 7th to the 8th day the young fry are transplanted to nurseries where they remain for 35 days. The natural food base is intensified with the application of chemical fertilizers. After the catfish fry have attained a weight of 3 to 4 g they are transplanted to rearing and foraging ponds at the rate of 50 to 100 fish per hectare in accordance with the necessary natural food availability. In the event of artificial feeding 5 to 10 young fish are planted per 1 m^2, which grow to a weight of 20 g toward autumn. The fry are fed twice a day with the following mixture having a food coefficient of 1.5 to 2.5: mince of fish, frogs, meat, waste, and flour. The food is given at shady places. The survival rate of the fingerlings is 80%.

Largemouth Bass (*Micropterus salmoides,* Figure 47) exhibits all the features recommended for additional fish in pond culture. It attains a weight of 3.2 to 8.0 to 10.0 kg and more; its meat is very tasty and contains about 19.4% protein and 1.0% fat. It is one of the important fish in pond culture in the USA. It was introduced in the USSR at the beginning of the present century and abounds in the Abrau and Limanchik lakes near Novorossiisk from where initial stock material can be obtained.

Largemouth bass is a predatory fish but not to the same extent as sander, pike, and catfish. Its food changes with age. In the earlier stages of development fry and fingerlings feed on the same foods as the fry and fingerlings of common carp. After attaining a length of 2.5 to 3.0 cm and a weight of 0.3 to 0.5 g, the fry feed on larvae and pupae of insects—Mayflies, bugs, etc.

Figure 47. Largemouth bass.

At the age of three to four months (16 to 20 g weight) largemouth bass changes over to feeding on small fish and, having grown to 25 to 30 g, vigorously feeds on tadpoles. With further growth it feeds to a greater extent on small fish which, toward the end of the first year of the bass's life, comprise 50% or even up to 86% of the diet. In its second year of life largemouth bass consumes small fish, frogs, and tadpoles. Hence additional stocking is meaningful if trash fish, tadpoles, frogs, aquatic insects and their larvae, etc. are present in the pond.

As an additional fish largemouth bass is valuable since it significantly increases fish productivity by improving the conditions for growth of the main object of fish rearing—carp. Stocking the fry of largemouth bass in rearing ponds containing carp fry increases fish productivity by 20 to 30 kg per hectare with the weight of the fingerlings of largemouth bass reaching 30 to 50 g or more toward autumn.

The combined raising of fingerlings of largemouth bass and two-year carp under normal stocking has given better results when the density of stocking of the former is 1,170 fish per hectare. The fingerlings attain a unit weight of 37.4 g toward autumn and their fish productivity has been recorded as 40.8 kg per hectare (24.5% of the natural productivity of carp).

The yearlings of largemouth bass are planted with those of carp at the rate of 10% of the carp stock. In the southern regions yearlings weighing 30 g are used for this purpose. Here the unit weight of two-year fish toward the end of the vegetative growth period reaches 250 to 280 g. The output in relation to stocked yearlings is 90%. Fish productivity in the foraging ponds increases by 30 to 50 kg/ha. If the quantity of trash fish is sufficiently high, the stocking of fingerlings of largemouth bass can be increased up to 250 fish per hectare. The following norms are recommended: average weight of fingerlings not less than 20 g, growth of two-year fish over the summer period 250 g, output of fingerlings 75% of planted number of fry, and output of two-year fish 80% of the total number of yearlings stocked. In the case of simultaneous raising of two-year largemouth bass with two-year carp and a stocking of each species at the rate of 400 fish per hectare, fish productivity toward autumn in the case of largemouth bass will be 53 kg/ha (25% of the productivity of carp).

RAINBOW TROUT (*Salmo irideus*). A number of fish species from the family of salmonids are of great value as additional stocking material in foraging carp ponds. These fish are very valuable for their food qualities and among them the rainbow trout occupies first place. However, not all carp ponds are suitable for this purpose. Only the better managed ponds with less silt deposition, devoid of overgrown aquatic vegetation, with a good oxygen regime, and depths up to 2 to 3 m at several points are suitable. The area of the carp ponds used for additional stocking of trout is also important since in large ponds catching trout is difficult and economically less rewarding. Schäperclaus (1961) has recommended the use of shallow carp ponds with a hard bottom. Additional stocking of rainbow trout is meaningful only when forage fish are plentiful, especially "verkhovka," which is preferred by trout over other fish species (ide, perch, ruff, loach, etc.). Studies on additional stocking of rainbow trout in carp ponds conducted in Belorussian SSR have shown that its fry feed on plankton and small larvae of insects, with the role of insects increasing with further growth. The feed of two-year fish comprises mainly aquatic and aerial insects and their larvae, but fish occupy a significant place, especially during autumn and spring. The minor similarity in the spectrum of feeding of carp and trout (chironomid larvae) is reduced by the fact that carp use food organisms living in plant thickets, in the water column, and on the bottom of the pond. The consumption of "coarse" fauna and trash fish by trout creates better conditions for the growth of carp. Water temperature is an important factor in the simultaneous raising of trout and carp. At 7 to 15°C and an oxygen content above 5 mg/l, two-year trout feed actively (Borobik, 1968); above 20°C food consumption decreases even if the oxygen content is high, and the growth rate declines. Growth also slows down under conditions of a low oxygen content, less than 3 to 4 mg/l.

Under favorable temperature, hydrochemical, and food regimes rainbow trout grow to a weight of 100 to 120 g toward autumn when the weight of the yearlings is 5 to 7 g, and reach 250 to 300 g when the weight of the yearlings is 25 to 40 g.

Trout are very susceptible to fish louse. Hence their stocking in carp ponds infected with this parasite is not permissible. Rainbow trout are more resistant to furunculosis but suffer (especially the young fish) from whirling sickness or myxosomatosis. Usually the rate of additional stocking of rainbow trout yearlings in foraging ponds is kept at 10% of the carp yearlings stocked. The stocking density of carp should never exceed three times the norm.

Taking into account the supply of artificial feed to carp ponds and the presence of trash fish, however, the norm of stocking may be increased under good conditions of fish production. A good output can be obtained by stocking up to 300 trout per hectare of carp pond under favorable conditions of feeding; the rate of additional stocking should be 150 to 200 trout per hectare when trash fish are minimal. The output of two-year trout toward autumn is 70% of the total quantity stocked. In ponds with an area of 50 hectares the output is 75 to 80%, and above 50 hectares 65 to 70%.

PELED (*Coregonous peled,* Figure 48) is an important and promising object for additional stocking in carp ponds and pond and lake fisheries. It is a typical planktophagous fish. The daily rhythm of feeding varies in peled. According to Sycheva (1955) the last feed occurs in the predawn hours. At daybreak peled feed intensively, mainly on *Bosmina* and other planktonic organisms, and toward noon feed twice more. In the second half of the day up to 6:00 p.m. the intensity of feeding reduces with an increase in water temperature; around 6:00 p.m. another period of intensive food consumption commences.

Figure 48. Peled.

Peled is a cold-loving fish. For its rearing better managed ponds, devoid of mud and overgrown higher aquatic plants are suitable, where the water heats up to 24 to 26°C, the oxygen content does not fall below 45 to 50% of normal saturation, and pH is maintained between 6.0 to 9.0 to 9.2. In winter the oxygen content should not be less than 3 mg/l. With three and four times the stocking density of carp and a reduction in normal oxygen saturation to below 30% peled mortality occurs. This species is less susceptible to parasitic diseases but its fry are attacked by fish louse, resulting sometimes in a significant loss. Peled reared in ponds may also suffer from diplostomatosis. The fingerlings are susceptible to the parasitic disease causing cataract of the eyes. When rearing is done in the first year together with carp, depending on the density of stocking and presence of zooplankton, peled attain a unit weight of 120 to 170 g, adequate enough for marketing. Fish productivity of under-yearlings is 100 to 200 kg/ha.

The density of stocking of peled young in ponds depends on the level of natural fish productivity. Up to a fish productivity level of 150 kg/ha stocking may be done at the rate of 3,000 fish per hectare; at a productivity level of 150 to 200 kg/ha stocking at the rate of 3,500 fish per hectare; and at a productivity of 200 to 250 kg/ha, 4,000 fish per hectare.

The growth of fingerlings in ponds from different zones is characterized by the following data (from Golovkov, 1968): Leningrad district—16.7 to 44.0 g; Velikoluksk district—29 to 81; Latvia—24.5 to 120.0, Kharkov district—18 to 121; and Belorussian SSR—25 to 40 g. Under conditions of simultaneous raising with carp results differ in different zones and with different methods of farm management. For example, in the foraging ponds of the fish farm "Skrunda" (Latvian SSR) two-year peled attained a weight of 402 g and 1,260 kg of fish were harvested from one hectare, i.e., 21% of the total fish productivity. In the Kharkov district, even with carp stocking three to four times above the norm, natural fish productivity was increased by 150 to 250 kg/ha as a result of additional stocking of peled. In Belorussian SSR with a stocking norm of 350 to 400 peled per hectare, productivity increased by 70 to 100 kg/ha and the output of two-year fish reached 80 to 85% of the stocked material. Sukhoverkhov (1966) believes that a plantation of peled yearlings in foraging ponds at the rate of 500 fish per hectare will yield a unit weight of two-year fish toward autumn of 300 to 400 g, and under favorable conditions even up to 500 to 700 g. At this age and weight peled is already considered a high grade food.

The output of peled fingerlings in relation to fry stocked in rearing ponds is taken as 50 to 60% and the output of two-year fish from foraging ponds 85 to 90%, with an average increase in productivity of two-year fish of 100 to 150 kg/ha. In individual cases under fairly favorable conditions for peled raising productivity may be even higher.

The stocking norm for peled fry in foraging ponds is 5,000 to 7,000 per hectare. In the case of rearing peled fingerlings as stock material in rearing ponds under monoculture the recommended norm of plantation under conditions in Belorussian SSR is 22,000 to 23,000 per hectare, with an output of 60% of the stocked material and a fish productivity of 170 kg/ha. Additional stocking of peled in carp foraging ponds is recommended if carp stocking is not greater than three times.

The simultaneous rearing of peled is possible not only with carp, but also with broad whitefish and Lake Chudskoe whitefish, which are more important fishes for the northern regions than carp.

Broad Whitefish (*Coregonus nasus*, Figure 49), is fairly valuable in terms of food (fat content 9 to 13%). It is a cold-loving fish and the largest of all the whitefishes (reaching 14 to 16 kg in weight). This fish is characterized by rapid growth and feeding throughout the year, except during the brief

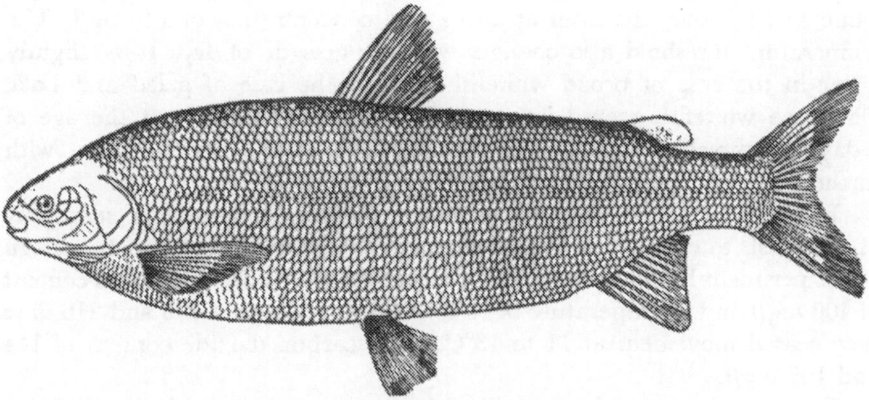

Figure 49. Broad whitefish.

period of spawning. Interest in rearing this fish in ponds and lakes is increasing every year. Broad whitefish from the Pyasina River grow better than whitefish from other waters in Siberia and hence are preferred for fish rearing. The Pyasina broad whitefish breeds from the beginning of October up to the 20th to the 23rd and mass spawning occurs from October 8th to the 15th. The average fecundity of the fish is 61,100 eggs with a fluctuation range of 45,000 to 152,000 eggs per female. The feeding of broad whitefish larvae under natural conditions has not been studied. Studies on feeding in aquaria have shown that juvenile broad whitefish with a weight of 5.3 g change over to feeding on benthos (chironomid larvae) in the first half of June, although small larvae of chironomids (*Cricotopus*) are found in the digestive tract even earlier (end of April), when the main food of broad whitefish is cyclopoid copepods. Still earlier (mid-April) rotifers and cyclopoid copepods constitute the main food items for young fry weighing 7 mg. In the middle of May fry weighing 31 mg mainly consume cyclopoid copepods, to a lesser extent daphnids and diaptomids, and rarely chironomids.

The food most relished by broad whitefish is small mollusks (*Sphaerium, Valvata, Pisidium*). In the absence of the required benthos and under obligatory feeding on zooplankton the growth of broad whitefish is retarded. Broad whitefish does not consume artificial feeds in ponds. Thus, broad whitefish is benthophagous and should not be raised together with carp; it is better to cultivate this species with peled or Lake Chudskoe whitefish. Broad whitefish is quite demanding with regard to conditions of external environment, especially the oxygen content in the water. When the oxygen content is 6.6 mg/l, seven-day-old fry become visibly disturbed, just as they do with an increase in water temperature to 25°C. This has also been established that day-old larvae at temperatures during transportation close to the required (4°C) can withstand a reduction in oxygen up to 0.9 mg/l, while at 12°C they die even at an oxygen concentration of 1.9 mg/l. The temperature threshold also changes with the growth of fry. It is slightly lower in the case of broad whitefish than in the case of peled and Lake Chudskoe whitefish reared in ponds in Ukrainian SSR. If at the age of 7 days this threshold is 22.8°C, at the age of 164 days it is 29.5°C, but with further growth (age 243 days) slightly reduces to 26°C.

Juveniles are very resistant to carbon dioxide. However, the sensitivity of these fish to carbon dioxide content in the water increases with age. In one experiment larvae were kept for a short time at a carbon dioxide content of 400 mg/l and a temperature of 13 to 15°C; at the age of 125 and 216 days they ceased movement at 11 to 13°C and a carbon dioxide content of 144 and 147 mg/l.

The growth rate of broad whitefish is not the same in different zones. Fingerlings under the conditions of the Leningrad district range in weight from 14 to 135 g, while in the ponds of the Kiev district (fish farm "Pushcha-Voditsa") where the most relished food of broad whitefish—small mollusks—is plentiful, the average weight is 200 g. Two-year fish in the Leningrad district have a unit weight of 171 to 420 g, while the average weight at "Pushcha-Voditsa" is 1,012 g; the unit weight of three-year fish is 771 and 1,850 g respectively. The initial sexual maturation of males and females of broad whitefish in Ukrainian SSR commences at age 2+, in the ponds of the Leningrad district at age 4+, and in its native waters at the age of 7 years (Yenisey broad whitefish).

In the autumn catches, with the commencement of water discharge, peled (if raised) and broad whitefish move out of the pond first. When collected in fish traps, broad whitefish concentrate on the bottom of the trap, while peled remain in the upper layers of the water. Thus, automatic segregation takes place which makes this process easier.

VENDACE (*Coregonus albula,* Figure 50) and its fast-growing form—ripus Swedish cisco (*Coregonus albula morpha vimba,* Figure 51)—are cold-loving fishes and widely used as material for additional stocking in carp ponds not only in the central regions, but also in the southern regions of the country. They mainly feed on zooplankton. Vendace prefers crustaceans and to a lesser extent feeds on rotifers, aerial insects, and algae. It feeds throughout the year except during the spawning period. The ripus feeds mainly on zooplankton (*Daphnia, Bosmina,* and other crustaceans); however, in some Ural lakes and in the whitefish ponds of the farm "Pushcha-Voditsa" it changes over to partial feeding on benthos (larvae of chironomids). This should be taken into consideration at the time of rearing ripus, especially in carp ponds. Experience has shown that competition for food between carp and ripus occurs only under the conditions of high stocking density and a deficiency in the typical natural food of the latter. Ripus feeds intensively with a reduction in temperature to 10 to 7°C.

Figure 50. Vendace (*Coregonus albula*).

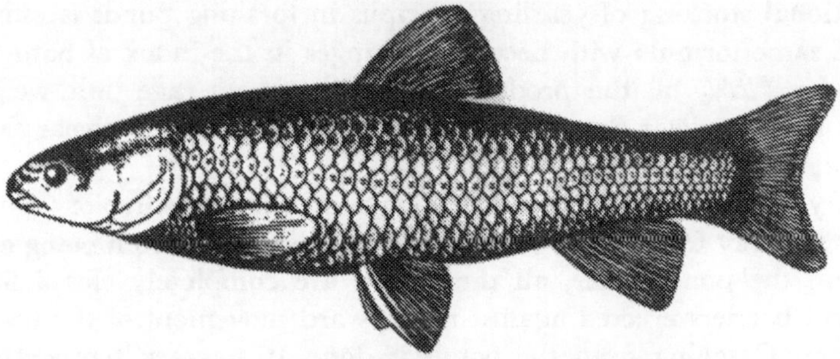

Figure 51. Ripus (*Coregonus albula morpha vimba*).

For the additional stocking of vendace and ripus only better managed carp ponds are suitable, namely those which are fully drainable, not overgrown with higher aquatic plants, devoid of silt deposits, with a good oxygen regime (not less than 35% of saturation point), and with an inflow of high quality water at the rate of 1.5 to 3.0 l/sec/ha. When the oxygen concentration falls below 30% the fishes die. In the northern and northwestern regions of the USSR ripus can be cultured in ponds with an average depth of 1.0 to 1.2 m, containing areas up to 3 m deep, and in the southern regions ponds with still deeper areas. In the Poles'e and forest steppes of Ukrainian SSR good results have been obtained from raising ripus in ponds with an average depth of 1.5 to 2.5 m and a maximum depth of 3.5 m. Fish ponds made in river beds are not suitable for additional stocking of vendace, when the waters are used for meeting the requirements of electric power stations, grinding mills (flour mills), etc., or if the waters are released through spillways. Ponds with an oxidizability greater than 15 mg O_2/l and ponds from which catching of fish with traps is not possible are also not suitable.

Vendace are stocked in carp ponds using fry, juveniles, and yearlings. Fry are released at the rate of 3,000 to 6,000 fish per hectare; in the case of juveniles this norm is increased by one-third. After stocking with juveniles the unit weight of vendace reaches 30 g toward autumn. Wintering ponds are stocked at the rate of 1,000 to 2,000 fish per hectare. Vendace feeds in winter and adds more than 16 g to its body weight. In Poland vendace yearlings are stocked in deeper ponds at the rate of 500 to 1,500 fish per hectare.

Spring-spawning Bauntian vendace from Lake Buryatin are recommended as a new object for additional stocking.

Ripus are stocked in carp ponds using fry or yearlings. The rate of stocking is determined on the basis of the formula used for carp stocking with the necessary modifications in indexes of natural fish productivity (15 to 20% of the productivity of carp) and planned unit weight of fingerlings toward autumn (on the average 60%, ranging from 50 to 65%).

Additional stocking of yearlings of ripus in foraging ponds is estimated using the same formula with necessary changes in the index of natural fish productivity (25% of the productivity of carp), average unit weight of yearlings and two-year fish, and percentage of output of two-year fish (on the average 85%).

The fry of both ripus and vendace are released in areas free of vegetation and located away from water outlets. To prevent the fry from being carried away from the pond water, all the outlets are completely closed for one month and barriers erected against the upward movement of the fry along the current. Catching from the ponds is done at a water temperature of 4°C, with a maximum of 6 to 8°C. Fish traps installed behind the water outlets are best since vendace and ripus are able to move from the ponds to the traps before the arrival of carp. Ripus reaches 60 to 70 g in body weight (maximum 100 g) in one summer period and is nearly ready as a marketable fish. If the fingerlings are kept for a second year of feeding, they are transplanted to wintering ponds with carp fingerlings for the winter period at the rate of 200,000 to 250,000 of carp fingerlings and 5,000 to 7,000 fingerlings of ripus.

In wintering ponds ripus can tolerate a reduction in oxygen concentration up to 1.1 ml/l. However, one summer, as a result of increased temperatures, under the conditions of the Savvinskii fish farm in the Moscow district, fish died at 29°C even when the oxygen tension was maintained at 4.7 ml/l. In deep water ponds with areas of lower temperatures, however, this species can survive even at a surface temperature up to 30°C. In the course of hibernation of fingerlings in the northern regions and under a prolonged winter losses toward spring have reached 5 to 10% with a maximum of 15%. Catching from wintering ponds ceases when the water temperature falls below 4°C. Ripus of two summers can attain a body weight of 300 g.

Lake Chudskoe Whitefish (*Coregonus lavaretus maraenoides,* Figure 52) is widely used for additional stocking in carp ponds in the USSR as well as abroad. This is a cold-loving fish which requires a clean and unpolluted pond with spring water, continuous water exchange, and a depth of 1.5 to 2.0 m. Lake Chudskoe whitefish is characterized by soft, tasty, and fatty meat, and better growth in larger, deeper ponds with an oxygen content of 6 to 7 ml O_2/l. It reaches 60 cm in length and a weight of 2.0 to 3.5 kg; the average weight of marketable two-year fish is 200 to 250 g. In the northwestern regions two-year fish reach 400 g and in Ukrainian and Moldavian SSR, 600 g. In the southern regions of Ukrainian and Moldavian SSR Lake Chudskoe whitefish attains sexual maturity earlier than in its native water—at age 2, and even at age 1 in individual cases. It has also been noted that for an earlier sexual maturation the weight of fingerlings toward autumn is significant: at a body weight of not less than 40 g sexual maturity is attained in two complete years, while in Lake Chudskoe (the native waters of the species) females achieve sexual maturity in the fifth year of life and males in the fourth. At the same time, under the conditions of pond culture, especially in the first three years, the growth rate of Lake Chudskoe whitefish is twice faster than in its native waters.

Figure 52. Lake Chudskoe whitefish.

The nature of feeding of Lake Chudskoe whitefish is not the same at different stages of development. In the ponds of Moldavia the larvae start feeding on the second day of life, consuming nauplii and copepodite stages of copepods. In the process of growth of the larvae *Cyclops, Daphnia,* and *Bosmina* comprise a greater proportion of their food. Fingerlings and adults primarily feed on zooplankton: *Cyclops, Vicinus, Acanthocyclops vernalis, Daphnia longispina, Bosmina longirostris,* and larvae of chironomids, as well as on their own eggs (especially during November, in the period of mass spawning). Pieces of higher aquatic plants are often found in the gut contents and out of the phytoplankton more often colonies of *Microcystis aeruginosa,* together with mineral components in the form of silt or sand. In the ponds of Ukrainian SSR the larvae change over to active feeding on the seventh to the eighth day of life and continue feeding in winter. The main food in winter (when carp does not feed) consists of chironomids and oligochaetes. Because of this some researchers consider Lake Chudskoe whitefish a benthophagous fish. The daily rhythm of feeding is not the same in each age group. Fry raised in ponds feed (predominantly on planktonic crustaceans) during the daytime, while fingerlings feed in the evening hours, and all the adult groups are characterized by a reduction in gut filling at night.

Lake Chudskoe whitefish feeds in winter also and hence should be released in the foraging ponds in autumn. If this is not possible under the conditions of a particular farm, whitefish is stocked in autumn in rearing ponds of sufficient depth, or (although less desirable) in wintering ponds where it undergoes hibernation with little loss if carefully handled at the time of catching and transplantation. The ponds are cleaned before autumn stocking, vegetation overgrowth removed, and lime applied. The stocks of natural foods are increased with live food. Whitefish is planted at a depth up to 0.5 m, preferably at places with a sandy bottom free of vegetation overgrowth. To prevent fish egress from the ponds formidable barriers are erected in the water flow near the outlets. Whitefish yearlings weighing 15 to 20 g are stocked in foraging carp ponds. In the process of rearing fingerlings the weight of whitefish in replenishing stocks increases to 25 to 30 g and more. The rates of stocking are decided by considering the average productivity 50 kg/ha with a survival rate of 85 to 90%.

Fish productivity of Lake Chudskoe whitefish under monoculture in the northwestern regions reaches 150 to 160 kg/ha and in the central belt 150 to 200 kg/ha. Taking into account that these fish respond poorly to transplantation from one category of pond to another, it has been recommended in Moldavian SSR that fish be raised in drainable ponds up to one year of age, and in nondrainable ponds up to two or three years of age.

An interesting object for additional stocking is the hybrid of Lake Chudskoe whitefish and peled. In Ukrainian SSR (Kiev district) when these fish were stocked with common carp and grass carp, the two-year hybrids reached 295 to 400 g, and some individuals even 560 g, in the foraging ponds.

STERLET (*Acipenser ruthenus,* Figure 53) is a very valuable fish for additional stocking. Large ponds with flowing water and good aeration are most suitable for its feeding. The pH level should be 6.4 to 7.6. However, sterlet can live not only in flowing ponds with cold water, but also in carp ponds. Sterlet should never be stocked as a marketable fish in silt-covered ponds or those overgrown with aquatic vegetation, since its meat in such conditions takes on a muddy taste that is very difficult to remove. Even keeping such sterlet in running water for several days is not always effective. Parts of the

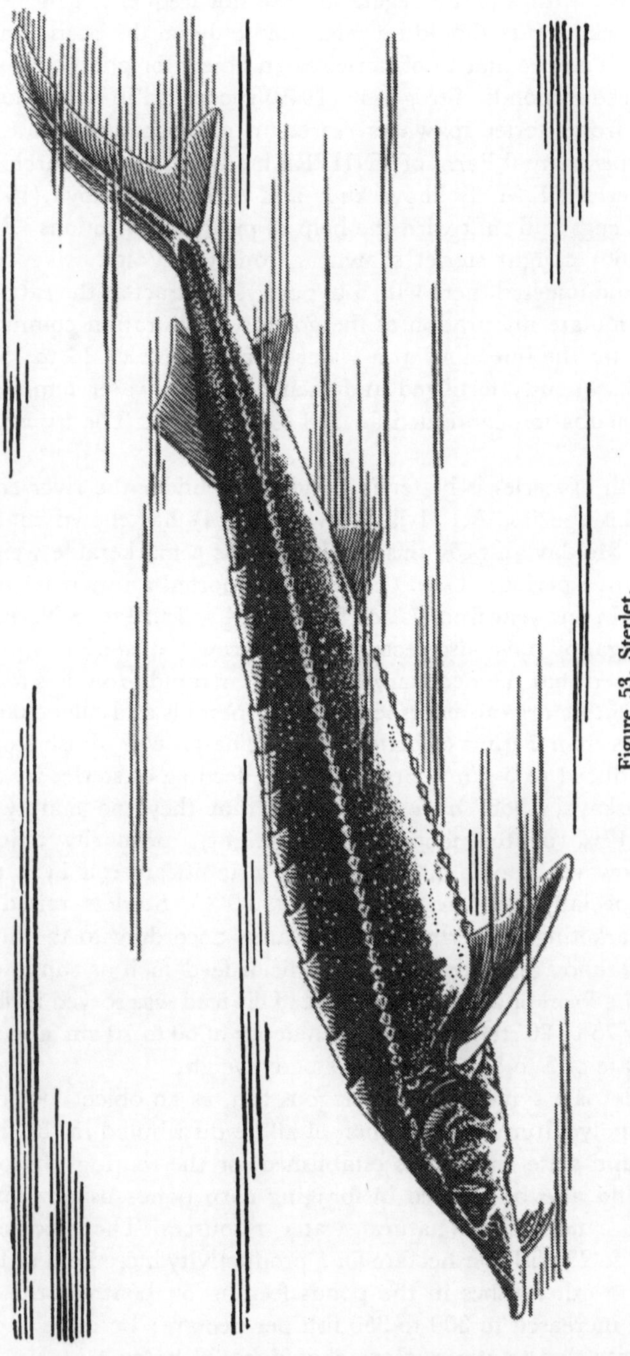

Figure 53. Sterlet.

pond overgrown with aquatic vegetation are not frequented by sterlet and hence their stocking rate should be calculated only on the basis of areas free of vegetation. The peculiarity of sterlet as an object for pond culture is that it does not breed in ponds. Stroganov (1949) succeeded in obtaining mature sex products from sterlet spawners reared in experimental ponds. At the Savvinskii Experimental Farm of VNIIPRKh (All-Union Research Institute of Pond Fisheries), F.M. Sukhoverkhov and A.S. Pisarenkova (1956) have also obtained eggs and milt with the help of pituitary injections. Rozhdestvenskaya (1966) caught sterlet spawners from the Volga River weighing 800 to 900 g and injected them with a hypophysis extract at the rate of 10 mg per fish to stimulate maturation of the gonads. Maturation commenced 24 to 36 hours after the injections at a water temperature of 14 to 15°C. The eggs were subsequently fertilized and incubated at a water temperature of 17 to 20°C; incubation continued for 108 to 120 hours. The fry were raised in ponds.

The growth of sterlet is better in ponds than under the river conditions natural for this species. A.I. Irikhimovich (1954) has shown that for the conditions of Moldavian SSR this species attains a marketable weight even in two vegetative periods. O.A. Grimm has reported an increase in weight of sterlet even in one year from 77 to 154 g to 614 to 1,024 g. S.N. Skadovskii and N.S. Stroganov have also recorded good growth of sterlet in ponds and have emphasized that the necessary condition for rapid growth is a sufficient quantity of benthic organisms, especially chironomids and oligochaetes, at a rate of not less than 7 g/m^2 of bottom area. The growth of fish stops if this quantity is reduced to 5 g/m^2. Studies on the feeding of sterlet fry in ponds (Rozhdestvenskaya, 1966) have also shown that they too mainly feed on benthos (52.12% of the food content weight), primarily chironomids (32.21%). However, zooplankton also plays a significant role in its nutrition (47.88%), especially *Daphnia longispina* (45.89%). Sterlets raised for the purpose of marketing in autumn will consume, according to the findings of F.M. Sukhoverkhov and others, even artificial feed such as sunflower meal prepared in the form of a doughlike mass. This feed was served in feed pans (one for every 75 to 80 sterlet) placed at a depth of 60 to 70 cm, and supplied daily at the rate of 3 to 5% of the fish's body weight.

Since sterlet is a typical benthophagous fish, as an object for additional stocking and polyculture it should first of all be distributed in the fish farms of collective and state enterprises established for the rearing phytophagous fishes. It should also be stocked in foraging carp ponds using yearlings or two-year fish caught from natural water resources. The stocking norm should be 200 to 250 fish per hectare for a productivity increment of 15 kg/ha. If there are no other fishes in the ponds feeding on benthos, the stocking norm may be increased to 300 to 350 fish per hectare; here the increase in fish productivity due to sterlet alone should be 20 kg/ha.

The hybrids of beluga and sterlet, which grow to 800 g toward the end of the second vegetative period, ought also to be seriously considered as additional stocking material. The "khramulya," a cyprinid fish, especially the Samarkand "khramulya" (*Varicorhinus heratensis stendachneri*), a subspecies of the Asiatic "khramulya" distributed in the basin of the Amu-Darya and Zeravshane Rivers, appears promising. Another subspecies (*Varicorhinus capoëta sevangi*), from Lake Sevan, has already become an important object of fishery. "Khramulya" species unfortunately grow slowly, a disadvantage in additional fish in pond fisheries. However, Asiatic and Samarkand "khramulya" grow more rapidly than Lake Sevan "khramulya". All of the "khramulyas" are phytophagous, feeding mainly on lower algae. They are delicious fatty fishes from which cured "balyk" is prepared. At a weight of 50 to 100 g they are considered marketable. "Khramulyas" could easily be cultured in carp ponds with a good oxygen regime.

MULLETS have been successfully cultured in recent years as an additional fish in carp ponds in the USSR and some foreign countries (China, India, Israel, Japan, etc.). First place among this group is held by the largest species, the striped or gray mullet (*Mugil cephalus*, Figure 54). In the USSR this fish is commercially fished in the Black and Azov Seas. The leaping gray mullet (*Mugil saliens*) is another candidate for culturing.

A number of studies have been done on the introduction of mullets for additional stocking. Babayan' has paid special attention to the use of water resources in the southern regions of the country located in the direct vicinity of the sources of mullet fry. According to him there are 240 estuaries in the Kuban region with a total area of 150,000 hectares; in the Odessa district the area under flood lands of the Danube and Dnestr together with the estuaries near the sea total 200,000 hectares. Similar water resources for rearing mullets are also available along the southern and northern coasts of the Caspian Sea, in Turkmenia along the sea coast of Azerbaidzhan SSR, Dagestan ASSR, and other places. All these coastal water resources are brackish and should be primarily used for mullet culture. However, even ponds are suitable for this purpose, although the biotechniques of mullet

Figure 54. Striped or gray mullet.

rearing in them are not yet fully developed. Experience abroad has shown that rearing mullet in monoculture is not profitable but that its introduction as an additional fish is quite economical. Furthermore, gray mullet, although a marine fish, can also live in fresh waters and grows more rapidly than in the sea. For example, in India it grows to a weight of 2.5 kg in the second year. In China the ponds for mullet rearing are 2 to 3 m deep but in some places even shallower. The optimum density of plantation is 1.5 to 3.0 fishes per m^2 in Japan and 10 to 12 fishes per m^2 in China.

In the countries listed above stocking material is either collected from the seas or from estuaries and transferred to the place of rearing. In China raising gray mullet with grass carp is considered a better combination; in Japan its combination with Japanese eel is preferred. In the process of mullet raising in the Odessa district (weight of stocking material in the spring 0.25 to 1.20 g) in nondrainable carp foraging ponds with an area of 160 hectares and different depths (maximum 6 m at some places) at a stocking rate of about 438 fish per hectare, the unit weight of the fish reached 100 g in the second decade of October. In the experiments of A.K. Chizhik (1968) two-year striped mullet reached 450 to 620 g during combined raising with carp and whitefish in the conditions of southern Ukraine. A tentative plan for raising these fish has been proposed for ponds under the following polyculture:

(a) carp yearling—1,000 fish per hectare with a unit weight of 25 g;

(b) grass carp yearlings—500 to 1,500 fish per hectare with a unit weight of 20 g;

(c) silver carp yearlings—1,000 fish per hectare with a unit weight of 20 g;

(d) mullet fry—200 fish per hectare with a unit weight of 1 to 2 g.

The anticipated unit weight toward autumn should be: carp—500 g, grass carp and mullet—400 g, and silver carp—450 g, with an output of carp and grass carp of 80% of the stocked material and of silver carp and gray mullet, 50%. The projected natural fish productivity is: mullet—40 kg/ha, carp—400 kg/ha, silver carp—360 kg/ha, and grass carp—32 to 160 kg/ha.

Detritus occupies an important place in the food of mullets. Spawning of mullets in ponds has not been successful to date. Let us mention here that V.K. Vinogradov and others have conducted experiments in the VNIIPRKh (1967 to 1968) on the artificial breeding of striped mullet. As a result of these studies it is now possible to use spawners of striped mullet for artificial breeding. In addition methods for their maintenance after pituitary extracts have been determined; even fertilized eggs have been obtained and observations conducted on the development of fertilized eggs and free embryos.

The acclimatization of the Pacific mullet (*Mugil soiny*, Figure 55) is of great interest in the European part of the USSR (Azov-Black Sea region); this fish can tolerate a reduction in water temperature to almost 0°C, which is lethal for Black Sea mullet.

Figure 55. Pacific mullet.

It should be noted that the fishes mentioned above are not the only potential candidates for additional stocking and polyculture.

Shemaya (*Chalcalburnus chalcoides*) and vimba (*Vimba vimba*) can also be used for additional stocking. Ponds for culturing these fishes should have a slow water exchange, be overgrown with higher aquatic plants, with coarse plants occupying not more than 15 to 20% of the pond area, and with a silt layer of up to 20 cm. Under favorable conditions two-year vimba attain a body weight of 100 to 150 g and fingerlings 6 to 8 g. According to Knyazeva (1960) fingerlings of shemaya and vimba enjoy a similar diet. Under the conditions of raising in ponds, daphnids, larvae of mosquitoes, and other insects were recovered from their guts. Two-year vimba feed on the larvae of mosquitoes and other insects, various kinds of worms, and the seeds of weed plants. Two-year Shemaya feed on beetles and insect larvae. Such a spectrum of feeding in shemaya and vimba makes their combined raising with planktophagous fish desirable, but necessary standards have yet to be worked out.

Omul (*Coregonus autumnalis*), muksun (*Coregonus muksun*), and Caspian inconnu (*Stenodus leucichthys*), depicted in Figures 56, 57, and 58 respectively, are also good candidates for additional stocking.

Suckers (family Catostomidae) likewise appear quite promising for pond culture. Of the 30 species of this family, most live in the waters of North America, from Mexico to the Arctic Ocean. Three species of the genus *Ictiobus*—bigmouth buffalofish (*I. cyprinellus*), smallmouth buffalofish (I. *buffalus*), and black buffalofish (*I. niger*)—have been used in pond culture abroad (Figure 59). All species of buffalofish are found in shoals.

These fishes are cultured in the USA in ponds, in the delta of the Mississippi River, and in rice fields. Bigmouth buffalofish predominates in rice

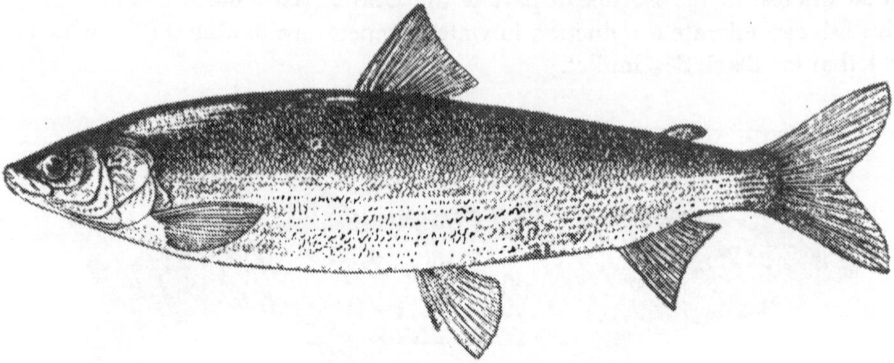

Figure 56. Omul.

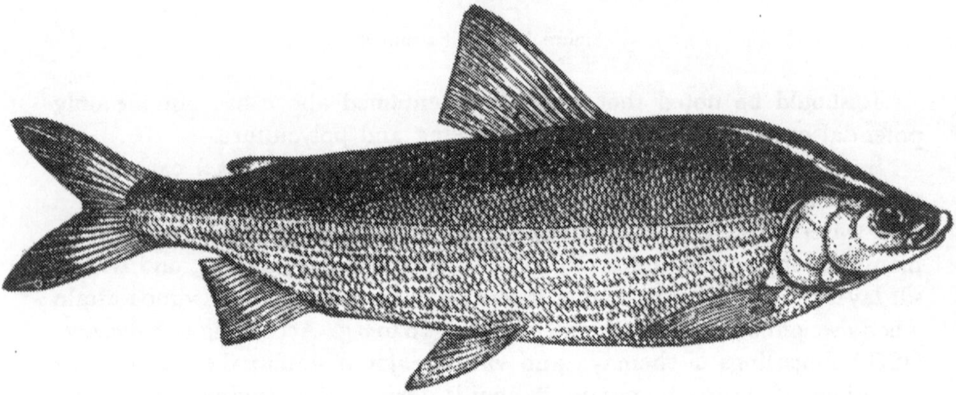

Figure 57. Muksun.

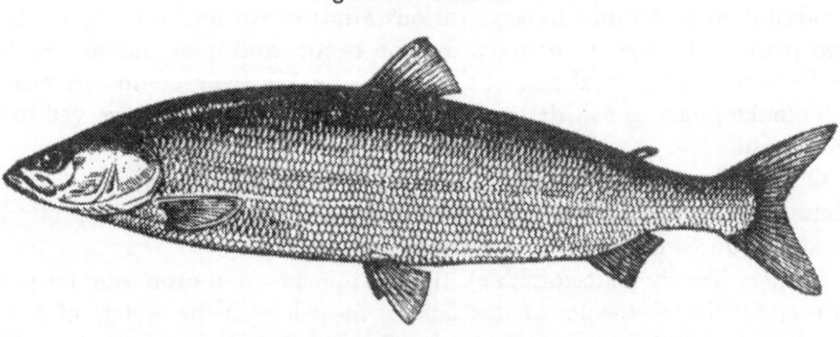

Figure 58. Caspian inconnu.

field culturing, although in taste qualities smallmouth buffalofish is considered the most valuable. In the USA these fishes are valued three times higher than carp. All of the aforementioned species of buffalofish spawn in the first half of summer, from the end of March to June, at a water temper-

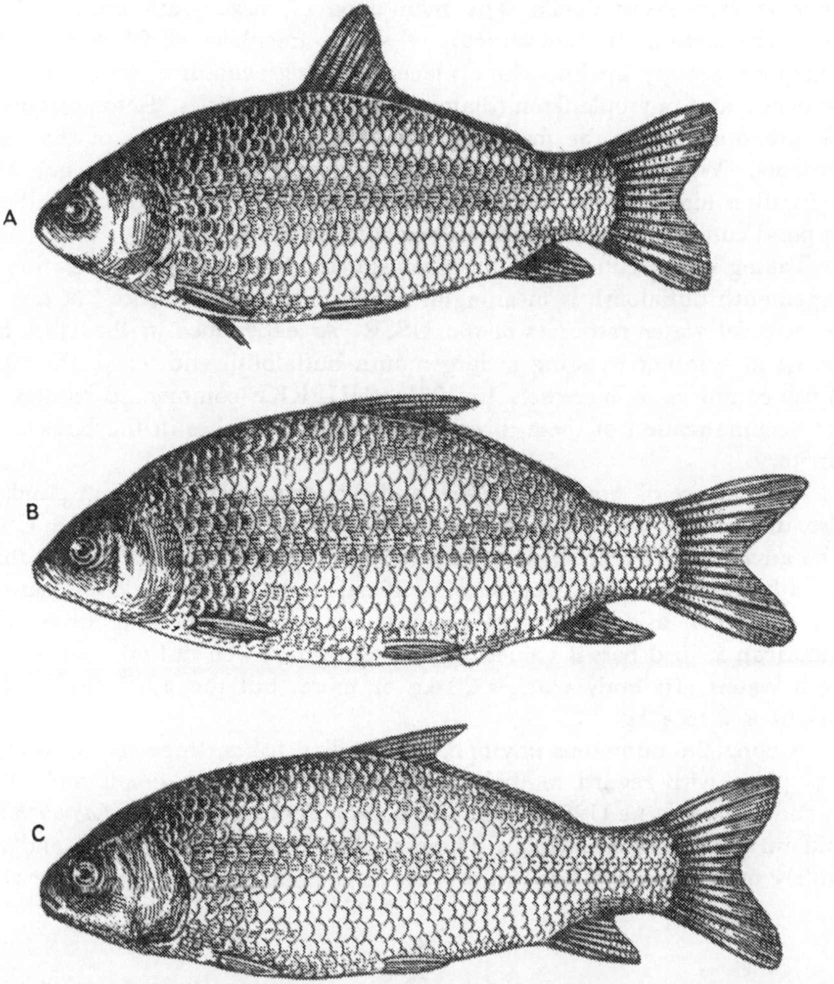

Figure 59. Under-yearlings of buffalofishes.
A—largemouth buffalofish; *B*—smallmouth buffalofish; *C*—black buffalofish.

ature of 14.4, or in the range of 15.6 to 16.7°C. At the time of spawning the males exhibit a spawning livery in the form of a brighter coloration and horn-like tubercles on their head. Males are generally smaller than females. Sexual maturation starts in the fourth year, or later in the northern regions. The eggs of buffalofishes are quite small and adhesive; they attach to aquatic plants. At 17°C hatching takes place after 9 to 10 days. Some species of buffalofish attain high weights. Bigmouth buffalofish reaches 15 and even 45 kg, smallmouth 15 to 18 kg, and black buffalofish 7 kg. All buffalofishes are carnivorous but docile. The main food of largemouth buffalofish is lower crustaceans (Entomostraca), of which members of Cladocera and Copepoda occupy an important place. Yearlings consume water beetles, ostracods, and phytoplankton (diatoms) in small quantities. Entomostracans are predominant in the food of older fish (comprising 75% of the gut contents). With further growth these fish prefer chironomids but still maintain a higher total weight of crustaceans in the total food composition. In pond culture they readily consume artificial feeds.

Taking into account what has been stated above, the acclimatization of largemouth buffalofish is meaningful not only for pond culture but also in the natural water resources of the USSR. As experience in the USA has shown in combined raising of largemouth buffalofish with carp, the yield of fish considerably increases. In 1971 VNIIPRKh commenced studies on the acclimatization of these three species of buffalofishes in the Krasnodar territory.

The culture of American catfish and American striped bass in ponds is also interesting. Both species are widely cultured in ponds in the USA. The advantage of rearing American catfish is not so much its food quality, as its highly effective utilization of artificial feed (1.8 kg per kg increase in body weight), aside from the fact that its rearing is not laborious. The American striped bass is valued as food and grows well in both marine and fresh waters. Its body reaches 20 kg or more, but the usual commercial weight is 2 to 4 kg.

Among the numerous phytophagous Indian fishes, three species occupy first place with regard to their prospects for rearing in ponds and other inland waters in the USSR: catla (*Catla catla,* Figure 60), rohu (*Labeo rohita*), and mrigal (*Cirrhina mrigala*). These species grow rather quickly and are widely used in fish culture not only in India, but also in other countries of

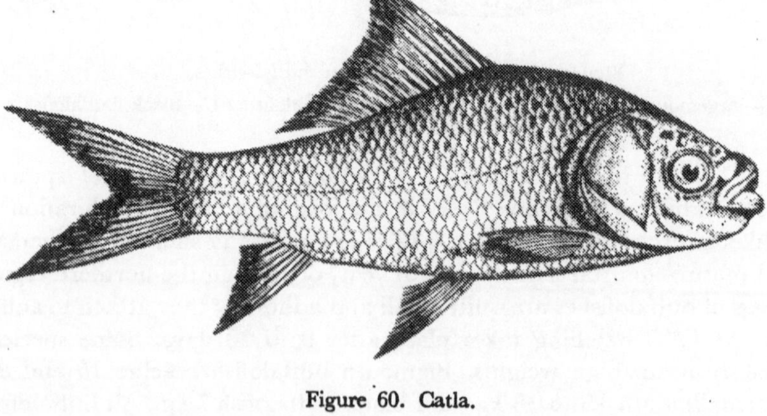

Figure 60. Catla.

Southeast Asia. Sufficient experience has been accumulated on their maintenance. Based on the nature of their feeding, these fishes can be cultured in a single pond: catla mainly consumes vegetation and decomposing higher plants; rohu feeds on growing and decomposing vegetation (50% of its ration) and large colonial algae, zooplankton, and detritus and silt; mrigal consumes blue-green algae, diatoms, filamentous algae, higher plants, and detritus to a larger extent than rohu (Table 15).

Table 15. Characteristic peculiarities of growth, weight, and fish productivity of catla, rohu, and mrigal (from V.V. Verigin)

Names of Indian phytophagous fishes	Maximum length, m	Maximum weight, g			Fish productivity by culturing in ponds, kg/ha
		Yearlings	2-year fish	3-year fish	
Catla (*Catla catla*)	1.8	1,125–4,100	—	6,750	1,100–2,200
Rohu (*Labeo rohita*)	0.9	900	3,600	5,400	1,000–2,000
Mrigal (*Cirrhina mrigala*)	—	650–1,800	2,600	4,000	1,100

Catla is the most rapidly growing and largest fish among Indian carp. All the species mentioned above live in rivers, flood lands, and lakes; they are capable of living not only in fresh waters but also in brackish. Under the climatic conditions of India they reach sexual maturity at the age of two years. Spawning takes place in rivers from June to September and the eggs develop in the water column.

The acclimatization of the three species mentioned above in the USSR is meaningful in regions with mild climatic conditions: Central Asia, Krasnodar, southern Ukrainian SSR, and Moldavian SSR. Warm industrial waters can also be used in culturing them.

In addition to the phytophagous fishes of the Far East complex, the black or Amur carp (*Mylopharyngodon piceus,* Figure 61), should not be overlooked. This fish lives in the rivers of continental China and Taiwan as well as in

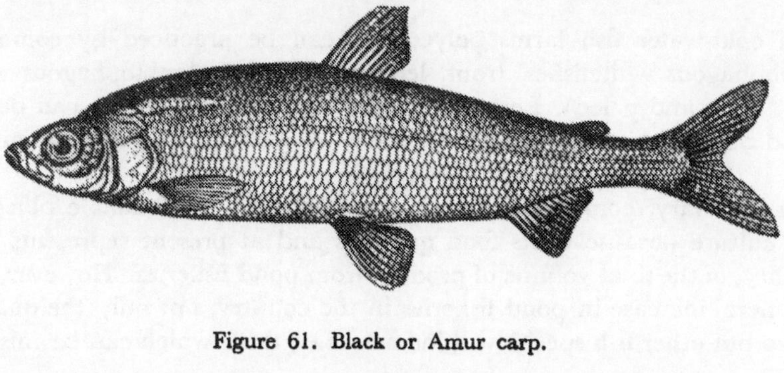

Figure 61. Black or Amur carp.

the Amur River. It feeds on gastropods, mainly *Viviparus,* and is quite large, reaching a length of 80 cm.

The fertilized eggs of ayu (*Plecoglossus altivelis,* family Plecoglossidae; Figure 62) were imported from Japan to the Chernorechenskoe trout farm (near the city of Gudauty). This fish is highly valued in Japan because of its good taste qualities. It can also be successfully reared in the USSR, especially in the subtropical regions of the Black Sea. Spawning takes place from August to October in rapidly flowing transparent streams and rivulets at deeper places. Fecundity reaches 10,000 eggs. The eggs are small and adhesive; they attach firmly to submerged objects. The fish spends the earlier period of its life in the estuarine areas of the sea at the confluence of fresh and marine waters and feeds on animal organisms. In March–April, by which time it is 7 cm long, it moves up the rivers and sexual maturation occurs anytime between August to October. In rivers ayu feeds on diatoms and blue-green algae. The length of an adult fish is 25 cm and its weight about 30 g. Larvae are seen in water temperatures of 15 to 20°C after 14 to 20 days. Ayu lives for one year, rarely two to three, since most of the fish die after spawning. Under the conditions of pond culture in Japan it grows better than in natural waters. Ayu is very sensitive to the presence of irson salts in water (more than 5.5 mg/l).

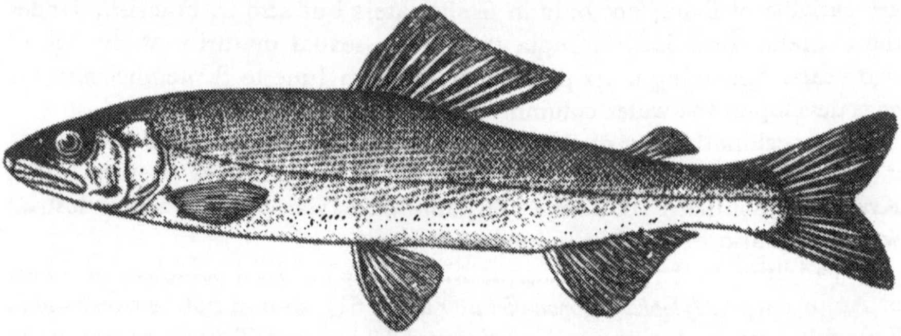

Figure 62. Ayu.

In cold-water fish farms polyculture can be practiced by combining benthophagous whitefishes, trout, lenok, char, and planktophagous whitefishes (ripus and peled). Lenok, due to its omnivorous nature, can become a good object for cold-water pond fishery, especially because, like rainbow trout, it too is a spring-spawning fish.

In summary, common carp is the main and most valuable object for pond culture because of its food qualities and at present represents more than 90% of the total volume of produce from pond fisheries. However, with the general increase in pond fisheries in the country, not only the quantity of carp but other fish species will increase, i.e., those which can be raised in additional stocking or in "polyculture". The examples of "polyculture" given above have proven that raising common carp together with grass carp, silver carp, and bighead increases fish yield two to three times. In the RSFSR in 1968 phytophagous fishes constituted 15% of the produce from pond fisheries, and in Uzbek SSR 50%. Carp culture together with one species of planktophagous fish and one species of predatory fish should be practiced more widely, since such polyculture increases fish productivity of the foraging ponds significantly compared to the monoculture of carp. In polyculture a combination of fish species should be given first preference.

A wider introduction of polyculture in the practice of pond fisheries and fish farms of collective and state enterprises would make it possible not only to increase the quantity of pond fishes and fish productivity, but would also ensure a wider assortment of fishes. Furthermore, such produce can be obtained with lower investments in food and for higher profits.

References

Agapov, I.I. and V.N. Abrosov. 1967. Ob izberatel'nosti pitaniya ryb i rybovodnom znachenii shchuki (Selective feeding of fish and the fishery importance of pike). *Vopr. Ikhtiol.*, 7, 1, 42.

Anisimova, I.M. 1963. Kachestvo potomstva serebryanogo karasya pri skreshchivanii s samtsami karpa zolotogo karasya i linya (Quality of the progeny of goldfish after crossing with male carp, Crucian carp, and tench). *Doklady TSKhA,* no. 85.

Babayan, K.E. 1960. Vyrashchivanie kefali kak dobavochnoi ryby v karpovykh prudakh (Raising gray mullet as an additional fish in carp ponds). Sbornik *Nauchn.-tekhn. Inform. VNIRO*. Moscow, vol. 6.

Babayan, K.E. 1963. Vyrashchivanie kefali v presnovodnykh vodoemakh (Raising gray mullet in fresh waters). *Trudy VNIIPRKh*. Moscow, vol. 12.

Babayan, K.E. 1965. Kefal' (The Gray Mullet). Pishchepromizdat, Moscow.

Balan, A.I. 1968. Som kak novy ob"ekt prudovogo rybovodstva (Catfish as a new object of pond fisheries). Avtoref. Kand. Diss.

Borovik, E.A. 1968. Raduzhnaya forel' kak dobavochnaya ryba v karpovykh prudakh (Rainbow trout as an additional fish in carp ponds). *Materialy XIV Konf. po Izucheniyu Vnutrennikh Vodoemov Pribaltiki,* vol. 1, pt. 2.

Burmakin, E.V. 1958. Razvedenie shchuki vo Frantsii (Pike rearing in France). Sbornik *Rybnaya Promyshlennost' za Rubezhom.* Izd. VNIRO.

Chizhik, A. and L. Abramovich. 1968. Rybnitstvo zoni zroshuvanogo *zemlerobstva* (Fisheries in the Zone of Irrigated Agriculture). Odessa.

Dombrovskii, V.K. 1964. Pishchevye atnosheniya karpa i serebristogo karasya pri sovmestnom vyrashchivanii v prudakh (Food interactions of carp and goldfish during their combined raising in ponds). Sbornik *Biol. Osnovy Rybn. Khoz-va na Vnutrennikh Vodoemakh Pribaltiki.* Izd. Nauka i Tekhnika, Minsk.

Dombrovskii, V.K., I.A. Krasheninnikov and G.F. Malysheva. 1964. Pelyad' kak ob"ekt razvedeniya v karpovykh prudakh (Peled as an object of rearing in carp ponds). Sbornik *Biol. Osnovy Rybn. Khoz-va na Vnutrennikh Vodoemakh Pribaltiki.* Izd. Nauki i Telhnika, Minsk.

Golovkov, G.A. 1966. Perspektiva ispol'zovaniya chira v ozernom i prudovom rybnom khozyaistve (The prospects of using broad whitefish in lake and pond fisheries). *Sobeshch. po Biol. Produktivnosti Vodoemov Sibiri.* Irkutsk.

Golovkov, G.A. 1968. Novye adresa pelyadi (New locations of peled). *Rybovodstvo i Rybolovstvo,* no. 4.

Golovkov, G.A. et al. 1967. Chir (*Coregonus nasus* Pall.) i perspektiva ego ispol'zovaniya v rybovodstve (The broad whitefish *Coregonus nasus* Pall. and the prospects of its utilization in fisheries). *Izv. GosNIORK,* vol. 35.

Hochman, L. 1967. Eerfahrungen mit der aufzucht der welsbrut in kleinen behältern, *Acta Union Agricult.,* vol. 15, no. 4.

Horoszewicz, L. 1972. Hodowla sumow stawach, *Gospodarca Rybna,* nos. 1–3.

Instruktsiya po razvedeniyu i tovarnomy vyrashchivaniyu gibridov belugi so sterlyd'yu (Instructions on Breeding and Commercial Rearing of Hybrids of Beluga and Sterlet). Izd. VNIRO.

Kanaev, A.I. 1952. Biotekhnika razvedeniya shchuki v karpovykh prudakh (Biotechniques of rearing pike in carp ponds). *Rybn. Khoz-vo,* no. 7.

Kharitonova, N.N. 1964. Serebryanyi karas kak dobavochnaya ryba (Crucian carp as an additional fish). *Rybovodstvo i Rybolovstvo,* no. 6.

Kharitonova, N.N. 1968. Pitanie i pishchevye vzaimootnosheniya karpa i serebryanogo karasya v prudakh (Feeding and food interactions of the common carp and Crucian carp in ponds). *Trudy UNIRKh,* vol. 15.

Knyazeva, E.S. 1960. Vyrashchivanie rybtsa i shemai v kolkhoznykh prudakh (Raising vimba and shemaya in fish ponds of collective enterprises). *Sel'sk. Khoz-vo Severnogo Kavkaza,* no. 7.

Kononov, V.A. and Z.A. Makina. 1952. Vyrashchivanie tovarnykh segoletkov shchuki v nagul'nykh karpovykh prudakh (Raising marketable pike fingerlings in foraging carp ponds). *Trudy UNIRKh,* no. 8.

Kozhin, N.I. 1948. Razvedenie siga v prudakh (Rearing of whitefish in ponds). *Rybn. Khoz-vo,* no. 10.

Lavrovskii, V.V. 1967. Vyrashchivanie dvukhletkov raduzhnoi foreli v polikul'ture (Raising two-year rainbow trout in polyculture). *Trudy VNIIPRKh.* Moscow, vol. 15.

Lyakhnovich, V. 1963. Serebryanyi karas' v prudakh Belorussii (Goldfish in the fish ponds of Belorussia). *Rybovodstvo i Rybolovstvo,* no. 1.

Mel'nikov, E.F. 1956. Nekotorye e'kologicheskie osobennosti vyrashchivaniya linya v prudovykh khozyaistvakh Ukrainskoi SSR (Some ecological peculiarities of raising tench in the fish farms of Ukrainian SSR). Sbornik *Rabot Ukr. N.-i. Stantsii Rybovodstva,* no. 2.

Mikheev, P.V. and E.V. Meisner. 1966. Razvedenie sudaka v prudakh (Sander Culture in Ponds). Pishchevaya Promyshlennost', Moscow.

Nikanorova, E.A. 1965. O biologicheskoi raznokachestvennosti sudaka (Biological variability of sander). *Rybn. Khoz-vo,* no. 12.

Nikanorova, E.A. 1968. Vyrashchivanie v prudakh razlichnykh e'kologicheskikh grupp sudaka (Pond culture of different ecological groups of sander). *Trudy Karel'skogo Otd. GosNIORKh.*

Nishkov, N. 1966. Otglezhdane i razv''zhdane na soma (Farming and rearing of catfish). *Ribno Stonanstvo,* no. 3.

Nosal', A.D. 1957. Biotekhnika razvedeniya i vyrashchivaniya chudskogo siga (The biotechnique of breeding and rearing Lake Chudskoe whitefish). Sbornik *Obmen Peredovym Tekhnicheskom Opytom v Rybnoi Promyshlennosti.* Moscow.

Nosal', A.D. 1961. Opyt razvedeniya shchuki na Ukraine (Experience in pike culture in the Ukraine). *Rybovodstvo i Rybolovstvo,* no. 4.

Nosal', A.D. 1963. O dobavochnykh rybakh (Additional fishes). *Rybovodstvo i Rybolovstvo,* no. 3.

Nosal', A.D. 1967. Chir v prudakh Ukrainy (Broad whitefish in the ponds of the Ukraine). *Priroda,* no. 7.

Petrov, S. 1961. Razvedenie sigov v Kurgonskoi oblasti (Rearing of whitefish in Kurgan district). *Rybovodstvo i Rybolovstvo,* no. 5.

Poltavchuk, M.A. 1965. Biologiya i razvedenie Dneprovskogo sudaka v zamknutykh vodoemakh (Biology and Rearing of the Dnieper Sander in Captive Waters). Kiev.

Razvedenie americanskogo soma v prudakh (Rearing of American catfish in ponds). *S.-Kh. E'kspress Informatsiya,* no. 9.

Romanycheva, D.D. 1966. Vliyanie rezkikh kolebanii temperatury vody na razvivayushchuyusya ikru sudaka (Influence of sharp fluctuations in water temperature on the developing eggs of sander). *Trudy Azovsk. In-ta Rybn. Khoz-va,* vol. 8.

Rozhdestvenskaya, A.D. 1966. Vyrashchivanie sterlyadi v prudakh (Raising sterlet in ponds). Sbornik *Nekotorye Voprosy Osetrovogo Khozyaistva Kaspiiskogo Basseina.* Izd. TSNIORKh i VNIRO.

Schäperclaus, W. 1961. *Lehrbuch der Teichwirtschaft.* Berlin.

Shkorbatov, G. and G. Kudryavtseva. 1961. Sigovye v Kharkovskoi oblasti (Whitefish in the Kharkov district). *Rybovodstvo i Rybolovstvo,* no. 5.

Shpet, G. and A.I. Balan. 1967. Razvedenie soma v prudakh (Rearing of catfish in ponds). *Rybovodstvo i Rybolovstvo,* no. 2.

Stroganov, N.S. 1949. Sozrevanie sterlyadi v prudakh i akvarial'nykh usloviyakh (Maturation of sterlet in ponds and aquarium conditions). *Vestn. Mosk. Un-ta,* no. 5.

Sukhoverkhov, F.M. 1966. Biologicheskie osnovy i e'ffektivnost' polikul'tury v prodovom rybovodstve (Biological Basis and Efficiency of Polyculture in Pond Fisheries). Pishchevaya Promyshlennost', Moscow.

Sukhoverkhov, F.M. et al. 1956. Osobennosti vyrashchivaniya i zimnego soderzhaniya sterladi v prudakh (Peculiarities of raising and winter maintenance of sterlet in ponds). *Trudy VNIIPRKh.* Moscow, vol. 8.

Sycheva, A.V. 1955. O sutochnom khode pitaniya pelyadi (The daily rhythm of feeding of peled). *Vopr. Ikhtiol.,* no. 4.

Tomnatik, E.N., V.I. Kralova et al. 1968. Rol' sudaka v povyshenii ryboproduktivnosti nagul'nykh karpovykh prudov (Role of sander in increasing fish productivity of foraging carp ponds). *Materialy XIV Konf. po Izucheniyu Vnutrennikh Vodoemov Pribaltiki,* vol. 1, pt. 2.

Verigin, V.V. 1966. Indiiskie rastitel'noyadnye ryby kak vozmozhnye vselentsy v vodoemy SSSR (Indian phytophagous fish and their potential for adaptation to waters in the USSR). Sbornik *Rybokhozyaistvennoe Osvoenie Rastitel' noyadnykh Ryb.* Nauka.

Vinogradov, V. and L. Erokhina. 1970. Perspektivny ob''ekt aklimatizatsii (A promising object for acclimatization). *Rybovodstvo i Rybolovstvo,* no. 4.

Vovk, P. 1962. O razvedenii soma v Vengrii (Rearing of catfish in Hungary). *Rybovodstvo i Rybolovstvo,* no. 2.

Zambriborshch, F.S. 1954. Opyt vyrashchivaniya kefali v karpovom nespusknom prudu (Exprerience in raising mullet in a nondrainable carp pond). *Rybn. Khoz-vo,* no. 10.

Zelenin, A., V. Korolev et al. 1965. O polikul'ture ryb v prudovom khozyaistve Moldavii (Polyculture of fish in the pond fisheries of Moldavia). *Rybovodstvo i Rybolovstvo,* no. 5.

Chapter IX

PRODUCTION PROCESSES IN THE CULTURE OF GRASS AND SILVER CARP AND BIGHEAD IN PONDS

In recent years, especially in the southern regions of the USSR, phytophagous fish have been introduced more and more into pond culture, primarily grass carp (*Ctenopharyngodon idella,* Figure 63) and silver carp (*Hypophtalmichthys molitrix,* Figure 64). In addition to these species, and in combination with them and common carp, bighead (*Aristichthys nobilis,* Figure 65) has also been included in pond culture. All three fishes belong to the cyprinid family and occur in rivers along the Pacific Coast of Asia from Amur to China. These species have been acclimatized in the European part of the USSR, western Siberia, Central Russian Republics, and other regions of the country, and are widely used in pond culture, especially in the southern regions. All three species are more heat-loving than common carp. In Turkmenia the females of grass and silver carp reach sexual maturity in their third to fourth year, in Uzbekistan in the fourth year, in Krasnodar in the fourth to fifth year, and in Moscow district in the seventh to eighth year. The males usually mature one year earlier than the females.

These species differ significantly in the nature of their feeding habits. Grass carp feeds on higher and lower plants and prefers soft young plants. When these food items are in short supply or absent, large fish will consume coarse plants (cattail, reed, etc.). Of the aquatic plants, those preferred are filamentous and narrow-leaved pondweed, hornwort, water thyme, and duckweed; from land plants, cereals, clover, and lucerne. This species utilizes grass mixtures better and digests them well. The daily ration of two-year fish consists, on the average, of 40% of the fish weight in the form of land plants, with a food coefficient of about 30. Grass carp changes over to feeding on plants during the first year of its life at a length of about 3 cm, and grows better if its ration includes about 30% animal food. It mainly feeds during the day and stops feeding at a temperature of 10 to 12°C. With a deficiency of aquatic plants in the ponds, grass carp consumes the food provided for common carp.

Silver carp feeds on phytoplankton and detritus. It prefers diatoms and green algae but consumption of blue-green algae is poor, as it avoids many of these species. The fish change over to feeding on algae at a length of

Figure 63. Grass carp.

Figure 64. Silver carp.

Figure 65. Bighead.

about 1.5 cm, but here, as in the first period of life of the fry, zooplankton also plays an essential role. The daily ration consists of about 17% of the fish weight.

Bighead feeds on phytoplankton and on zooplankton; the total weight of the two constitutes 50% or more of the diet. It can compete with common carp in the consumption of zooplankton, a fact to be taken into consideration in their combined culture.

The weight ratios of edible and inedible parts of the body of grass and silver carp and bighead are given in Table 16.

The usual carp ponds are used for culturing these fishes.

Table 16. Weight ratios of body parts of grass carp, silver carp, and bighead (% of live weight)

<table>
<tr><th>Type of fish</th><th>Average weight, g</th><th>Edible part (meat)</th><th>Head</th><th>Bones</th><th>Fins</th><th>Scales</th><th>Viscera</th><th></th></tr>
<tr><td>Grass carp</td><td>882.4</td><td>59.4</td><td>13.2</td><td>4.0</td><td>3.5</td><td>3.9</td><td>11.0</td><td rowspan="2">From A.P. Sivertosov</td></tr>
<tr><td>Silver carp</td><td>782.6</td><td>62.9</td><td>19.8</td><td>3.2</td><td>3.4</td><td>2.0</td><td>8.7</td></tr>
<tr><td>Bighead</td><td>703.0</td><td>51.1</td><td>23.2</td><td colspan="2">14.6</td><td>3.5</td><td>7.6</td><td>From F.M. Sukhover-khov</td></tr>
</table>

Broad Fish of Phytophagous Fish and their Maintenance. Grass and silver carp and bighead do not breed in ponds and the only method for obtaining eggs is pituitary injections. Male gonads, although they mature without pituitary injections, nevertheless produce a larger quantity of milt with artificial stimulation.

According to the report of the Tropical Research Institute of Fisheries in Malaysia (Malacca), the quantity of feed influences the rate of maturation of phytophagous fish under pond conditions. Planktophagous silver carp and bighead mature more rapidly than grass carp which feeds on aquatic plants. During the period of maturation these fishes should be provided feed consisting of prawn meal, rice bran, and wheat flour.

The number of brood fish to be raised is not the same for different species. Taking into account that the production of silver carp in southern regions is, according to approximate estimates, 85 to 90%, and that of grass carp only 10 to 15%, the number of brood fish should be adjusted accordingly. The stock of female fish should not be less than 50%. Artificial egg fertilization makes it possible to reduce the number of males; no more than three or four males are required for every five females. The annual replenishment of the brood fish stock, due to the rejection of old and unproductive spawners and injuries caused during breeding work, is about 20% of the number of brood fish kept at a particular farm. For maintenance in summer ponds the stocking of brood fish should not be more than 100 fish in the case of silver carp and grass carp, and no more than 50 fish per hectare for bighead. In rearing the progeny of an artificial breeding program, natural productivity is taken as not more than 400 to 500 kg/ha, with the following approximate ratios: grass carp (without feeding) not more than 100 kg/ha, bighead 100 to 150 kg/ha, and silver carp 200 to 300 kg/ha. The growth of brood fish during summer should not be less than 1 kg. In this process measures are taken for the conservation of the food base by applying manures to the ponds and, in the case of grass carp, when there is a deficiency in higher aquatic plants, land plants (clover, lucerne, barley, etc.) should be provided. Special fields are maintained for sowing these crops, and several crops are grown since feeding this fish just one type of plant for any length of time leads to a drop in consumption. The food coefficient for grass carp is, on the average, taken as 30 in calculations of food consumption.

In rearing grass carp the provision of higher aquatic plants is mandatory. No formula feed, or cereal, or other well-known feed of carp has proven a proper substitute for such plants. Experience has shown that a deficit in such plants causes serious functional disruptions leading to retardation in growth and maturation and, even, mass-scale mortality.

In stocking brood fish for hibernation the following norms are observed: fingerlings at the rate of 200,000 to 300,000 fish per hectare, and older age groups at the rate of 100 to 200 quintals per hectare of wintering pond. The brood fish of each species are kept in separate wintering ponds. Other age groups of different species serving as replenishing stock need not be kept in different wintering ponds.

ARTIFICIAL FERTILIZATION OF EGGS AND PROGENY OF PHYTOPHAGOUS FISH. When the temperature of the water has stabilized at 19 to 20°C, the water is discharged from the wintering ponds, the brood fish collected, and an inventory taken. The fish are grouped according to species, sex, and level of condition for spawning. Those with injuries, symptoms of sickness, and deformations are rejected. Individuals just attaining maturity, which have not reached the stage of spawning, are also not used for breeding work. They are planted in foraging ponds. Depending on their stage of gonadal maturation, female fish are divided into three groups or classes and males into two (Table 17).

After grading the fish according to species, sex, and class, the brood fish are stocked in hatching ponds (Figure 66). These should be small (0.05 to 0.20 ha), with a well-planned bottom, depths of 1.5 to 2.0 m, and a constant water discharge which prevents overheating of the water. The fish are held in these ponds until breeding work commences. To avoid overmaturation in this period they should not be kept in well-heated ponds and breeding should not be stretched over more than 25 to 30 days. The norm for prespawning maintenance of brood fish is 1,000 fish per hectare of pond (not more than 150 quintals/ha). Pond filling with water should be accomplished in 2.0 to 3.0 hours and discharge completed within 1.0 to 1.5 hours.

The periods of work on artificial breeding for individual species of phytophagous fish are not the same because maturation of the sex products occurs at different periods. Work commences with silver and grass carp, followed 7 to 10 days later by bighead. Commencement of artificial breeding depends on climatic zones. In Turkmenian and Uzbek SSR breeding is done in the first ten days of May; in Krasnodar and other regions of similar climatic conditions, in the second half of May; in Moldavian SSR, at the beginning of June; and in the Volgograd and Astrakhan districts, in the second half of June. All these are approximate periods and may change

Table 17. Classification of males and females

Group (class)	Females	Males
1	Fish with mature, ripe gonads, ready for spawning.	Milt readily released when body pressed. Inner surface of pectoral fins rough or serrated. In silver carp large number of sharp spinules felt, which are less sharp in bighead and in form of tubercles. In grass carp spinules smaller and surface of pectoral fins rough.
2	Fish with less mature gonads; used in a given season in second place and much later (after 15–20 days of constant warming of water up to 19–20°C in morning hours).	Milt either not released when body pressed or discharged in small quantity and with difficulty. Such males used in case of absolute necessity at end of breeding work or not used at all.
3	Fish with immature gonads, slender in appearance, almost indistinguishable from males. Not used in current year and after grading returned for foraging in stock ponds.	

Figure 66. Pond for prespawning maintenance of brood stock of phytophagous fish.

according to the peculiarities of a given year. Progeny of phytophagous fish in the central belt of the RSFSR have not been obtained on a commercial scale to date due to temperature conditions. In this region it is recommended that the warm waters of hydroelectric stations be used, which would considerably reduce the period of rearing of brood fish, and ensure the corresponding temperature conditions required for an earlier maturation of the gonads and initiation of artificial egg fertilization.

To stimulate the maturation phenomenon in females, especially in the case of those fish which are spawning for the first time, phased administration of pituitary injections is best as it accelerates the process in the ovaries during the fourth stage of maturation. In the prespawning period changes take place in the ovaries in two stages—the preovulation stage in which the eggs ripen, and ovulation when the eggs are liberated from the follicular membrane holding them in the ovaries.

The first stage proceeds under the influence of very small doses of the gonadotropic hormone (the primary injection) and the second stage through larger doses (the final injection). For stimulation of the maturation process pituitary glands obtained from sexually mature sazan are dried and defatted in acetone and kept ready for the prespawning period. The spawners are given injections in the form of a water suspension prepared immediately before the injection. The requisite quantity of pituitary glands is poured into a small porcelain mortar (30 to 50 cm^3) and thoroughly ground with a pestle. A few drops of water are added and the resultant paste thoroughly ground again, taking care that no granules remain which could block the needle of the syringe. Boiled and cooled or distilled water or isotonic physiological solution (6.5 g of common salt per liter of water) is then added to the paste to obtain a homogeneous suspension. The volume of the suspension

Figure 67. Grass carp being given pituitary injection.

for the preliminary injection should be 0.5 to 1.0 ml, and for the final injection 1.0 to 1.5 ml.

"Record" Syringes with a 5 to 10 ml capacity are recommended for these injections. The latter are administered in carriers containing water, and the suspension injected into the muscles of the back in the first third of the body, above the lateral line and below the base of the dorsal fin (Figure 67). The needle should penetrate rather deeply into the tissues at an acute angle under the scales, without piercing them, and the place of injection manually pressed and massaged to prevent a backup of the suspension on withdrawing the needle.

Before the females are injected their weight is recorded, measurements noted, and tagging done according to need as indicated in the register (Table 18).

Table 18. Register for recording data characterizing spawners

Date					
Sl. No. of female fish					
Fish species					
Fish weight					
Marking*					
Preliminary dose					
Final dose					
Sl. No. of fish enclosure					
Date and time of egg extraction					
Weight of eggs obtained					

*Marking is done by cutting the tips of some of the soft rays of the anal fin in the order of their arrangement with a pair of scissors.

The preliminary injection for females constitutes only one-eighth to one-tenth part of the specified dose of gonadotropic hormone; the final dose is given 24 hours later. It is important that the correct dose of pituitary extract be ascertained at the time of the injection since the results desired will not be obtained if it is low. With low doses eggs are either not obtained or only partially laid.

The quantity of pituitary extract in the final dose usually varies in the range of 3 to 6 mg/kg weight of female fish. In the process of dosage determination it is essential to take into account not only the body weight of the fish, but also the level of its fecundity. The greater the fecundity, the greater the quantity of pituitary required for total ovulation. In practice the dose is determined by multiplying the weight of the female fish by the quantity of the pituitary gland, within the limits of 3 to 6 mg, and corrections made in subsequent batches depending on how fully the females discharge the eggs. For convenience it is recommended that the preliminary injection consist of 3 mg of dry hypophysis for females weighing 5 to 7 kg, and 5 to 6 mg for larger fish. Milting males are generally used for this work. However, to obtain a sufficient quantity of milt the males are also given a pituitary injection. For males weighing 5 to 7 kg 4 to 6 mg of pituitary gland suffice. For larger quantities of milt larger males (weighing up to 10 kg) should be injected at the rate of 12 to 15 mg pituitary gland per fish. Males are given the injections one hour before the final injection is administered to the females. The male : female rate is 1 : 2–3.

To determine the final dose of pituitary gland a nomogram can be used, which gives the relationship between the pituitary dose and body girth (Figure 68). For example, if a female fish has a weight of 10 kg and a span of 54 cm, as can be seen from the correlation curve 5 mg of pituitary gland per kg of weight is required. Hence in this example 50 mg of pituitary would be the proper dose.

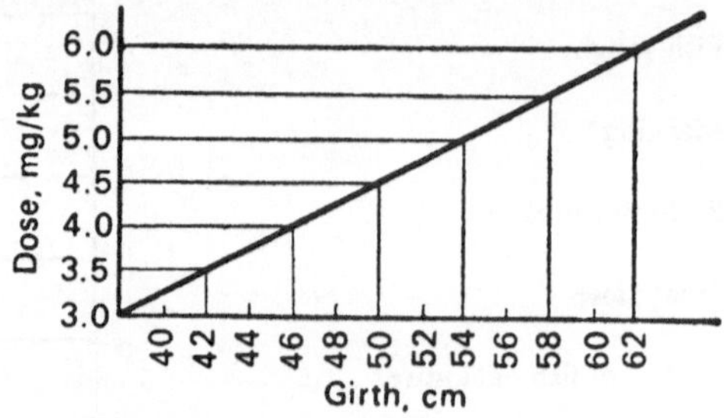

Figure 68. Nomogram showing relationship between dosage of pituitary gland and body girth.

Injected spawners require a temperature close to the spawning temperature and a favorable oxygen regime to ensure maturation of the sex products. These conditions are achieved in small (20 to 30 m²) spawning ponds with a constant water exchange, a depth of 0.8 to 1.0 m, and provided with discharge and replenishment of water within 30 minutes. Such a spawning pond can hold about ten spawners. After injection spawners may be kept in baths prepared from tarpaulin, fiber glass, or any other suitable material. When the spawners are kept in such containers a water exchange of 3 to 4 l/min must be ensured. After the final injection females become ripe at a water temperature of 20 to 22°C within 10 to 12 hours; at 23 to 25°C, after 9 to 11 hours; and at 26 to 28°C, after 7 to 10 hours. It is convenient to inject in the evening 9:00 to 10:00 p.m.) sine all the subsequent elaborate operations (stripping of eggs, their fertilization, distribution in apparatus, and quality control) can be done the next day. Stripping of eggs should be done on alternate days for a better and more complete utilization of the hatchery ponds and apparatus, and to ease personnel workloads. Before stripping, the water is drained from the spawning pond so that personnel can move freely in it. Catching is done with hand nets made from jute or some other material, ranging in length from 1 to 3 m. The opening of the net has a metal hoop 30 to 35 cm in diameter. The hand net is maneuvered in such a way that the fish slips in headfirst (Figure 69) while another laborer simultaneously grips its caudal peduncle with a muslin cloth. The fish thus held is quickly turned abdominal side up in water to avoid loss of sex products, and the vent pressed shut with a finger. It is carried in this position to the shore and carefully cleaned of water and slime with a dry muslin cloth.

Figure 69. Catching spawner with a hand net.

The process of egg stripping is shown in Figure 70. The eggs from each female are collected in separate containers, separately fertilized with milt by the dry Russian method, and separately measured by weight or volume (see Chapter XV). Egg stripping and fertilization is done in shade or under a tent. Dry, clean, noncorrosive containers in which the eggs can slide smoothly down the sides should be used. Ripe eggs stream from the vent of the female easily and range in color from grayish-sky-blue to bright orange; overripe eggs are dull white.

Usually one gram of nonfertilized eggs contains 800 to 1,000 eggs of grass carp, 900 to 1,200 eggs of silver carp, and 600 to 800 eggs of bighead. A

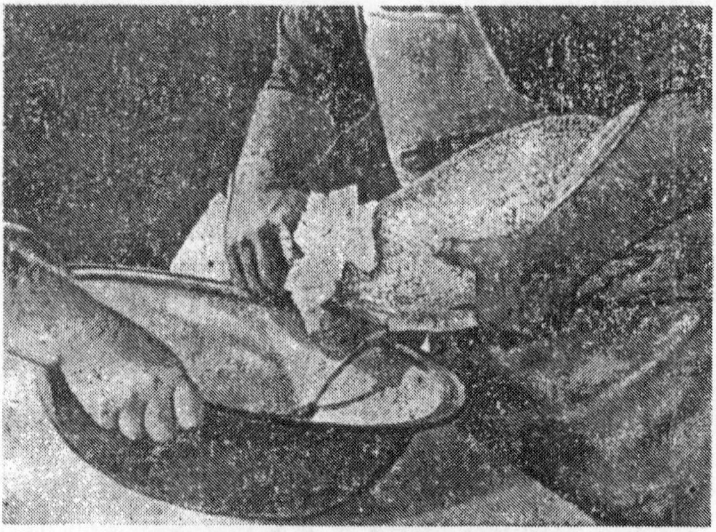

Figure 70. Stripping of eggs in a silver carp.

similar quantity of eggs generally occurs in one gram of nonfertilized roe. Depending on the size and age of the fish, the fecundity of grass and silver carp varies from tens of thousands to one-and-a-half to two million eggs; females weighing 7 kg produce about 500,000 eggs on the average.

The fecundity of fish is estimated on the basis of the number of eggs in the ovary. Usually, the larger the number of eggs, the smaller their size.

To determine fecundity the ovaries are weighed at the prespawning stage. Then one gram of eggs is taken (in the case of large eggs more), and their number counted. The number of eggs per gram is multiplied by the weight of the ovary and the *absolute fecundity* index obtained.

Absolute fecundity is quite distinct from *practical fecundity*. In nature not all of the eggs are fertilized; under artificial breeding the index of fertilization is higher. In the process of stripping some eggs remain in the ovaries and body cavity of the female; the quantity actually obtained from the female is known as *practical fecundity*. The magnitude of practical fecundity provides a basis for calculating the number of female fish required.

Not only the age and size of the fish, but also water temperature and oxygen influence the number of eggs which will develop. The feeding of brood fish is also significant since it exerts a considerable effect on gonadal development. In experiments with brook trout (*Salmo fontinalis*) 910 ripe eggs were obtained from a single fish after intensive feeding, 520 eggs when the ration was cut in half, and 405 eggs when the ration was reduced to one-fourth.

However, overfeeding of fish is harmful because fattening leads to sterility and ultimately to degeneration of eggs. Degeneration may also set in due to improper maintenance and unsatisfactory conditions.

It is better to use part of the milt collected from two to four males to fertilize the eggs of a single female, for the purpose of introducing selectivity in fertilization. The milt may be collected in advance or the males caught and stripped over the eggs. For the fertilization of one liter of eggs 5 ml of milt suffice.

Milt collection should only be done 30 to 60 minutes before the eggs are stripped from the female. For this purpose the males are caught, their flanks thoroughly cleaned with a dry muslin cloth, and the fish gently stroked from upside downward. The milt which oozes out is collected in dry, clean test tubes (the most convenient size is 15 cm long and 3 to 4 cm diameter) or glass jars, with a separate container for each male fish. Care is taken during milt extraction to see that it is clean and not accompanied by blood, mucus, gut contents, or water. The first few drops of milt should not be collected. When the test tubes or jars are filled, they are closed with corks or cotton plugs and stored in thermos flasks containing ice until the time of egg fertilization. Broad-mouthed thermos flasks with a capacity of 0.5 l are best; one-third of the thermos is filled with ice and covered with a thin layer of cotton or a muslin cloth folded over four or five times.

The quality of the milt is also very important and assessed beforehand. High quality milt is characterized by a white color and a creamy consistency; poor quality milt is more watery with a greenish or pale sky-blue tinge.

In the process of egg fertilization that part of the milt which is not used is left in the test tubes and stored again in the thermoses. Warming the test tubes of milt is not advisable. Storage in a thermos of ice for 10 to 12 hours does not reduce the viability of the spermatozoa. They actively move in water for 15 to 30 seconds and sometimes even longer. In fact, the higher the water temperature the more rapid the mortality of the spermatozoa.

Milt is carefully spread over the eggs (Figure 71) with a bird feather. A small quantity of water is then added, the eggs gently stirred with the feather, and the container held in the hand and gently tipped from side to side. The water will become milky in 2 to 3 minutes. The milky water is poured out and fresh water added, the process being repeated until the eggs are fully cleansed of milt, mucus, blood clots, and scales.

Fertilized eggs swell considerably (their diameter increases from 1.2 to 5.0 mm and more) and become quite buoyant. They are placed in the incubation apparatus not later than 5 to 10 minutes after fertilization. Extreme caution should be exercised in the course of water replacement to

Figure 71. Pouring of milt over eggs.

avoid loss of fertilized eggs during water displacement. The Weiss apparatus is often used for the incubation of eggs of phytophagous fish. It consists of a cylindrical glass jar which tapers downward. An apparatus with a volume of 8 liters can hold about 50,000 small eggs (Figure 72). For better coverage of each egg with water, which moves from the bottom upward under pressure, the jar is positioned vertically. The lower opening is corked with a plug in which a metal pipe has been inserted for supplying water under pressure. The water rises upward along the sides of the jar and surrounds the eggs in the apparatus. Depending on the pressure and consumption of water at one or the other height, the eggs stop rising and settle until circulated by the water once more. Fertilized eggs are visible in the apparatus and their continuous movement ensures thorough cleansing. In the laboratory of acclimatization of VNIIPRKh (All-Union Research Institute of

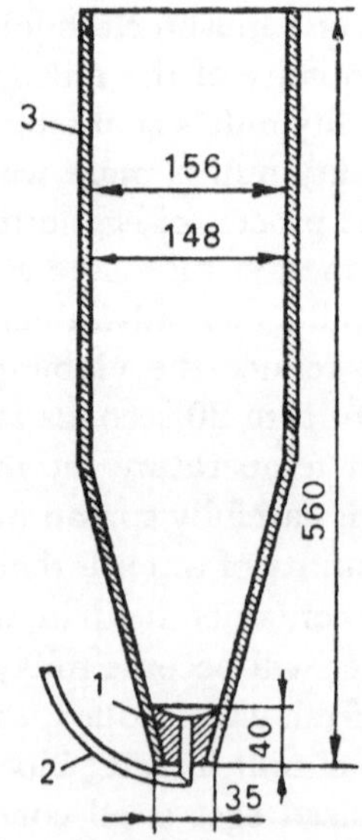

Figure 72. Weiss apparatus commonly used for incubation of small eggs; capacity 8 liters.

Pond Fisheries) apparatus of larger capacity (from 50 to 200 liters) have been constructed for the incubation of eggs of phytophagous fish. An apparatus of 50-liter capacity can hold 350,000 eggs, of 100 liters up to 750,000 eggs, and of 200 liters, 1.5 million eggs (Figure 73).

Figure 73. Improved apparatus of VNIIPRKh and stand for incubation of eggs of phytophagous fish; capacity 100 liters.

INCUBATION OF EGGS OF PHYTOPHAGOUS FISH AND MAINTENANCE OF THEIR LARVAE. The incubation of eggs of phytophagous fish is carried out in a hatchery located near the spawning ponds. Special attention is paid to the quality of the water supplied to the hatchery (Figure 74). When water is heated in the hatchery, it should be supplied to the apparatus through tanks where the air bubbles are removed, since these adhere to the eggs and carry them out of the apparatus. The water must be clean, unpolluted, and free of mechanical suspensions. To prevent the entry of predatory cyclopoid copepods into the incubation apparatus together with the water, a filter made of plastic (capron) net with a mesh size not less than No. 46, should be fitted on the water collection tube. Before loading the apparatus with fertilized eggs the hatchery is prepared as follows. The entire fish rearing apparatus

Figure 74. Hatchery for phytophagous fish.

is thoroughly washed and, if necessary, disinfected, the regime of water consumption adjusted, the necessary containers prepared (enameled tubs, buckets with lids, polyethylene mugs, etc.), the stand-by incubation apparatus and thermometers checked, and the register for recording progress in the development of eggs and free embryos readied (Table 19).

Table 19. Register for recording incubation data

Date of loading apparatus with eggs	Species and serial number of female	Quantity of eggs	Number of apparatus	Time of fertilization of eggs

Percentage of fertilization	Quantity of egg losses	Number of hatched eggs	Number of larvae	Remarks

Using polyethylene mugs, the incubation apparatus is loaded with eggs directly from the containers in which artificial fertilization was done, taking care not to injure the eggs. The mugs should never touch the sides of the container while collecting eggs and to avoid egg loss through water overflow in the apparatus during loading, the water level in them is reduced one-third beforehand. After loading the incubation apparatus the water supply must be so regulated that the eggs, which are still not fully swollen, are gently circulated in the lower half of the container (water consumption at the rate of 0.4 to 0.5 l/min). As the eggs swell and increase in volume the flow of water is increased to 0.7 to 0.8 l/min. To load an apparatus of larger capacity the eggs are poured directly from the containers (Figure 75). The water flow in a 50-liter VNIIPRKh apparatus is 3.4 l/min, in a 100-liter apparatus 7 l/min, and in a 200-liter apparatus up to 10 l/min.

Figure 75. Loading eggs of phytophagous fish from container into VNIIPRKh incubation apparatus of large capacity.

The percentage of fertilized eggs from each female fish is determined separately in the Bogorov counting chamber at the cleavage stage of four to eight blastomeres to morula (examining 100 eggs each time). The data obtained are recorded in the respective columns of the hatchery register. If the quality of eggs has been high, the fertilization percentage will be 90% and the output of free embryos 70 to 80% of the total number of eggs loaded.

The extent of loss in eggs during the period of incubation may increase with faulty incubation conditions, incorrect installation of incubation apparatus, untoward fluctuations in water temperatures, and variations in quality of sex products. The minimum water temperature for the development of eggs into free embryos is 18 to 19°C; a reduction in temperature to below 17°C not only slows down the development of the eggs, but also causes mortalities and the hatching of underdeveloped and deformed embryos. At a temperature of 21 to 25°C incubation requires 23 to 33 hours and at 27 to 29°C reduces to 17 to 19 hours. A high level of waste in eggs in incubation apparatus and of free embryos in fish tanks also occurs when the optimum temperature maintained during the day is allowed to drop to 15 to 16°C at night.

The conditions of maintenance of brood fish during the prespawning period also influence the percentages of fertilization and hatching of the eggs. In protracted spawning periods and extended retention of brood fish at high temperatures the quality of the sex products deteriorates and the percentages of loss at different stages of development increase.

Care and maintenance of the eggs in the apparatus during the incubation period include regulation of the conditions of water consumption to exclude the possibility of the eggs being carried away through excess water supply, or their settling to lower levels because of a reduction in water supply. Both these situations lead to infective conditions and egg mortality, necessitating the troublesome task of siphoning dead eggs out through a thin rubber tube. Dead eggs are dull white and float on the surface layers; they are readily discernible and easily removed. They should be removed 8 to 10 hours after the commencement of incubation, when the stage of gastrulation is complete.

To determine the expected output of prolarvae, the percentage of dead eggs and eggs with deformed embryos is estimated a few hours before hatching.

Mass-scale hatching of embryos usually continues for 1 to 3 hours, but sometimes extends over a longer period, even up to 24 hours. Uniform hatching can be achieved by reducing the water supply in the apparatus three to five times compared to the normal water consumption. The larvae are transferred to enclosures fixed in cement tanks with a depth of 1 m. The enclosures are made of capron net Nos. 18 to 20 measuring 60 cm × 60 cm × 45 cm or 70 cm × 70 cm × 45 cm. They are placed in the water at a depth of not less than 30 cm and care is taken that their bottom is level, i.e. does not slant. About 250,000 larvae can be placed in each enclosure. The maintenance of these enclosures includes provision of the required oxygen regime, for which purpose the water is aerated. This is especially important during the quiescent period when most of the larvae settle to the bottom of the enclosure. Eggshells and dead embryos are removed by means of a glass tube inserted in a rubber syringe. The larvae are kept in such enclosures until they change over to mixed feeding and, depending on the temperature, transferred as follows: at 18 to 20°C after 90 to 100 hours, at 20 to 23°C after 80 to 85 hours, and at 26 to 27°C after 48 hours. The percentage of loss is low in the fish enclosures under favorable temperature conditions and water exchange. The survival rate from fertilized eggs up to larvae which have changed over to mixed feeding should not be less than 50%. Counting is done visually, for which some kind of lightcolored container is used (pan, basin, etc.). A certain number of larvae are counted and subsequently, on the basis of their concentration in the container, calculations made and compared with existing standards.

In farms without special facilities for the development of phytophagous fish simple equipment is used, requiring less extensive accommodation. Brood fish are kept in ponds with an area of up to 1.5 hectares and a depth of not less than 1.5 m. Spawners given pituitary injections are maintained in flowing water tanks. For incubation of the eggs and maintenance of the larvae a special mobile unit developed by VNIIPRKh is used, which consists of an incubation apparatus of 50 to 100-liter capacity and a metal tank with three enclosures for larval maintenance. Water is supplied to the unit through a network of tubes attached to the bottom. The size of the larval enclosures is 50 cm × 50 cm × 45 cm. The water is exchanged in the tank for larval maintenance at the rate of 4–5 l/min, and in the incubation apparatus according to its capacity.

After maintenance of the larvae in tank enclosures either they are transplanted for further rearing to other farms, or reared to more viable stages, or transplanted directly into rearing ponds. The empty tank enclosures are scrubbed thoroughly with a brush and running water to remove overgrowth, the frames removed and washed in a detergent, and both the nets and frames dried in shade, away from direct sunlight. The enclosures are then stored until needed the following year.

There are several methods of rearing larvae before they commence mixed feeding. Generally, they are reared in nursery (transplantation) ponds in carp farms or well-managed ponds of other categories with a well-planned bottom and an average depth of 0.5 to 0.7 m. Dirt accumulators are fixed in the water supply units and fish screens fitted in the water outlets. The food regime is extremely important in rearing larvae. The concentration of food organisms should not be less than 1,000 to 1,500 food items per liter. The animal organisms should exceed the plant matter, and the major components in the initial days should comprise zooplankton of small forms, and in the second half of larval rearing larger forms of varying quantities. However, larger forms of zooplankton (cyclopoid copepods and daphnids) are not readily available throughout the entire period of larval development of silver carp. Many species of invertebrates are predators of fish larvae; the most dangerous are cyclopoid copepods, beetles, bugs, larvae of dragonflies, etc. To prevent the entry of predatory forms from the source of water supply special screens of capron net No. 32 are used. Shortening the period between filling the ponds with water to releasing fish in them also delays the development of predatory forms. The period of rearing is determined from the time the larvae attain viability, i.e., have changed over to the consumption of all or most of the smaller and larger organisms, including those which constitute threats. For most species the changeover occurs when the larvae are 11 to 12 mm long and weigh 15 to 20 mg. The growth period under the conditions of Krasnodar average 10 days, which makes it possible to use the same pond twice in a season. Well-prepared ponds with good soil-climatic conditions can be stocked with larvae at the rate of up to three to four million per hectare in the first rotation and two to three million per hectare in the second. These norms may be raised to six to seven million larvae per hectare under conditions of manuring. Water discharge from the ponds and catching of larvae are done at night when the water temperature in the surface layers is low. The larvae settle into the deeper layers and move more rapidly with the flow of water into the fish traps. They are collected from the traps in dip nets and transferred to pans or other containers. Since larvae accumulate at the bottom of the dip net, a pan or mug filled with water is placed under it while transferring the fish. The output of grown larvae is not less than 60 to 70%.

Rearing of Fingerlings. Fingerlings of phytophagous fish are reared with fingerlings of common carp in small rearing ponds with a well-planned bottom, preferably with an area of not more than 10 hectares. The larvae are stocked in these ponds not later than six to seven days from the time of their filling with water. Screens with a mesh size of not more than 1 mm should be placed earlier at the source of water supply.

In stocking larvae directly into rearing ponds it is essential that the water temperature be equalized throughout the pond for 30 minutes. For larvae transported in polyethylene bags, prior to their release in the ponds the packets are opened and the temperature equalized. For control, 200 to 300 larvae are placed in fish enclosures made of capron net Nos. 18 to 20 positioned in the ponds into which the larvae are released.

The survival rate of larvae which have changed over to mixed feeding up to the stage of fingerlings is not less than 40% in the southern regions and not less than 30% in the central belt. The output of fingerlings from fry has recently increased and is now 70 and 50% respectively in the two regions.

In addition to carp, the following norms of stocking of phytophagous fish larvae in rearing ponds are recommended: for southern regions—50,000 to 70,000 per hectare, for the central chernozem zone—40,000 to 50,000 per hectare, and for regions in the central belt—30,000 to 40,000 per hectare. The stocking of larvae of grass carp reared without additional plant feed should not exceed more than 10,000 per hectare for all zones, and in the case of bighead 30,000 per hectare. The productivity of phytophagous fish cultured additionally in carp ponds has tentatively been estimated as follows:

in the southern regions—6 to 10 q/ha from fully drainable rearing ponds, and in the central belt—3 to 5 q/ha. The projected average weight of fingerlings in the central belt is 15 to 20 g, and in the south 25 to 30 g.

Periodic control catches are made throughout the vegetative growth period, as is done in carp culture. The results of control catches are compared for each fish species separately. Autumn catching commences when the water temperature drops to 14°C. The water is discharged from the ponds gradually to avoid sharp fluctuations in the water level. In both the autumn and control catches the peculiarities of phytophagous fish must be taken into account. They are more easily injured than common carp, especially silver carp. Turbidity while catching silver carp must be avoided since the gills become clogged with silt particles. Catching must be done quietly since noise frightens the fish and, as mentioned above, without sharp fluctuations in the water level. A frightened silver carp will jump out of the water and die. Fish traps made of netting and installed in the water outlets are best for catching. This method avoids the possibility of injury to the fish. It also makes grading then and there much easier since the use of grading tables is not recommended. In the process of catching silver carp move first into the fish trap, followed by bighead. This makes their catching easier with an anchored seine. These species are followed by grass carp and carp which can be caught from the flow of fresh water. Fingerlings caught by one or the other method are counted and graded according to species and size. The use of dip nets is not recommended for collecting the fish from the bag (bunt) and the wings of the seine; it is better to use a bucket filled with water.

The fingerlings of phytophagous fish are stocked for hibernation either in separate wintering ponds or with the fingerlings of carp when the former predominate in the stock material. If the predominance is reversed, one runs the risk of injury to the yearlings of silver carp at the time of winter catches.

According to the data of the fish breeding farm "Yakot" in the Moscow district, all the adult groups of phytophagous fish hibernate better under water temperatures higher than 0.5 to 1.0°C.

The output of yearlings[1] from the wintering ponds is 80 to 90% of the fingerlings stocked, depending on the zone.

Raising of Two-Year and Three-Year Fish. Two-year and three-year phytophagous fish are cultured in rearing ponds together with carp. The norms of productivity under combined rearing with carp are as follows for individual phytophagous species: in the southern zone for grass carp—1 q/ha and for silver carp and bighead—5 to 9 q/ha; and in the central belt 1 to 2 and 2 to 3 q/ha respectively. The stocking norms for individual

[1]Damage to the swim bladder was noted in silver carp at the fish farm "Trud" in Krasnodar for the first time in the winter of 1969–1970.

species in the south vary as follows: (a) silver carp—from 500 fish per hectare (if the carp are not fed regularly) to 2,000 to 2,500 fish per hectare (under the conditions of abundant phytoplankton), and an average of 1,000 fish per hectare; (b) bighead—average of 500 to 700 fish per hectare; and (c) grass carp, according to experience in Krasnodar, not more than 50 to 100 yearlings per hectare of drainable foraging ponds, even with the provision of abundant aquatic plants.

In the central belt, where the temperatures are lower and hence the intensity of feeding on plants by grass carp also less, stocking in foraging ponds with adequate vegetation can be increased to 300 to 500 yearlings; silver carp up to 1,000 fish per hectare; and bighead 500 to 700 per hectare. The average weight of two-year fish in these two regions has been determined as follows: (a) in the south for silver carp not below 500 to 700 g, and for bighead and grass carp 800 to 1,000 g; (b) in the central belt for grass carp and bighead 300 to 400 g, and for silver carp not more than 200 g.

The output of two-year fish for all the species of phytophagous fish should be 80% of the quantity of yearlings stocked.

Taking into account that under the climatic conditions of the central belt it is not possible to grow two-year fish up to a higher unit weight, and that weights obtained (200 to 300 g) are not of high food quality, it is more appropriate in this region to change over to a three-year rotation program. The weight of three-year fish should be as follows: silver carp 500 to 600 g, and grass carp and bighead 700 to 1,000 g. Three-year fish consume more coarse food, utilize higher plants and phytoplankton more fully, and are more tolerant compared to two-year fish. In the southern regions there is no need for a three-year rotation program since two-year phytophagous fish grow to a sufficiently high marketable weight.

The weight of three-year phytophagous fish from foraging ponds is not less than 85%. In stocking fish for hibernation the norms are fixed in accordance with the age of the carp.

Thus, the use of phytophagous fish in polyculture increases the fish yield by approximately 50%. The production processes hardly differ from those involved in the monoculture of common carp. When the density of stocking of phytophagous fish is increased to more than 50% of the total fish production, separate measures should be implemented to accommodate the biological peculiarities of the phytophagous species and its food requirements.

The prospects of using phytophagous fish in nondrainable natural water resources such as lakes, large reservoirs, and irrigation systems are definitely good.

References

Il'in, V.M. and L.M. Solov'eva. 1965. Vyrashchivanie i zimovka segolatkov rastitel'novadnykh ryb (Raising and hibernation of phytophagous fish). *Trudy VNIIPRKh,* vol. 13.

Problemy rybokhozyaistvennogo ispol'zovaniya rastitel'noyadnykh ryb v vodoemakh SSSR (Problems in Fishery Utilization of Phytophagous Fish in the Water Resources of the USSR). Izd. Akad. Nauk TurkSSR. Ashkhabad.

Rukovodstvo po biotekhnike razvedeniya i vyrashichivaniya rastitel'-noyadnykh ryb (Handbook on Biotechniques of Breeding and Raising of Phytophagous Fish). VNIIPRKh. Moscow.

Rybokhozyaistvennoe osvoenie rastitel'noyadnykh ryb (Adaptation of Phytophagous Fish for Fish Farming). Nauka, Moscow.

Sivertsev, A. 1965. Vesovoe sootnoshenie chastei tala prudovykh ryb (Weight ratio of body parts of pond fish). *Rybovodstvo i Rybolovstvo,* no. 6.

Vinogradov, V.K., L.V. Erokhina et al. 1966. O metodakh iskusstvennogo rasvedeniya rastitel'noyadnykh ryb (Methods of artificial rearing of phytophagous fish). In *Rybokhozyaistvennoe Osvoenie Rastitel'noyadnykh Ryb.* Nauka, Moscow.

Chapter X

MELIORATION OF FISH PONDS

Biological and physicochemical changes take place in the process of exploitation of a pond for the purpose of pisciculture: formation of a silt layer, bog conditions, and the appearance of higher aquatic plants. Associated with such changes are a deterioration in the hydrochemical regime and sanitary conditions and an increase in the acidity of water. Such conditions ultimately lead not only to a considerable reduction in natural fish productivity, but also to a deterioration in growth and development of fish and, under extremely unfavorable conditions, the impossibility of using the pond for pisciculture. Experience in foreign countries has shown that with proper management practices it is possible to preserve ponds in a good state of exploitation for hundreds of years (for example, ponds have been maintained in the German Democratic Republic for 200 years and even more). Melioration[1] refers to a system technical and organizational farm measures directed toward a radical improvement of unfavorable conditions and, in the present work, improvements in ponds for the purpose of bettering their fish productivity.

The famous Soviet scientist Academician A.S. Zernov wrote: "Water can and often does yield the same quantity of meat per unit area and unit of time as the land surface." The value of water resources for the purpose of food production is underscored by that statement.

Melioration with regard to pond fisheries is applied not only to the pond itself but also to the territory surrounding the fish farm. The latter involves measures for improving the hydrochemical conditions, the quality of the water supply for fish ponds (see below), and the prevention of siltation of ponds as a result of turbid water supply carrying suspended particles which settle in the ponds. At the time of floods, high waters, and rains, waters saturated with particles from eroded ground may also enter the ponds from the surrounding areas and ultimately cause an increase in siltation. To prevent these undesirable situations necessary measures must be undertaken in the catchment area: construction of sedimentation areas, installation of filters, equipment for purification of spring waters, afforestation and grass plantation on slopes and banks, etc.

[1]From Latin: *meliorare*—to better or improve.

Measures directed toward the improvement of ponds of different categories include the creation of hydrochemical conditions essential for the fish under culture, control of aquatic plants to preclude harmful overgrowth, checking of excessive siltation, estivation, lime application, and technocultural measures.

Melioration of fish ponds is directed along two lines—care of water and care of soil. The close interaction between water and soil determines the qualitative assessment of the entire complex of biological and physicochemical phenomena occurring in a fish pond.

Meliorative Measures for Pond Water. Oxygen deficiency is often associated with a low concentration in the source of water supply as well as an unsatisfactory preparation of the pond before fish stocking. To increase oxygen concentration the water is aerated. This is especially important for wintering ponds when they are covered with ice and enrichment with oxygen from the air blocked. The high density of fish stocked for the winter period, the decomposition process of organic matter, although less at low temperatures, and the accumulation of a large quantity of metabolic products—all these tend to lower the oxygen content in the water, very often to a critical level. Even in summer the oxygen content may decrease notably,

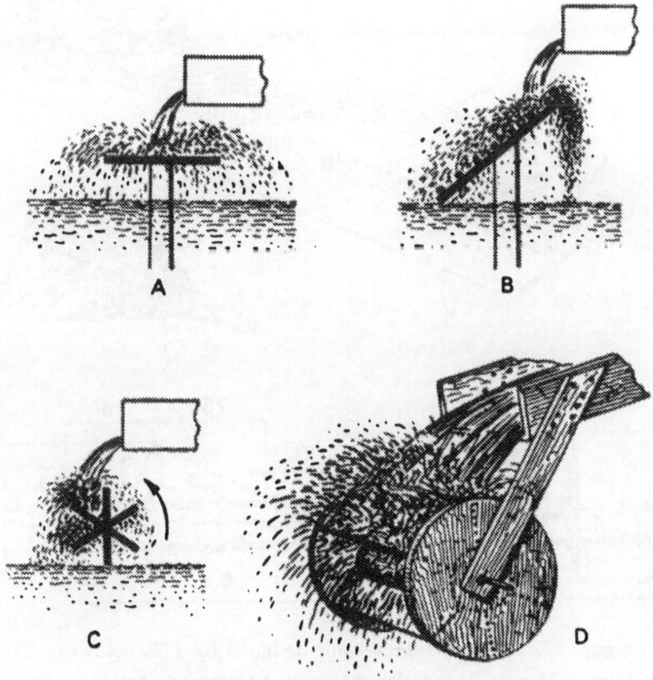

Figure 76. Simple aerating equipment.
A—Table; *B*—Ladder; *C* and *D*—Impellers (rotators).

especially in densely stocked ponds, with the addition of large quantities of artificial feed and the accumulation of metabolic products. The oxygen regime deteriorates particularly during the night and predawn hours when the photosynthetic activity of aquatic plants is absent.

Water can be aerated with the help of aerators, devices designed for enrichment with oxygen. One type of aerator involves sprinkling the water; the minute drops become enriched with oxygen through coming into contact with air. Aerators of simple construction (Figure 76) and different kinds of mechanical sprinklers used for irrigation purposes (DDN-45, DNN-50, DDA-100M, KDU-55M, etc.) are effective in fish farms. The propeller type aerator designed by T.T. Solov'ev (Figure 77) has given good results. P.M. Reshetnikov has invented a pneumatic aerator for frozen waters during winter (Figure 78). Aerators of other types are also available, especially in the USA, where a floating aerator which churns the water at any desired depth is quite popular.

The mechanical aerators used in the USA are quite effective, increasing the oxygen content to such an extent that fish production has been improved

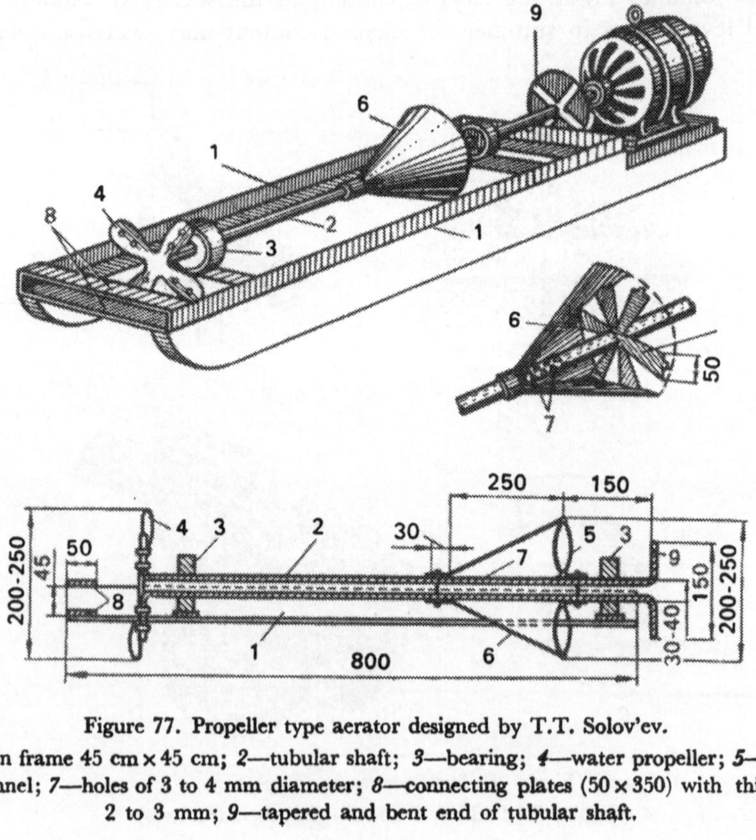

Figure 77. Propeller type aerator designed by T.T. Solov'ev.

1—iron frame 45 cm × 45 cm; *2*—tubular shaft; *3*—bearing; *4*—water propeller; *5*—vane; *6*—funnel; *7*—holes of 3 to 4 mm diameter; *8*—connecting plates (50 × 350) with thickness 2 to 3 mm; *9*—tapered and bent end of tubular shaft.

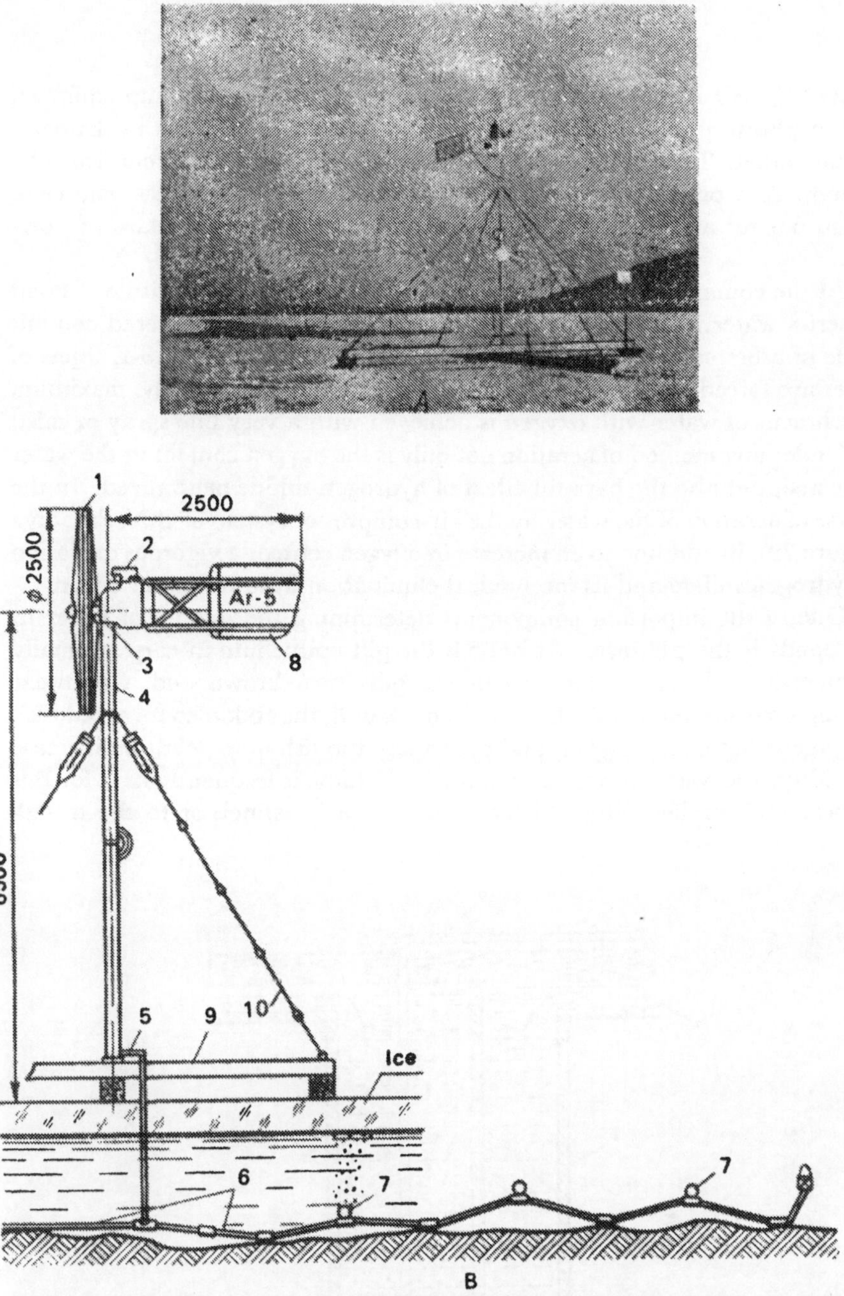

Figure 78. Pneumatic wind-operated aerator for wintering ponds (Reshetnikov system). *A*—General view; *B*—Diagram.

1—wind wheel; *2*—cylinder compressor; *3*—cavity of mast from where air is sucked in by compressor and forced into tube; *4*—air; *5*—rubber tube continued down to water bottom; *6*—air pipes; *7*—diffusers; *8*—flat steering tail for orientation of wind wheel; *9*—wooden frame for fixing of mast; *10*—extension for holding mast.

by 29.7%. Air is supplied to the ponds using a centrifugal pump equipped with a plastic pipe 2.54 cm in diameter and two pipes 0.6 cm in diameter for air outlet. The plastic pipe is located at a distance of 15 cm from the bottom. At a pressure of one atmosphere and temperature 23°C a volume of 6.2 to 6.9 m^3 air enters the pond every minute for each hectare of pond area.

At the commercial fish farm of the All-Union Research Institute of Pond Fisheries water is supplied to the wintering ponds along a covered conduit made of asbestos and cement pipes. Using pumps 4-NDV and 3-K-9 jets of water are forced out through an opening 5 mm in diameter. The maximum enrichment of water with oxygen is achieved with a very fine spray or mist.

Under any method of aeration not only is the oxygen content in the water increased, but also the harmful effect of hydrogen sulfide neutralized. In the course of aeration of the water by the air-compressor system of G.N. Paradiya (Figure 79), in addition to an increase in oxygen content a vigorous oxidation of hydrogen sulfide and its mechanical elimination from the water occur.

One of the important components determining the quality of water in fish ponds is the pH level. At pH 5.0 the gill epithelium in carp gradually disintegrates (the outer margins of the gills turn brown and a granular coating develops on them), the skin becomes dull, the abdomen turns red, fish movement becomes sluggish, and eventually the fish dies. With an increase in acidity, the water has to be neutralized. Lime is frequently used for this purpose and applied either to the water supply channels or to the trough

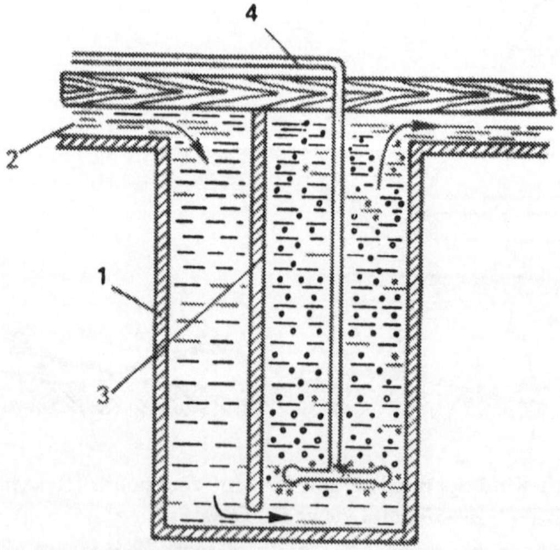

Figure 79. Air compressor of system devised by G.N. Paradiya.
1—block; *2*—water supply trough; *3*—partition in cistern which does not reach bottom; *4*—tube supplying air from compressor with net cover.

using a lime mill (Figure 80). The mill is installed in a water supply channel and automatically releases lime in the water. Several types of lime grinding machines have been constructed, but their basic principle of operation is the same. The water wheel, turned by the water current, propels the chain drive connected to a hammer, which strikes the outside of a conical lime container, causing a certain quantity to fall out with each blow. The container has a grinder inside. The lime powder is uniformly released in the water through an opening 1 cm in diameter at the lower end of the container. The opening is equipped with a shutter valve to regulate the rate of lime

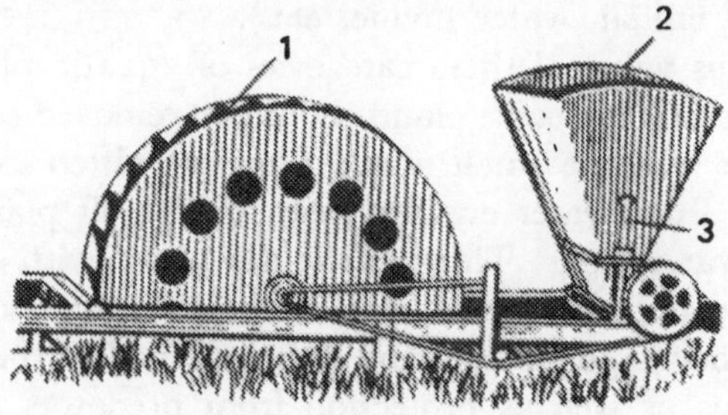

Figure 80. Lime mill.
1—blade wheel; *2*—conical container with grinder inside; *3*—hammer.

supply. The output of the lime mill is 50 kg in 7 days, which makes it possible to notably reduce the acidity of water within a week or two.

Excessive ferrous oxide (FeO) in water (more than 4 to 5 mg/l) has a harmful effect on fish. Under the influence of increased aeration, the iron bonds are changed and ferrous oxide converted into insoluble ferric oxide (Fe_2O_3); calcium (present in the lime) facilitates this reaction. Ferric oxide, precipitated in the form of red-brown sediment, accumulates in the gills of fish, making respiration difficult, which results in death by asphyxia. It also acts as a catalyst, enhancing the harmful acid reaction of water. To remove excess ferrous oxide from water and to protect fish from the effect of insoluble iron sediment, shallow ponds with lime piled in heaps on the bottom make excellent sedimentation tanks. Under the influence of lime and aeration ferrous oxide is precipitated and the water can then be used in the rearing ponds.

Meliorative Measures for Control of Excess Aquatic Vegetation. Excess overgrowth of aquatic plants in fish ponds considerably deteriorates the hydrochemical regime, especially the oxygen regime (usually during the night and predawn hours), increases acidity of water, creates shade over the surface which prevents the penetration of light and heat, and causes difficulties in the normal exploitation of the pond, especially in catching fish. Left unchecked, plant overgrowth can lead to mass fish mortality and decomposition, deteriorating the conditions of environment for fish culture.

Higher aquatic plants (macroflora) occurring in ponds are usually categorized as follows:

(a) Emergent coarse plants; stems, leaves, and inflorescences located directly above the water surface (bullrush, sedge, cattail, sweet flag, reed, "trezubka", manna grass, arrowhead, water plantain, etc.);

(b) Semi-emergent plants; leaves and flowers float on water surface (duckweed, frogbit, yellow water lily, white water lily, perennial buttercup, water lady's thumb, floating pond weeds, etc.);

(c) Submerged soft plants; do not protrude above or on the water surface (hornwort, water milfoil, water thyme, etc.).

The plus-minus value of these categories of aquatic plants varies from pond to pond. Emergent coarse plants should be removed completely except in foraging ponds where a small number are permitted in a large area to protect the dikes from water erosion. Submerged soft plants, in a limited quantity, are advantageous. They enrich the water with oxygen, serve as a place for the concentration of food organisms, and exert a significant influence on the development of such. Fish find food in the thickets of submerged soft plants as well as protection from predators and unfavorable weather conditions. Ponds from which coarse plants have been removed and a small quantity of soft vegetation retained are most productive.

In the conditions of Ukrainian SSR the biomass of food zoobenthos collected from plants per 1 m^2 of pond area is more than the quantity collected from 1 m^2 of the bottom. A good concentration of larger chironomids and other organisms occurs in the root part of the plants. The average biomass of zoobenthos among soft plants is 25 to 27 times greater than in coarse plants (16.4 g/m^2 versus 0.62 g/m^2). Food organisms congregate especially in pondweeds (*Potamogeton*), water milfoil (*Myriophyllum*), hornwort (*Ceratophyllum*), and aquatic smartweed (*Polygonum amphibium*). Of these, large-leaved pond weeds, which do not form dense overgrowths, are easily thinned at the time of mowing or weed cutting. Considering the contribution of soft plants to the food base of ponds, dense outgrowths of vegetation should be thinned with weed cutters in the form of longitudinal and transverse corridors, which cover no more than 25% of the total pond area.

Semi-emergent plants must be kept to a minimum since they create shade on the ponds and hinder the penetration of light and heat.

The methods for clearing ponds of coarse and semi-emergent aquatic plants and any excess quantity of soft plants range from biological to mechanical and chemical, depending on the plant species.

Aquatic weeds should be removed from the ponds before flowering, but not so early that the young shoots have time to propagate again.

Biological Control Measures entail the direct or indirect destruction of some species of plant or animal by another species. In the USSR grass carp is used to check the growth of higher aquatic plants. The phytophagous species *Tilapia zillii* and *T. mossambica,* imported from Africa, are currently being tested for this purpose. The latter species has already proved useful in checking growth in cooling tanks of electric power stations, etc. The growth of aquatic plants can also be controlled with domestic ducks (see Chapter XIV).

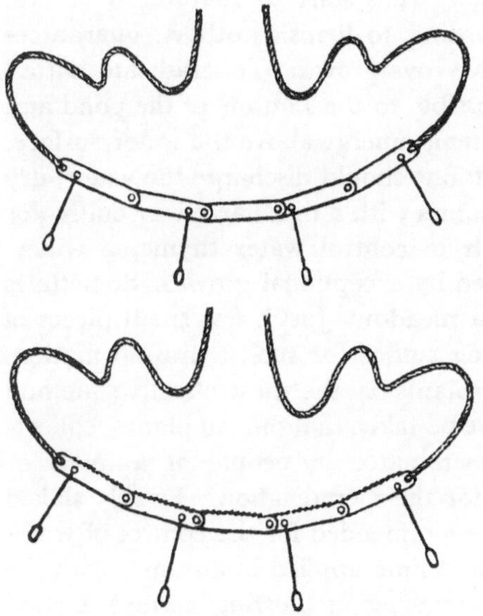

Figure 81. Chain weed cutter.

To suppress plant growth in rearing and foraging ponds 400 yearlings, 200 two-year, and 50 three-year grass carp per hectare of pond area have proved effective norms.

Mechanical Methods of Controlling aquatic weeds vary but mainly involve cutting the plants with the usual or chain-operated cutter (Figure 81), or cutters of different types (Figure 82), depending on the pond size. According to the reports of British

Figure 82. Weed cutter "Libella" (German Democratic Republic). Cutting depth 0.75 m, draft 0.2 m.

pisciculturists, repeated cutting over the vegetative growth period has a deleterious effect on cattail and water lily. It should be noted here that cattail, reed, and some other plants are most vulnerable immediately after maximum flowering and before the new seeds start to mature. A second cutting four to six weeks later, according to British authors, guarantees the complete destruction of vegetative overgrowth. To eradicate cattail the first mowing is done as close as possible to the bottom of the pond and subsequent mowings done before the stems emerge above the water surface.

To effectively control aquatic plants one should discharge the water, dry and plow the bottom, and cut the rhizomes with a disk harrow or cultivator and remove them. It is more difficult to control water thyme, a widely distributed plant in ponds characterized by exceptional growth. Sometimes its proliferation converts a pond into a meadow. Just a few small pieces of stem left during the process of clearing suffice for such a transformation. Pond estivation and removal of the plants is the most effective method of controlling water thyme. Care must be taken that the cut plants, collected in heaps, are not accidentally disseminated by people or animals or the implements and equipment used for their eradication. Freshly slaked lime at the rate of about 4 tons/ha is recommended for the control of water thyme. The ponds should be drained and lime applied in autumn, with the ponds left dry throughout winter. Harrowing or plowing ensures a good mixture of lime in the soil.

Another method of controlling water thyme is pond estivation during the summer (for one or two years) and planting of kitchen garden crops on the bottom which need continual tillage and earthing.

Duckweed often covers the water surface in the form of a continuous green mat and prevents the access of oxygen, light, and heat. It is removed from ponds with seines, push nets, and rakes, and its growth further controlled by clearing (before seed formation and better at the time of flowering) such plants along the edge of the pond as bur marigold or beggar's ticks (*Bidens tripartitus* and *B. cirnua*). The flat seeds of the latter plant species have three hard awns which terminate in recurved hooks. If eaten by fish (especially fry), these seeds stick in the oral cavity, causing ulceration of the mouth and making food intake difficult. Horse sorrel has an adverse effect on the development of zooplankton in ponds—not so much the young plants but the dying ones, which release acid compounds. These plants should be uprooted.

Filamentous algae in excess are also harmful, especially for fry which become entangled in the thin filaments. Such algae is also an obstacle in netting out fish. Most often filamentous algae develop in small ponds with stagnant or mildly flowing water, or in ponds in which the proportions of phosphorus and nitrogen salts used as chemical fertilizers have not been correctly ascertained. The best method for getting rid of filamentous algae is to dry the ponds for one vegetative period and use them for the cultivation of agricultural crops (oats, vetch-oats mixture, potato, etc.). If this is not possible, filamentous algae are removed with push nets or rakes. Extreme care must be taken in collecting filamentous algae, especially from rearing ponds containing juvenile fish, since a large number of fry may be entangled in them. The application of chemical fertilizers, which facilitates the development of phytoplankton, also reduces filamentous algae. Phytoplankton can absorb solar rays without giving rise to the development of filamentous algae.

Chemical Methods of Control should be used with extreme caution since many substances (weed killers which contain arsenic, and others) are toxic not only for aquatic organisms but also man and animals. Some of the very important problems associated with further prospects of using different chemical substances, such as their effect on natural fish foods, have yet to be fully studied. The effect of the accumulation of different natural and biologically active synthetic substances in the food items of both plants and animals must be analyzed. Considerable caution must be exercised in the application of such substances in ponds when the water is used by people and animals for drinking purposes. Toxic chemicals should be used for controlling aquatic plants only after careful and comprehensive studies have been done on their toxicity and effectiveness under various conditions. The recommended preparations should also conform to the requirements for the conservation of nature.

A search should be instigated for chemical substances which are less stable. Detoxication proceeds more quickly in them and they are dangerous neither to man, animals, nor the surrounding environment.

Prevention of Siltation in Fish Ponds. Silt accumulates in fish ponds through the water supply, especially from the catchment area at the time of spring floods, overflows, and rain, and as a result of the processes taking place in the pond itself. Depending on the pond area and nature of the catchment area, siltation in ponds takes place at different rates. To reduce the possibility of siltation a belt 100 m wide should be left unplowed around the pond's edge. Forest belts 10 to 20 m wide and located at a distance of not less than 20 to 30 m from the pond constitute a good protection from siltation (and are even more effective at a still greater distance). Silt enters ponds located in ravines from those parts of the ravines directly adjacent to the ponds. To prevent siltation in such ponds trees and shrubs should be planted in the dry part of the ravine right up to its bottom in the form of interrupted belts. Behind (or rather, above) the trees grass stands can be developed, forming a belt not less than 70 to 100 m wide. These measures facilitate the clearing of surface waters by removing the suspended particles and reduce siltation in the ponds. Water flowing down the slopes at the time of high floods is absorbed by both the forest and grass belts and through this process the supply of ground water to the ponds increases.

The decomposition of higher and lower plants, aquatic invertebrates, etc., contributes notably to the pond silt.

Carp ponds are characterized by a soft silt layer, which is rich in readily mineralized organic substances. All the beneficial chemical processes leading to natural fish productivity occur in the active silt layer, which is usually 30 or even 40 cm thick. Nitrogen-fixing bacteria (nitrification) add an active productive property to the silt layer. Silt deposits consisting of an excessive amount of coarse cellulose residue, however, produce adverse effects (increase in acidity, deterioration in the hydrochemical regime, and reduction in fish productivity).

An unwanted superficial silt layer can be removed by periodic pond drying. More radical measures are needed however, for checking thick silt depositions in which coarse aquatic plants with strong root systems have already developed. Shallow plowing after drying the pond is effective but tillage which loosens the soil and destroys the rhizomes of coarse plants is better (with a disk harrow, cultivators, etc.). Deep plowing should never be done as the less productive lower layers are brought to the surface and the more productive layers buried. This type of soil exchange reduces natural fish productivity. However, when the rhizomes of coarse plants have developed into a strong, highly intertwined mesh, deep plowing is unavoidable.

Power-operated hoes are used for plowing the bottom of fish ponds, covering up to 5,000 m^2 per hour. In old neglected ponds in which the silt layer is 2.5 m or more thick cultivation with agricultural implements is not possible because the equipment bogs down. Such ponds are cleared with hydraulic dredges. Left dry throughout winter the silt layer in the pond freezes, is broken into pieces, and transferred to fields (it serves as a good manure). A layer of 15 to 20 cm of silt should be left in the pond, however, at the time of desiltation.

In terms of nitrogen content, pond silt is almost comparable to dung. But silt from forest and park ponds often contains harmful organic acids (humic and ulmic acids), sulfur compounds, and a high quantity of iron oxide. Hence the silt from such ponds should be analyzed before it is used as manure.

The production processes in warm-water carp culture are so managed that ponds of different categories are not under water throughout the year. They should be drained some time during the year to permit necessary repairs. Furthermore, drying is an effective measure against siltation. Under the influence of air, light, and heat silt deposits in dry ponds mineralize, and the enemies and parasites of fish die. Food organisms likewise die but, especially plankton, are restored fairly rapidly (approximately 10 to 14 days) after filling the ponds with water. The rate of restoration of different planktonic and benthic species depends on their respective dynamics of development. Summer ponds should be left dry during winter, and winter ponds kept dry during the spring-summer period. Hatching ponds should be left dry for 10.5 to 11.0 months, rearing ponds for 6.5 to 8.0 months, foraging ponds for 6 to 7 months, and wintering ponds for 5 to 6 months. In the southern regions summer ponds (with the exception of hatching ponds) may be kept under water for a longer period and wintering ponds, contrarily, for a shorter period. In the northern regions just the opposite is better: a longer dry state for summer ponds and a shorter dry state for wintering ponds.

Estivation of Fish Ponds. The annual periodic drying of ponds, the time of year specified by production processes, makes it possible to undertake necessary reclamations. Not even one reclamation operation is possible under the conditions of prolonged exploitation of fish ponds. Removal of excess silt and the enrichment of bottom soil with substances facilitating maintenance and increase in fish productivity make estivation essential, i.e., leaving the fish ponds (mainly rearing, foraging, and summer nurseries) dry for longer periods—one year or more—during which time different agricultural crops can be cultivated. Depending on the conditions of the ponds such an alternation of culture (fish—agricultural crops) is done every four years, i.e., four years of fish culture and one year of agricultural crops. From the fishery point of view the entire gamut of sanitary measures requisite for fish culture is effected in estivating ponds by the cultivation of some agricultural crop, total drying of the bottom, cleaning and drying of the irrigation network, lime application, and repair of hydroengineering installations.

Vetch and oats are the two most commonly sown crops (oats utilizes nitrogen from the deeper layers of the soil and vetch enriches the soil with nitrogen); lupine, maize, pea, bean, sugar beet, cabbage, carrot, cucumber, etc. can also be cultivated. These crops give good yields and simultaneously augment the destruction of aquatic flora, increase soil aeration and friability, neutralize soil acidity, and provide nitrogen enrichment. In Ukrainian SSR the following yields have been recorded from one hectare of estivating ponds with a thin layer of silt and few higher aquatic plants without the application of manures: 600 to 650 quintals of sugar beet, 1,000 quintals of cabbage, 450 quintals of carrot, and 200 quintals of cucumber. Grain crops are more advantageous, however, as the postharvest residue with decomposition creates favorable conditions for the development of chironomid larvae. Hence stubble is never plowed under after harvesting. The maximum productivity of ponds is achieved by sowing maize; insects lay eggs on the postharvest residue and the larvae of chironomids utilize the decomposing stubble for food. The reported yields of grain crops in the fish farms of the central chernozem zone and southern regions of the RSFSR are: 50 quintals of maize, 30 to 40 quintals of spring wheat, 35 to 40 quintals of barley, 30 quintals of bean, and 30 quintals of sunflower per hectare.

As far back as 1785, A.T. Bolotov in his article entitled "Sowing of food grains in ponds," recommended that in the first year of estivation winter and spring rye should be sown, followed by barley in the second, and oats in the third. In ponds with peaty soils spring crops should be sown but lowlands containing considerable moisture (especially in a moist summer) should be left under grasses. These recommendations are still valid today in spite of the wider assortment of crops available for cultivation.

The "rejuvenated" pond yields a higher output of fish the following year. However, a significant increase in fish productivity is realized only when the bottom of the pond has been well dried. It has been proved that a correct follow-up of estivation in the first year increases natural fish productivity two or more times. The highest natural fish productivity of the ponds after their utilization for cultivation of agricultural crops is observed in the first two years; in the third year it reduces by about 40%, and in the fourth year reaches the level of productivity which existed before the ponds were placed under estivation. If artificial feeding of fish under dense stocking is combined with the simultaneous rearing of ducks, pond estivation should be done every two years. In general, the periodicity and duration of estivation depends on the conditions of the ponds and the peculiarities of the geographic zones of the USSR. Fish productivity of ponds under different periods of rotation is shown in Table 20.

Table 20. Fish productivity from pond with an area of 300 hectares with different periods of fish rotation (from F.M. Sukhoverkhov)

Parameters	Periods of alternation of fish rearing and cultivation of agricultural crops				
	Without cultivation of agricultural crops	After one year	After two years	After three years	After four years
Area of pond under fish culture, ha	300	150	200	225	240
Natural fish productivity, q/ha	2.5	5	4.8	3.8	2.05
Annual output of fish, q	750	750	960	720	600
Area of pond under cultivation, ha	—	150	100	75	60
Grain yields, q/ha	—	4,500	3,000	2,250	1,800
Output of fish as a result of feeding, q	—	1,500	1,000	750	600
Total output of fish, q	750	2,250	1,960	1,470	1,200
Fish productivity per 100 hectares of land, q	250	750	680	490	400

Neglected and marsh-type ponds are kept under estivation for two years, not one. In the first year they are dried, the rhizomes of aquatic plants removed, and the ponds plowed in July–August; in the second year the pond bottom is utilized for cultivation of agricultural crops. Drying and cultivation of agricultural crops ensures mineralization of organic matter. This neutralizes acidity or prevents acidification of the pond soil and encourages aeration, as a result of which not only the chemical properties but also the physical features of the pond improve. Soil is an important factor in fish productivity and its fertility cannot be maintained by manuring alone. The physical properties of the soil play an equally important role in maintaining and improving the level of natural fish productivity in ponds.

When the ponds are annually filled with water, even with the application of organic manures and inorganic fertilizers, estivation is nonetheless important—just as crop rotation is in agriculture—because soil structure is very significant. Besides, estivation is an effective prophylactic measure against possible epizoons.

For quite some time in Western Europe, but only recently in the USSR, rotation of ponds under estivation has been done. For example, in the German Democratic Republic, Federal Republic of Germany, and Austria, many carp ponds are filled with water for three years (discharging the water every year in autumn), then sown with agricultural crops for two to three years. In France field areas meant for the cultivation of grain crops (usually not more than 5 hectares), are surrounded by small but sufficiently high dikes for holding enough water to carry out fish culture. The soil in such cases should be clayey and poor in water permeability. In the low-lying parts of such an area a water outlet is provided in the dike and channels made to drain water from the field. In the area itself the usual collection channel is made before draining the water, from which all the fish are collected at the time of water discharge. Fish are raised in such ponds for two years. Oats is sown in the third year and spring wheat in the fourth without applying manures (since the silt layer accumulated in the previous two years serves this purpose) and yields of nearly 40 q/ha are obtained. The rotation cycle, commencing with fish culture for two years, is then repeated.

In Hungary ponds with highly saline, unfertile soils, unsuitable for agriculture, and low in organic matter are placed under estivation every three years. In the fourth year even the most saline soil becomes suitable for the cultivation of different fodder crops useful to fisheries. Sorghum, maize, bean, and other crops are cultured, and the yields even under such unfavorable soil conditions from an agricultural point of view are sufficiently high (sorghum 36 to 63 q/ha and maize in terms of cobs 32 to 79 q/ha). The surface layer of the soil becomes thicker by 3 to 4 cm within one rotation. Thus after three years of exploitation of the area for fish rearing an accumulation of bottom deposits containing organic matter is obtained. In the course of annual water discharge a goodly portion of harmful salts dissolved in the water is removed from the ponds. The soil of such a silted pond, after three years of fish rearing, acquires a more friable structure and becomes suitable not only for effective fish culture but also for crop raising. After a year under fodder crops the natural fish productivity even in these unfavorable soil conditions reaches 395 kg/ha (versus 65 to 70 kg/ha) without using the rotation system and without applying manures. The total productivity exceeds 12 q/ha, the sanitary conditions improve, and incidence of disease in fish has not been observed. Under good soil conditions the cultivation of agricultural crops is economically profitable not only for one year, but even for two, especially when it is necessary to increase the food resources for culturable fish.

The foregoing data on the yields of agricultural crops from estivation ponds in Ukrainian SSR and European countries bordering the Soviet Union make it possible to recommend a wider utilization of this practice in our country. Alkaline soils occupy more than 100 million hectares in the USSR.

On the initiative of the Belorussian Agricultural Academy, a similar experiment is now being organized in the flood lands of the Golyshi River (Goretskii region of Mogilev district). The available dry area will be flooded with water and used for carp rearing for three years. Subsequently the water will be drained, the area plowed, and cereal-legume crops sown for two years. After that the flood lands will be filled with water again and used for fish rearing. Other forms of mixed farming, combining fish culture with other branches of agriculture, are also being considered.

Pond pisciculture is thus included in a general rotation scheme as an economically profitable aspect of agriculture.

In an experiment conducted at the fish farm "Konigwart" (German Democratic Republic) half of two ponds was moistened (in March) and the other half left dry. The results of this experiment are presented in Table 21. It can be seen that the biomass of benthos was several times greater in that half of the ponds left dry, and that chironomids predominated.

Lime application is generally considered a basic measure for improving the conditions of external environment in ponds and thus increasing fish productivity. The success of fish culture, especially in ponds, which are under continual exploitation for long periods of time, depends on regular applications of lime. Lime is utilized by the entire aquatic flora and fauna, including fish, and effects important changes in the pond soil, which radically improve the conditions of external environment. Deficiency in calcium salts, needed for the life of aquatic plants and animals, especially fish growth, is characteristic of less productive ponds, which are poor in nutrients. Calcium salts are required in the formation of skeletal bones, assist in the normal development of embryos and fish tissues, and regulate the functions of the neuro-

Table 21. Effect of drying ponds on their productivity

Date	Pond No. 20				Pond No. 7			
	Wet area		Dry area		Wet area		Dry area	
	Chironomids	Oligochaetes	Chironomids	Oligochaetes	Chironomids	Oligochaetes	Chironomids	Oligochaetes
June 12	0.22	0.15	23.93	0.51	0.69	2.40	16.94	0.58
June 28	2.89	0.02	11.88	0.80	15.58	4.86	35.16	3.86

muscular system. The quantity of calcium in a pond is depleted each year through consumption by various organisms and the total discharge of calcium-laden water; the calcium lost cannot be replenished by the quantity present in the water supply system. Lime applied to fish ponds forms compounds with all the acids present in the pond silt and soil (oxalic, tannic, sulfuric, humic, and others) and neutralizes the acidic medium. After combining with silicates, lime facilitates conversion of silicates into a soluble form, as a result of which the soil is enriched with mineral substances. Lime improves the physical properties of the soil by loosening it, which leads to better aeration. Lime becomes bonded with the carbon dioxide present in the water and forms less stable bicarbonate compounds. The deficiency in free carbon dioxide in the water, observed in summer as a result of intensive absorption by lower aquatic plants, is replenished by the bicarbonates. Other important properties of lime include detoxification of magnesium, sodium, and potassium; augmentation of nitrogen in the soil; development of nitrogen-fixing and nitrifying bacteria; and transformation of biogenic compounds (phosphorus, potash) into easily assimilable ones. "Phosphorescence," the result of lime application, is reduced during the summer and the gas regime at night improves.

It is essential to analyze the soil of the pond for calcium content before lime application. An excess quantity of calcium salts can cause harmful effects. Lime should not be applied to ponds with loose soil nor to those with a small quantity of humic compounds. In such cases lime reduces absorption and facilitates the dissolution of calcium salts, which are removed from the ponds at the time of water discharge and thus their benefit lost.

The simplest index to lime content in the ponds is the type of aquatic plants present. Sphagnum moss, horsetail, and sedge indicate a deficiency in lime; water thyme and stonewort signify an excess quantity. Pond soil samples containing lime "boil" under the action of hydrochloric, nitric, or acetic acid, releasing carbon dioxide ($CaCO_3 + 2HCl = CaCl_2 + CO_2 + H_2O$).

It is essential that lime be applied when the water reaction is acid, and more so if the acidity is high; lime applied to highly alkaline water is harmful. At neutral and mildly alkaline reactions lime application is beneficial, especially when, as a result of pond pollution or the entry of marshy waters, this reaction moves slightly toward acidity. Ponds with acidic, marshy, podsolic, loamy, sandy, clayey, and heavy-clayey soils have the maximum need for lime application.

Different kinds of lime are used, depending on the purpose of application: unslaked (CaO) or slaked $Ca(OH_2)$. The latter is transferred into calcium carbonate ($CaCO_3$) under the influence of carbon dioxide from the air. Powdered limestone is also used.

Unslaked lime (CaO) is used for the disinfection of pond soil, control of fish pests, including their eggs and embryos, and the destruction of small trash fish and predators. Used for such purposes, unslaked lime in many cases makes it possible to restore the productivity of the ponds (for example, in the case of infection), raise the output of fish (in the case of a significant development of fish pests), and to expedite preparations for pond tillage (when trash and predatory fishes are abundant).

The application of unslaked lime for prophylactic purposes should be done systematically. Hatching and wintering ponds are disinfected immediately after catching and transplantation of the fish to the respective category of pond (rearing and foraging). Disinfection is done using the same ratio of lime as used for fish-collecting pits, drying network, and marshy regions of the ponds. Fish farms located near sugar factories can substitute disinfected slurry for lime since this waste material mainly consists of lime, water, a small quantity of organic material, and phosphoric acid up to 1.5%. Since the action of slurry is weaker than that of lime it is used at a higher rate. The phosphorus and nitrogen content of slurry increase the value of the fecal matter.

The degree to which lime is ground is important. Academician D.N. Pryanishnikov has written: "The durability of large lime particles is indicated by the fact that undecomposed $CaCO_3$ has been found in soil six years after lime application." Data supporting this observation follow:

Diameter of applied particles of lime (mm)	Undecomposed $CaCO_3$ (% of applied quantity)
2.30–1.25	62
1.50–0.50	24
0.50–0.125	17
less than 0.125	5

It is better to apply lime to fish ponds when the bottom is moist, after the discharge of water in autumn and spring. In spring, if unslaked lime is used, lime application should be done 14 to 20 days before filling the ponds with water or releasing the fish. Calcium carbonate can be applied directly while filling the ponds and stocking fish.

There are different methods of lime application. Powdered lime is distributed over the bottom in a uniform layer. Lumps of unslaked lime are arranged in small heaps of 30 to 50 kg and covered with a layer of moist soil. If there are no rains, these heaps are periodically moistened with water. At the end of the slaking process the lime is pulverized and uniformly distributed over the bottom. For disinfection of small ponds lime milk (unslaked lime with water) can be uniformly sprinkled over the bottom (using a watering can, for example).

Lime, especially unslaked, should not be applied manually but with the help of some special equipment such as lime application drums on wheels (Figure 83), which are either horse-driven or provided with a mechanical

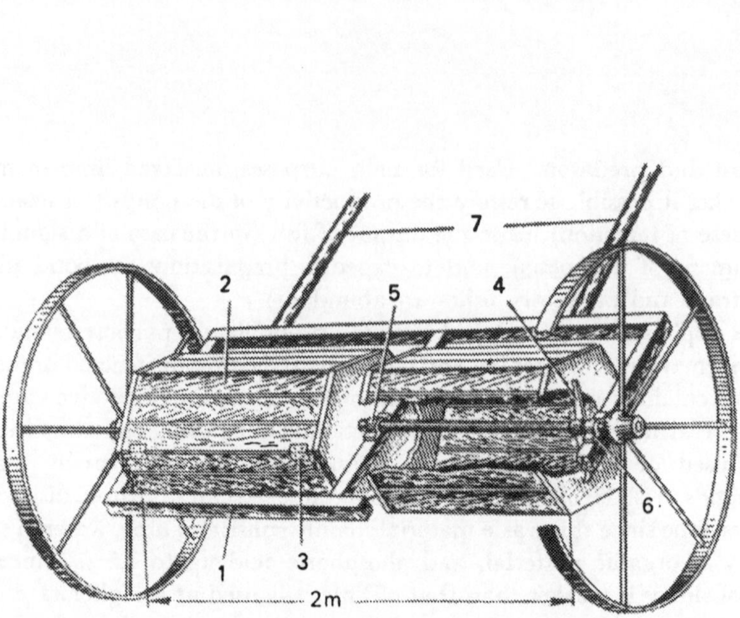

Figure 83. Lime drum.

1—wooden frame; *2*—hexagonal wooden drum for lime; *3*—face of drum adapted in form of hatch for lime; *4*—iron clamp attached to body of drum through metal or wooden semi-axle; *5*—second end of semi-axle inserted into bearing fitted on crossbeam of frame; *6*—bearing of frame; *7*—shafts for horse.

drive. It is not possible in manual application to spread the lime uniformly over the pond area, which reduces the effectiveness of this measure. Aerial broadcasting has been introduced just recently, which reduces the labor force and speeds up the process of lime application. Lime application from an airplane or helicopter is carried out on a windless day or when the wind intensity is low (speed does not exceed more than 6 m/sec) since under a strong wind much of the lime is wafted away.

In calm weather, with a head and lateral wind up to 2 m/sec, lime is applied from a height of 10 to 15 m; with a wind velocity of 2 to 4 m/sec, it is done from a height of 5 m. A mild rain causes no harm, but lime should be protected from moisture at the takeoff-landing platform.

The dusting of unslaked lime from aircraft AN-2 equipped with a special device is most advantageous from a height of 10 to 15 m. In this case the consumption of unslaked lime increases to 42, and that of chlorinated lime to 36 kg/sec. If it is necessary to increase the rate of lime application, the flying height of the aircraft should not be beyond 5 m.

Lime applied by an aircraft is uniformly distributed over the area; its application at different sections of the ponds can be regulated from a helicopter. The workers must wear protective apparel (overalls and gas masks).

The quantity of lime (CaO) required for soils with different pH levels can be estimated on the basis of data given in Table 22.

Table 22. Relationship between quantity of lime (q/ha) and mechanical composition and pH of soils

pH of soils	Type of soil		
	Thick clayey and loamy	Sandy-loamy	Sandy
Less than 4.0	42	22.0	14.5
4.0–4.5	32	17.0	14.5
4.5–5.0	27	14.5	12.0
5.0–5.5	17	12.0	7.0
5.5–6.0	12	7.0	4.5
6.0–6.5	7	5.0	2.0

The proper rates of lime application, depending on the zone and nature of the soil, conditions of the ponds, and purpose of the application, have yet to be precised. The following approximate norms are recommended, however:

(1) In annual systematic lime application of ponds, 1 to 2 q/ha[2];

(2) For high siltation and reducing the condition of pond soils, 30 to 40 q/ha of unslaked lime;

(3) To control gill disease (gill rot) and obtain sedimentation of decaying organic matter, which provides a favorable medium for disease development, 2 q/ha of unslaked lime every day for several days;

(4) For disinfection, 10 to 20 q/ha; in the case of carp suffering from rubella, dactylogyrosis, hydrodactylosis, etc., 25 q/ha of unslaked lime;

(5) To improve the quality of pond soils, 2 to 4 q/ha of unslaked lime;

(6) In the case of low calcium content in the ponds, 10 to 15 q/ha of unslaked lime;

(7) To control small trash fish and predators, 15 to 25 q/ha of unslaked lime.

For ponds which use night soil the rate of lime application is 300 q/ha.

According to the findings of the All-Union Research Institute of Lake and River Fisheries, under the conditions of the northern and northwestern regions of the USSR wintering ponds should be covered with 25 to 30 q/ha of unslaked lime in July–August for the purpose of general disinfection. In the case of large-scale infection of fish with external parasites and considerable mortality, lime application should also be done in the spring after the fish are caught. If the epizooty is rather severe, unslaked lime is applied twice. For sporozoan diseases (coccidosis,[3] myxosporidiosis[4]) the rate of lime

[2]In Bulgaria 2 to 4 q/ha unslaked lime are used in annual lime applications.

[3]The causal organism of coccidosis isCoccidia—sporozoans which lead a parasitic mode of life inside the cells of invertebrates as well as vertebrates. In fish they parasitize the epithelial cells of the intestines, liver, and kidneys.

[4]Myxosporidia are parasites which infect marine and fresh-water fish.

application is increased to 40 to 45 q/ha. In these same zones lime should be applied to the ponds at the rate of 5 to 25 q/ha to improve the quality of soil structure in the case of lime deficiency, poor mineralization of organic matter, and marshy development and acidification of some portions of the ponds. A much larger quantity of lime is applied to marshes and bogs. Shallow plowing of lime into the soil is not harmful. One application of lime is done before plowing in July–August after removing the aquatic plants. The plants are cut not less than twice in the summer to facilitate better soil heating.

To control small trash and predatory fish unslaked lime is better applied at the end of winter through holes made in the ice in a chessboard pattern. It is applied in equal parts to all the holes. In the farms of collective and state enterprises where the ponds are mostly nondrainable and often heavily populated with small trash fish and predators, this method of lime application is most convenient and facilitates a better preparation of the ponds for fish release. At the time of slaking the lime the water turns milky, but at the end of the slaking process it returns to a normal color.

Control of Trash and Predatory Fish. Small trash and predatory fish sometimes enter the ponds from the sources of water supply and through the network of water outlets; these affect the normal course of fish culture. Unwanted fish are competitors for food, both natural and artificial, and the predators among them pose a threat to the fish under culture. Hence adequate measures must be taken to prevent ingress of such fish into the ponds.

Different filter-fish traps are in use. The sectional, rubbleglass, filter-fish trap has been found effective at the Alma Ata fish farm (Figure 84). Its distinguishing feature is that it allows the passage of nearly 1,000 liters of water or more per second, which is very important during floods. This filter-fish trap consists of a rectangular frame (better constructed from angle iron) 1 m high, at the bottom of which metal net-grills are fitted (size of mesh no more than 1.5 mm), with wooden planks along the sides. On the bottom of the filter-fish trap, layers of rubble are laid up to 10 cm thick, then common glass pieces 5 cm, and so on; the diameter of the rubble in cross section should exceed that of the opening in the grill by 0.3 to 0.5 cm. Smaller pieces of rubble, although they increase the efficiency of the fish, reduce the water-letting capacity. The upper layer of rubble is covered with a metal net to prevent disturbance. After filling in rubble to one-third the height of the trap, the latter is installed in the channel at a depth of 25 to 30 cm. For installation it should be noted that the level of water in the water supply source and the channel should differ; hence a coffer dam with a wooden trough is fitted in the channel which supplies water in the fish trap where filtration takes place. At the time of determining the size of the fish trap that quantity of water expected to pass through it must be taken into account,

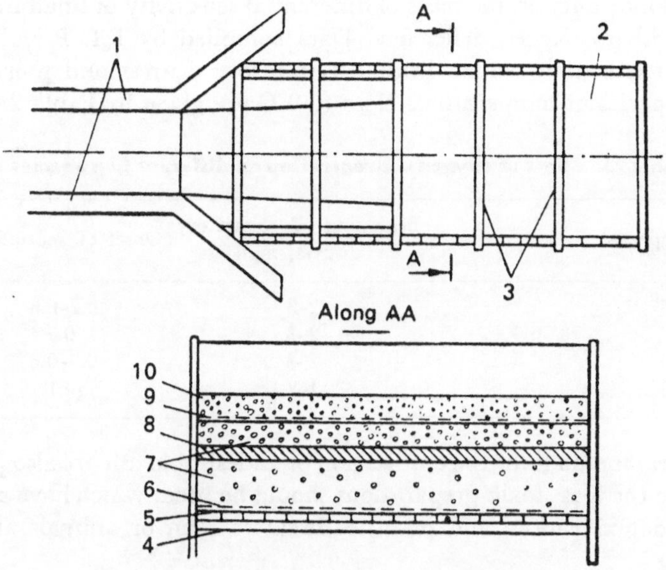

Figure 84. Filter-fish trap for unwanted fish.
1—concrete flaps; *2*—replaceable trough; *3*—section of screen; *4*—crossbeam; *5*—metal frame; *6* and *9*—metal net; *7*—layer of rubble; *8*—layer of glass; *10*—cobbles.

together with the intensity of its pollution and the diameter of the rubble. The average water-letting capacity should be 20 to 25 l/sec per 1 m^2 of filtration layer. The filter fish trap of this system is convenient for exploitation since, being sectional, it is easier to clean. When debris has accumulated in the first section, the water is directed into the second section by removing the replaceable trough. The efficiency of the filter-fish trap is determined by the complete retention of all unwanted fish 10 to 15 mm long and their further entrapment as the water passes through the rubble and glass layers.

Filters are also currently being used which are made of fibrous and nonfibrous materials or materials made of glass and fiberglass. The Institute of Hydrobiology of the Academy of Science of Ukrainian SSR has found that filters of glass wool are extremely effective. The filter is 5 to 6 cm thick and consists of individual fibers 10 to 20 μm in diameter. Such filters ensure a higher degree of water purification, are more convenient to use, and cheaper than a number of other materials currently used in pond fisheries.

An extreme measure for the eradication of small trash fish and predators is the complete drying of the ponds, plowing, and lime application. This measure can be implemented in totally drainable ponds. It is difficult to remove less valuable and predatory fish from nondrainable ponds. The only method which works in such ponds involves the creation of artificial conditions of fish mortality on the basis of differential sensitivity of small trash and predatory fish to oxygen deficiency. Data compiled by T.I. Privol'nev on oxygen concentrations which lead to respiratory distress and mortality of some fish species at temperatures close to 0°C are given in Table 23.

Table 23. Effect of oxygen concentration on different fish species

Fish species	Beginning of respiratory distress, mg/l	Onset of mortality mg/l
Perch	2–3	0.2–0.6
Roach	2–3	0.7
Pike	2–3	0.3–0.6
Crucian carp	1–2	0.1

Chemical methods for the eradication of undesirable fish are also possible. In this case the least toxic preparations should be used, which have a shorter period of detoxification, and pose no threat to man or animals and their food base.

TECHNICAL CULTURAL OPERATIONS. To ready fish ponds for exploitation they are cleared of stumps, stubble, shrubs, trees, stones, and other objects; hummocks and floating mats are removed; the bottom is levelled by filling in depressions; and the water channels and fish pits cleaned.

Water channels can be cleared of silt with the help of a canal cleaner, a tractor with a screw conveyor fitted on one side driven by the motor of the tractor. Moving sequentially along both sides of a channel, the screw conveyor sucks in silt and debris and throws it to one side. The output of the canal cleaner is sufficiently large (5 to 6 km per work day). Its working mechanism can be adapted to the general dimensions of the channels.

Sedge hummocks are cut up with a common joiner's saw in which the blade is fitted at a right angle to the frame; the latter consists of steel tubes. With such a saw 0.66 hectares can be cleared per work day.

Floating mats (vegetational layer with peat) prevent the penetration of light, heat exchange, and oxygen dissolution. At the time of water discharge from the ponds floating mats are removed, which otherwise would settle to the bottom, compressing the fish and clogging the water channels.

The main features of floating peat in different types of marsh land are given in Table 24.

Smaller floating mats are pushed to the edges and the larger ones cut into pieces with a special saw or hydromonitor, towed to the edges, and lifted out with a winch or windlass. A motor boat, launch, tractor, or winch is used in towing the floating mats to the edge of the pond.

As a prophylactic measure, places which are liable to develop floating mats are spread with sand or clay, pegs, and intertwined branches of trees.

Table 24. Main external features of floating peat

Type of marsh land	General characteristics of upper layers of marshy area	Floating capacity of peat
Low-lying, grassy	With decomposed peat. Extent of decomposition 40–50%. When pressed in hands passes through fingers and dirties hands. Upper nondecomposed layer not thick. Distinct traces of alluvial ground.	Does not float.
Low-lying, grassy	Peat poorly decomposed (up to 30%), with a clearly noticeable fibrous residue of root system. Usually brown in color.	Floats.
Low-lying, forest land	Well-decomposed peat (more than 40–50%), brownish or almost black, with a few small roots and herbaceous plants. Upper nondecomposed layer almost absent. Distinct traces of mineral detritus.	Does not float.
Transitional	Poorly decomposed peat (less than 30%) with clearly noticeable residue of hypnum mosses and grassy plants. Externally resembles fibrous mass poorly enriched with decomposed peat.	Capable of floating.
Upper layer	Poorly decomposed peat, consisting of moss residue (sphagnum).	Capable of floating.

Sowing of agricultural crops for several years or burning the surface peat layers are also effective.

Floating mats pose a great problem for pond fisheries. Controlling them is so difficult and costly that at the time of selection of an area for the construction of new ponds, one must make certain that the threat of floating mats does not exist.

References

Bank, D. 1960. Die dekämpfung von fadenalgen mit chemischen mitteln, *Alg. Fisch. Zeit,* no. 2.

Berezovskii, A.I. 1935. Melioratsiya v rybnom khozyaistve (Melioration in Fish Culture). Vses. Koop. Izd-vo, Moscow.

Bobrov, A. 1966. Aeratsiya vody (Aeration of water). *Rybovodstvo i Rybolovstvo,* no. 6.

Engashev, V. and A. Aleshin. 1969. Aviyatsiya na slzuhbe v rybovodov (Aviation in the service of pisciculture). *Rybovodstvo i Rybolovstvo,* no. 4.

Erokhina, L.V. and M.A. Brudastova. 1957. Primenenie dozhdevatelei dlya aeratsii rybovodnykh prudov (Use of sprinklers for aeration of fish ponds). *Rybnaya Promyshlennost',* no. 27.

Ivanov, M. 1965. Maksimal'na produktsiya ot edinitsa ploshchad (Maximum production from unit area). *Ribno Stopanstvo,* vol. 12, no. 2.

Jahnichen, H. 1967. Die aufnahme von wasser- und landpflanzen durch jnge graskarpfen (*Ctenopharyngodon idella*), *Dt. Fisch. Ztg.,* vol. 14, no. 15.

Khomchuk, A.A. 1959. Metodicheskie ukazaniya po letovaniyu prudov (Methodical instructions on estivation of ponds). Sbornik *Rybnaya Promyshlennost',* no. 52.

Kormilin, V., M. Kenzhalin and N.A. Bulanov. 1969. Dlya bor'by s sornoi ryboi (Control of trash fish). *Rybovodstvo i Rybolovstvo,* no. 1.

Lennon, R. 1967. Clearance and registration of chemical tools for fisheries, *Pogr. Fish. Culturist,* vol. 29, no. 4.

Maisuryan, N.A. 1957. Prudovoe rybovodstvo v polevykh sevoobortotakh Frantsii (Pond fisheries in field crop rotations in France). *Vestn. S.-kh. Nauki,* no. 12.

Martyshev, F.G. 1968. Bol'she ostorozhnoisti (Be more careful !). *Rybovodstvo i Rybolovstvo,* no. 6.

Merla, G. 1969. Einwandfreies trockenlegen der teichebedeutungsvoll für die teichfruchtbarkeit, *Dt. Fishrei. Ztg.,* vol. 3, no. 16.

Nikol'skii, G.V., B.V. Verigin, A.P. Makeeva, G.V. Popova and S.G. Soin. 1968. Issledovaniya rastitel'noyadnykh ryb i vnedrenie ikh v praktiku rybovodnogo khozyaistva i melioratsii vodoemov (Studies on phytophagous fish and their introduction into the practices of pond fisheries and the melioration of ponds). *Vestn. Mosk. Un-ta,* no. 2.

Privol'nev, T.I. 1954. Fiziologicheskie prisposobleniya u ryb k novym usloviyam sushchestvovaniya (Physiological adaptation in fish to new conditions of survival). *Trudy Soveshch. Ikhtiol. Komissii Akad. Nauk SSSR,* no. 3.

Pryanishnikov, D.N. 1946. *Agrokhimiya* (Agrochemistry). Sel'khozgiz, Moscow.

Shpet, G.I. 1954. Regulirovanie vodnoi rastitel'nosti, v rybovodnykh prudakh s tsel'yu uvelicheniya kolichestva benticheskikh organizmov kak kormovoi bazy dlya karpa (Regulation of aquatic vegetation in fish ponds with a view to increasing the quantity of benthos as a food base for carp). *Trudy Soveshchaniya po Rybovodstvu.* Moscow.

Shpet, G.M. 1962. O mekhanicheskikh sredstvakh aeratsii karpovykh prudov v letnii period (Mechanical means of aeration of carp ponds during summer). *Trudy Soveshch. Ikhtiol. Komissii Akad. Nauk SSSR,* no. 14.

PART III

INTENSIVE TYPE OF CARP CULTURE

The production of carp from ponds can be increased not only by the most exhaustive utilization of the natural organic food resources in the pond as such, but also by combining natural foods with additional fish feed. To achieve the maximum increase the methods of intensive fish farm management must be followed. With this objective in view, artificial feeds are added and the ponds manured. With the first measure the increase in productivity is direct, while in the second it is indirect, through increasing the stocks of natural food organisms. These basic methods of intensification for a given pond area can be used in conjunction with the intensification complex,[1] thereby further enhancing the total fish productivity.

[1]See Chapter XIII for a detailed analysis of these aspects.

Chapter XI

ARTIFICIAL FEEDS OF CARP

Carp, in general, are stomachless fish and, as such, consume small quantities of food, resulting in more frequent evacuations than observed in fish having a stomach. Each group of the basic components, whether organic or inorganic in origin, is significant in the nutrition of fish and must be provided in its ration. However, while considering the organism as a whole, care must be exercised that the groups of basic components provided in nutrition are interrelated and essential for normal physiological activity.

PROTEINS. PROTEIN RATION IN CARP FOOD. In fish cultured in ponds the protein content varies from 14.7 to 21.7% (Lake Seven trout) and fat content from 0.4 to 22.2% (common eel). Proteins play an important role in fish nutrition. They cannot be replaced either by fat or by carbohydrates. In the assessment of the food value of proteins their amino acid composition is also significant. At present, more than 20 amino acids have been identified in animals and plants. Some are nonessential and others essential. The former are synthesized in the animal organism, while the latter are not and must be obtained from nutrients. There is a special group of semi-essential amino acids, in the presence of which the nonessential ones are consumed more economically. Fish proteins comprise the total complex of essential amino acids, but their individual quantity varies in different species. Without knowing the requirements of pond fish for individual amino acids, it is impossible to compose a balanced feed, reduce the cost of feed per unit growth of the fish under culture, and to increase the winter-hardiness of fingerlings.[2]

In pond fisheries the problem related to providing the required ratio of essential amino acids in feeding carp and other fish species has been rather poorly studied. In foreign countries the amino acid content in the body of animals is expressed as a percentage of the content of the major limiting amino acid from plant sources—lysine—as a basis for composing rations balanced in terms of amino acids. It is believed that the requirement of animals for amino acids corresponds approximately to the amino acid composition of their body and in the process of determination of their requirement for amino acids one should commence with the aforementioned percentage base.

[2]For essential amino acid content in feeds, see I.P. Popov's *Amino Acid Composition of Feeds*, Rossel'khozizdat, 1965.

The following 17 amino acids have been isolated in the proteins of two-year carp from autumn catches: cystine, arginine, lysine, histidine, ornithine, aspartic acid, serine, glycine, glutamic acid, threonine, alanine, proline, tyrosine, methionine, valine, phenylalanine, leucine, and isoleucine. The qualitative amino acid composition of the proteins in under-yearlings, yearlings, and two-year fish is similar. The essential amino acids are arginine, histidine, threonine, leucine, isoleucine, valine, lysine, tryptophan, methionine, phenylalanine, cystine, glycine, and tyrosine.

The requirement for methionine may be met by consuming such algae as *Scenedesmus* and *Chlorella*, which contain up to 60% proteins in a dry state. Furthermore, the proteins of *Scenedesmus* contain 18 amino acids. Individual components of natural foods of animal origin are also quite important, not only for meeting the requirement for methionine, but also other amino acids (see below). The methionine content in natural foods and feed mixtures changes during the vegetative growth period. These changes as reflected in carp fingerlings are shown in Table 25.

The tryptophan present in fish proteins, according to Melikova (1969), is 1.3 to 1.5 times greater than in homoiothermal animals. This amino acid can be used as an index of the full-value proteins of the muscle tissues. The quantity of tryptophan in the proteins of carp fingerlings varies from 0.02 to 0.10% (Mukhina and Koroleva, 1964).

Deficiencies in other amino acids produce the following adverse effects: arginine—disruption in growth; histidine—anemia; leucine, isoleucine, and threonine—disruption in protein synthesis; phenylalanine—hormonal imbalance; and valine—injury to the nervous system.

It is known that essential amino acids should be present in a quantity not less than 6 mg/100 g of the total amino acids in proteinic food.

Fish raised in ponds consume not only extraneous feed but also natural food, which is physiologically complete and hence offsets the deficit of one or the other amino acid in artificial feed. In composing the feed mixture one must select individual components according to the quantity of amino acids present in them, while ensuring that the entire complex is represented. An excess quantity of essential amino acids in the feed, similar to a deficiency, has an adverse effect on metabolism and growth.

Any excess quantity of proteins in the feed is accumulated for energy metabolism, a process otherwise essentially met by a carbohydrate component. Such a utilization of proteins is irrational since their value as a source of energy is not very high. Here it should be kept in mind, however, that fingerlings, especially in autumn, should not suffer from a protein deficiency. The optimum protein content in carp food of good quality should be about 15% (according to Schäperclaus, 17 to 18%) and for yearlings stock material even higher.

An analysis of the amino acid composition of a number of planktonic and

Table 25. Changes in methionine content in fingerlings of carp during the vegetative growth period (from Mukhina and Koroleva, 1964)

Type of food	Methionine content, %							
	July 2		July 15		September 1		October 1	
	Total proteins	Total nitrogen	Total proteins	Total nitrogen	Total proteins	Total nitrogen	Total proteins	Total nitrogen
Natural food	1.30	0.73	2.71	1.52	2.07	1.21	1.86	1.10
Feed mixture	2.06–3.68	1.19–1.96	2.80–2.88	1.12–1.75	1.28–2.92	0.76–1.19	1.18–1.45	0.21–0.83
Feed mixture. green plants	3.28	1.94	3.00	1.76	2.23	1.10	0.67	0.39

Table 26. Percentage of amino acids in the proteins of some invertebrates (m g per 100 g)

Amino acids	Daphnia	Chironomidae	Caddis fly	Enchytraeidae	*Limneae ovata*	*Limneae stagnalis*	Gammaridae	Corixa
Arginine	10.92	4.75	5.36	5.62	4.44	4.92	4.76	4.50
Histidine	2.69	2.38	2.28	1.86	1.33	1.24	1.69	2.15
Tryptophan	3.62	2.06	2.98	1.79	—	—	1.72	1.64
Methionine	3.45	1.48	1.72	1.69	1.68	1.04	1.56	2.02
Cystine	1.17	1.05	1.21	1.05	1.17	1.02	0.99	1.92
Tyrosine	4.27	3.16	3.34	3.37	2.45	2.24	2.24	3.04

benthic organisms has shown that it varies in different species (Table 26).

The quantity of different amino acids in the proteins of food organisms is, on the average: arginine—6.6, histidine—2.7, lysine—6.0, tryptophan—1.2, phenylalanine—4.6, methionine—1.3, cystine—4.7, threonine—3.9, leucine—5.4, isoleucine—4.7, and valine—4.7.

In pond fisheries the protein ratio is considered a criterion for evaluation of feeds; in other words, the ratio of digestible nitrogen to digestible non-nitrogen substances in the feed.

A protein ratio up to 1:5 is considered narrow, and higher than this wide. It was long thought that the narrower the ratio the more valuable the feed, but in practice this is not always true. Sometimes a feed with a wider protein ratio (for example, 1 : 7) has the same effect as a feed with a narrow protein ratio (for example, 1 : 2). This is explained by the fact that the deficiency in assimilable protein in the feed is offset by valuable natural food. The values of natural food and feed are determined not only by this ratio, but by a complex of factors which create the conditions of the external environment. Vitamins must be included in any evaluation of fish nutrition; since they are obtained mainly from natural food by carp, factors affecting the external environment are automatically involved.

The protein ratic 'n the feed varies depending on the age of carp: for juvenile fish it should be 1 : 0.4 to 1.05, for two-year fish 1 : 5, and for three-year 1 : 8. The protein ratio anticipated may change in the course of the vegetative growth period depending on the temperature of water and the growth rate of carp. With an increase in water temperature, which steps up feeding and fish growth, the protein ratio in the feed of two-year fish may be narrower (for example, 1 : 2.5). In the pre-autumn and autumn periods, when the water temperature drops, the protein ratio should be broadened to 1 : 5. The same holds true with reference to stocking density. The lower the density, the wider the protein ratio and, conversely, the higher the density of stocking and hence reduction in quantity of natural food intake, the narrower the protein ratio. Here one must pay attention to the completeness of the amino acid composition in the proteins given.

Under the conditions of the Moscow district (Gerasimova, 1967) a well-balanced feed mixture with a protein ratio of 1 : 5 contains a sufficient quantity of proteins for two-year carp raised at 5N stocking.[8] This makes it possible to economize 16.5 to 35.2% protein in the form of concentrated feeds. Narrowing the protein ratio is permitted only at the end of July–beginning of August, when the natural food base of the pond has decreased.

Fats and Carbohydrates. Fats and carbohydrates are important for a normal discharge of the physiological functions of fish. Fats are primarily

[8]Stocking density calculated only on the basis of natural food is called normal density and designated by the letter *N*.

formed from the carbohydrates of the food consumed by fish and participate in plastic metabolism, constitute an important source of energy for the organism,[4] and serve as solvents for certain biologically active substances, in particular vitamins. Fat reserves in the body of fish constitute a most unstable component, varying according to season, age of fish, and conditions of feeding and rearing.

If the percentage ratio of protein in the body of fish is consistently maintained at the same level, considerable fluctuations in the fat content will be observed from season to season.

Under satisfactory conditions of fish rearing and supply of natural food and artificial feeds, the fat content in the body of fish increases with advancing age. Given the age-dependent peculiarities of metabolism, the level of energy metabolism is higher in younger fish than in adults. On the basis of fat content and intensity of fat accumulation it is possible to determine the biological and physiological condition of the fish; the dynamics of this content often give one an idea of the direction of fat metabolism in the organism. The quantity, and most probably, quality of fat formation is particularly significant in the process of rearing stock material. In winter, when the temperature falls, heat-loving fish are the first to consume accumulated fat reserves and in large amounts; the degree of winter-hardiness depends on the total quantity of fat present in the fish body. Furthermore, the quality of the produce is evaluated on the basis of the quantity of fat in the body. The quality of the fat is assessed on the basis of fatty acid composition. The most important of these acids are myristic, palmitic, arachidonic, oleic, linoleic, and linolenic.

However, not all of the fatty acids essential for fish can be synthesized in the organisms. In particular, carp need linoleic and linolenic acids, which can only be obtained from fats in feed.

In fisheries decisions are still based on the index of total fat content in the fish body. Such an index does not explain incidences of mortality of carp fingerlings in the process of hibernation in spite of a sufficient quantity of fat and the observance of other conditions necessary for successful hibernation. Obviously the characterization of hibernation cannot be restricted to the index of total quantity of fat; one must also assess the fatty acids it contains.

According to the findings of Bottesch (1958) the fat content in carp undergoes significant quantitative changes in the second year of life, depending on the time of year (Table 27).

Fat accumulation takes place in all the important organs, although the rate of deposition is not equal and the percentage varies. The moisture content of the body is inversely proportional to the quantitative changes in fat

[4]D.A. Lovern believes that synthesis of fat from carbohydrates occurs as an exception in marine fish, and that formation from proteins in food is very insignificant.

content. There is also a relation between total fat content and the fats present in carp food.

Table 27. Percentage of fat in whole fish and individual organs of two-year carp throughout the year

Month	Total fat content	Fat content in various organs, %				
		Muscles	Head	Liver	Testes	Ovary
May	3.47	1.71	5.77	6.39	—	—
June	3.52	4.14	5.75	8.16	—	—
July	5.14	—	—	—	—	—
August	6.49	7.44	13.59	8.72	—	—
September	10.13	7.75	10.59	8.89	2.94	12.52
October	14.61	13.07	13.39	10.44	2.98	25.88
November	12.76	12.49	13.90	10.54	5.56	26.01
December	12.74	12.48	14.13	7.35	7.83	13.68
January	11.70	11.75	16.29	5.40	5.25	27.53
February	—	—	—	—	—	—
March	—	—	—	—	—	—
April	8.17	10.52	16.21	3.95	3.67	30.43
Average	8.32	8.04	12.25	8.49	4.70	22.67

Changes in total fat content and moisture content in the body of two-year carp in different months are shown in Figure 85.

Seasonal changes in fat content, as a high energy reserve material, are also associated with gonadal development in fish and their maturation. The physiological commissioning of requisite substances in fish in the process of gonadal development ensures the production of sex products of a definite quality and quantity, a prerequisite for normal breeding. In this process the nature of fat accumulation and utilization in the process of gonadal development is specific for individual fish species.

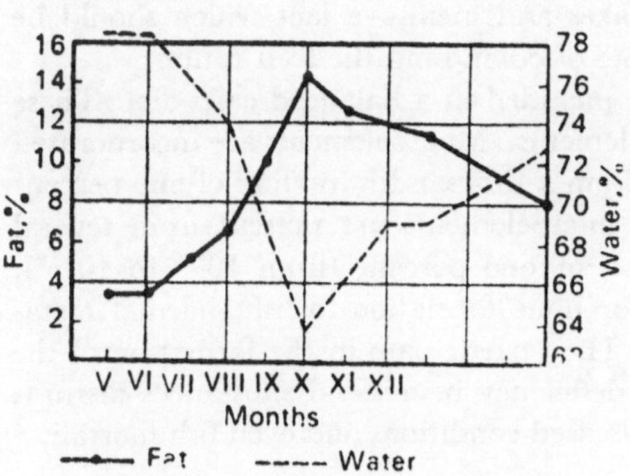

Figure 85. Monthly changes in total fat and moisture contents in two-year carp.

Studies conducted by Maslova (1968) show that in carp feeding on natural food the ratio between fat and protein is 1:2.0 to 1:2.5, which ensures a high growth rate and normal life activity. In the course of consumption of high quality food this ratio is expressed as 1:1.1 to 1:0.9, which justifies fat accumulation in carp. The intensity of fat accumulation considerably increases with high stocking density (especially notable when 10 and 15 repeated stockings were undertaken in an experiment). This phenomenon is probably explained by a reduction in the consumption of natural food and changes in the ratio between this food and artificial feed. The properties of the fat deposited in the fish body depend on the quality of the fat in the food. The composition of fatty acids is more variable in fish feeding on natural food than in fish consuming one kind of artificial feed.

Thus, the quantity and quality of natural food and artificial feeds, as well as the conditions of fish culture, influence the intensity of fat accumulation. The use of fertilizers, especially superphosphate, leads to a reduction in the intensity of fat accumulation in two-year carp (Maslova, 1968).

The carbohydrate metabolism of carp is quite interesting but has been poorly studied to date. When carbohydrates are increased in the ration, an increase in body fat content occurs, i.e., a distinct change in metabolism. Pond fisheries, especially when fingerlings are being raised and prepared for hibernation, usually increase the supply of carbohydrates from the second half of the vegetative growth period to augment fat deposition in the body, a factor related to winter-hardiness of fish. However, an excess quantity of carbohydrates in the ration is harmful to the protein content of the body, particularly in growing fish. It leads to an increase in the quantity of food consumed, overstrain on the digestive system, increase in the food coefficient, and protein starvation.

E'rman (1969) has shown that by adding 13.8% starch in carp feed, the quantity of nitrogen excreted by the fish decreased by 52 to 59%. This effect takes place under different protein contents in feeds of both animal and plant origin. On the basis of easily assimilable monosaccharide content, cereal feeds are more nutritional than cakes and meals—a fact which should be taken into consideration at the time of composing the feed ration.

MINERAL SUBSTANCES[5] must be included in a balanced carp diet. These are divided into macro- and microelements. Macroelements are incorporated in the body in quantities ranging from a thousandth fraction of one percent up to an integral percent, while microelements are present from several thousandths to a trillionth fraction of one percent (from 10^{-3} to 10^{-12}). Mineral substances are essential for bone formation in fish, normal metabolism, and other vital functions. They participate in the formation of the cells and tissues of the organs. A deficiency in mineral substances disrupts mineral metabolism and leads to diseased conditions and even fish mortality.

[5]For the macro- and microelement content in the feed of carp and other fish species of pond culture, see *Mineral Composition of Feeds* published by Kolos, 1958.

Their relationship with proteins, vitamins, and hormones has been established. Their participation in many enzymatic systems has been demonstrated. The extent of accumulation of trace elements by fish depends on the season of the year and is determined by intensity of feeding and physiological activity. The growth and development of fish depend not only on their presence but also on the ratio of mineral substances in natural food, feeds, and water of the ponds. The mineral substances dissolved in water enter the fish body through the gills and mucous membrane of the oral cavity, and are also absorbed by the body surface through the skin and scales. It has been established that magnesium, calcium, iron, zinc, strontium, yttrium, iodine, cobalt, and a number of other cations penetrate the tissues and organs of fish through the body surface. Anions of sulfur, carbon, and phosphorus also penetrate the gills and skin.

The intensity of penetration of the fish body varies for different elements and their compounds. It has been established that 69 to 88% of calcium (Ca^{45}) penetrates from water through the surface of the gill filaments, mucous membrane of the mouth, and the body, while penetration through the skin surface varies from 12% in scale carp up to 31% in mirror carp (which is covered with scales to a lesser extent). The intensity of penetration of Ca^{45} in fish reared in waters with different calcium contents is not the same. In the experiments of G.S. Karzinkin and M.P. Bogoyavlenskaya (1956) the lowest intensity was recorded when the quantity present in water was high (150 mg/l) and the Mg : Ca ratio 1 : 11.1; the maximum intensity occurred in fish kept in water containing 30 mg/l and the Mg : Ca ratio 1 : 2.2. Fish reared in water with different quantities of calcium exhibited differences in weight increase. Weight was best in water with 50 mg/l Ca and an Mg : Ca ratio of 1 : 3.7, and minimum in water containing 150 mg/l Ca and an Mg : Ca ratio of 1 : 11.1. Certain pH levels of water increase the calcium metabolism in fish and they lose a large quantity of this element. Low calcium content in water leads to mortality of carp fry and increased feed expenditures. Higher density of stocking also reduces the relative calcium content in the body of two-year carp.

Phosphorus ions (p^{32}) penetrate the mucous membrane of the mouth and even more effectively the surface of the gill filaments. Only about 1.5% of the total requirement for phosphorus ions is met by phosphorus dissolved in water; the greater part is assimilated from natural food and artificial feeds. The phosphorus content in the body of fish has an inverse relationship with fat accumulation; with an increase in quantity of fat in dry matter the phosphorus content decreases. The most significant accumulation of phosphorus is associated with a maximum increase in protein content. Fluctuations in organic and inorganic phosphorus in the blood are associated with the process of carbohydrate and fat metabolism.

Calcium and phosphorus absorbed from water are mainly deposited in the bones. A large percentage of the phosphorus goes into the muscles and a smaller percentage into the composition of "soft tissues" and body fluid. Under normal conditions, calcium and phosphorus are deposited in the organism in an approximate ratio of 2 : 1.

Similar to homoiothermal animals, inorganic phosphorus is absorbed in the intestines of fish, but most intensively in the anterior section. Irrespective of the mode of penetration of the fish body, phosphorus is quickly distributed throughout the organs and tissues.

The dynamics of phosphorus and calcium accumulation in the body of two-year carp (as percentage of dry weight) depend on the density of stocking. With an increase in density the percentage of deposition, especially that of calcium, drops, which is manifested toward the end of the vegetative growth period.

Cobalt influences blood formation (erythrocytes and hemoglobin increase) and accelerates general metabolism and synthesis of muscular proteins. Cobalt deficiency delays synthesis of vitamin B_{12}. In spite of the important role of cobalt in the life activity of fish, the available data are contradictory and the dynamics of its action not yet fully understood. Considering the lack of studies on a number of factors associated with its use, pond fisheries should avoid its inclusion in artificial fish feeds until these problems have been resolved.

Iodine deficiency causes goiter in fish. Iodine penetrates fish through the gills and body surface.

Copper, manganese, and zinc also exert an influence on fish.

The effect of zinc on eggs, embryos, and larvae of carp was studied by Vladimirov (1969). Doses of 0.05 and 0.50 mg/l of zinc sulfate improved the development of embryos and the vitality of larvae. Doses up to 5 mg/l reduced fecundity, disrupted the egg membranes, induced hatching of underdeveloped embryos, and diminished the vitality of larvae.

According to Sabotash (1970) the initial zinc content in eggs obtained from spawners of different age is not uniform. The zinc content is more in three-year fish, less in six- and seven-year fish, and still less in older fish (11 to 14 years).

The inclusion of a number of other trace elements in artificial fish feeds or for increasing natural productivity can only be recommended after scientifically based norms of application have been determined, taking into account the biological and zonal peculiarities of the ponds.

The addition of one or the other mineral salt to carp feed may be desirable if the quantity of trace elements in the medium surrounding the fish is known (including artificial feeds). The proportion of trace elements for a given complex of conditions should be determined accordingly.

Water-Salt Exchange in carp depends on the conditions of their rearing. The aquatic medium in which fish live exerts an influence on all the vital functions including water-salt exchange. In most of the fish species undertaken for pond culture (carp, sazan, tench, grass carp, silver carp, goldfish, pike, sander, catfish, etc.) the water content is around or above 70%; the water content is higher in young fish and gradually decreases with age. At a higher concentration of the metabolic products released by the fish in water, they tend to accumulate in the blood, leading to intoxication of the fish and ultimately to a reduction in water exchange. An inverse relationship is observed between the quantity of fat and water in carp. The extent of water and mineral accumulation in the body of carp, especially calcium and phosphorus, depends on the density of stocking; it decreases sharply at 10 and 15N stocking (Maslova, 1968).

VITAMINS have a high biological activity and are required for normal life activity. It is believed that fish need the same vitamins required by homoiothermal animals. A deficiency in vitamins in fish feeds does not cause such severe physiological dislocations as in homoiothermal animals, but it does retard growth and reduces the organism's resistance to diseases. Artificial feeds containing certain vitamins help in the assimilation of mineral and protein substances in fish. The total absence of vitamins, especially the B complex, poses serious consequences in fish.

Vitamin B_1 (thiamine) should be added to feed mixtures. A deficiency in this vitamin in carp results in convulsions, muscular atrophy, exudation, exophthalmos, increased respiration, and poor growth.

The vitamin requirements of carp per kg of body weight, according to the findings of a number of authors, are as follows: B_2—from 0.11 to 0.33 mg, B_6—0.15 mg, nicotinic acid—0.55 mg, pantothenic acid—1.0 to 1.4 mg, and vitamin A—100 to 500 units. Carp also needs vitamins C, D, and E.[6]

A feed containing vitamins A, B, C, D, and E is assimilated by carp better, their growth rate increases, and the food coefficient decreases considerably.

The addition of vitamins under highly dense stocking is especially important, when the possibility of obtaining vitamins by fish from natural foods decreases considerably or is almost totally excluded (for example, culture in enclosures and cages). Steffens (1969) has recommended (mainly on the basis of experiments by Japanese pisciculturists) the addition of the following vitamins per kg of feed mixture in the raising of two-year carp (in mg):

B_1	10	Choline chloride	50	
B_2	20	C	50	
B_6	0.82	E	20	
Nicotinamide	30	K_3	3	
Pantothenic acid	100	A	600	IU
Inositol	500	D	2,000	IU

[6]According to the data of British scientists the best growth of fry is observed when α-tocophenol (vitamin E) is present at the rate of 10 mg per 100 g of dry weight of feed.

Studies by M.R. Sidelnikov have shown that with the addition of B_{12} concentrate at the rate of 24 mg/ton of carp feed productivity increased by 27.7%, and the food coefficient decreased by 26.7%.

All the essential vitamins are received by fish from natural foods, the composition of which is quite variable in the food spectrum of carp. All species of crustaceans are rich in vitamin content (vitamin B_1 has been found in them); daphnids also contain vitamin B_2. A high percentage of vitamin A has been found in daphnids, amphipods (*Gammarus pulex*), and other invertebrates. All the larvae of insects are rich in vitamin B_2 (0.483 to 0.615 mg% of fresh weight) but less so in vitamin B_1 (0.180 to 0.361 mg%). These vitamins are minimally present in larvae of chironomids, and nil or almost nil in *Corixa* and caddis fly. Insects are a poor source for vitamin A. Worms have an even lower vitamin content than insects. None have been found in Enchytraeidae; *Tubifex* contains a little vitamin A and a negligible percentage of vitamin B_1 (0.087 mg%). Many microorganisms contain different vitamins which reach the fish through the medium of zooplankton and benthos and sometimes even directly through their ingestion. The Ural formula feed plant has recently started marketing vitamin granules for fish.

For the enrichment of feeds with vitamins a farina of such conifers as spruce and pine, and aquatic plants (duckweed, water thyme, pond weed, etc.) has been recommended. Currently, attention has focused on planktonic unicellu'ar algae which yield a high biomass and contain vitamins C, E, and provitamin A.

In addition to unicellular algae, filamentous algae are also a rich source for increasing the food value of prepared mixtures. They are no less valuable as a source of vitamins, proteins (up to 15%), and fat.

For the enrichment of artificial feeds with vitamin C and carotenes fresh filamentous algae is recommended; for reinforcement of the vitamin B complex dry algae is better. Algae are protected during drying from the action of direct sunlight to avoid destruction of the vitamins.

ANTIBIOTICS also act as stimulators in fish nutrition, aside from their use in medicinal preparations for certain infectious diseases. However, the addition of antibiotics to feeds of farm animals is somewhat controversial. It has been demonstrated in experiments that part of the antibiotics in feeds accumulates in the animal's body, and reaches man through meat consumption, reducing his resistance to different diseases. According to the findings of H. Mann (1967) antibiotics such as oxytetracycline, chloramphenicol, penicillin, and chlortetracycline added to the feed of trout did not produce the desired effect. Simultaneously they proved fairly effective in controlling the infectious diseases of fish. These two situations should not be treated as identical: in the first case the antibiotics systematically act on the organism and in the second are used for a short time. Experience in animal husbandry has revealed that after treatment of cows suffering from mastitis with penicillin, the milk is not suitable for processing into sour-milk products for some time. Such milk, in the opinion of veterinary experts, is not totally harmless for the human body. Thus further studies are required on the influence of antibiotics on fish after their addition to feeds.

The addition of biomycin in feeds in experiments conducted by E.M. Malikova (USSR) and in experiments carried out with carp in Bulgaria gave unsatisfactory results. Tests conducted with the by-products of the penicillin industry have shown that penicillin doses given to juvenile carp and salmon, which were several times lower than clinical doses, stimulated their growth. Very small doses of penicillin (200 mg/kg) and purified biomycin (50 mg/kg) considerably activated the enteric microflora of the fish. Studies on the action of tetramycin given with feed to carp showed that it stimulates growth and improves food assimilation.

Feeding of Carp. The food base is the most important aspect in the development of intensive pond fishery. Studies on the nutritional aspects of feeds and feeding of fish are steadily increasing. The characteristics of the feeds commonly used in pond fisheries are presented in Table 28.

Sorghum grain is also used for feeding carp. In the process of feeding juvenile fish broken grains of sorghum the food coefficient was found to be 4.5, while in feeding older carp the coefficient was 5.0, i.e., closer to the coefficient of barley and maize. Red gram (*Cajanus indicus*) is also of interest as the seeds contain proteins, fat, starch, and vitamins. A high yielding variety of this crop under moderate irrigation produces three tons of pods per hectare. In the steppes and forest-steppes, regions of Ukrainian SSR, southeastern regions of RSFSR, Azerbaidzhan, and Tadzhik SSR, and in other southern republics, marsh pea can be used in the form of a cereal-pulse concentrate with a high protein content. The dry matter of seeds of cultivated marsh pea contains from 25 to 30% proteins, 52 to 58% nonnitrogen soluble matter, about 3% water, and 4% cellulose. The waste materials from local food and fish industries are no less important, as are the trash materials from fish farms (small trash fish, tadpoles, frogs, mollusks, paste made from aquatic and terrestrial plants). Aquatic and terrestrial plants contain in addition to proteins, fats, and carbohydrates, vitamins and minerals, and hence constitute a valuable component in feed mixtures. The offal from the fish industry contains 20 to 50% dry matter, 14 to 18% proteins, 10% mineral matter, and a rich assortment of vitamins. The waste from meat factories can also be utilized. Of the waste material from the food industry, carp can be fed corn flour, sweepings, and maize cake (food coefficient 4 to 5), groundnut cake (food coefficient 3 to 5), malt sprouts (food coefficient 11 to 12), and brewery waste (food coefficient 20 to 26). Agricultural waste is also used for feeding carp. Locusts, earthworms, larvae of blowfly, protein paste from whale liver, and different kinds of food wastes are other food sources. The food value of dried locusts is characterized by the

Table 28. Characteristics of some carp feeds

Feed	Water	Raw protein	Fat	Carbo-hydrates	Ash	Cellulose	Protein ratio	Food coefficient
Leguminous crops:								
yellow lupine, split	14.0	33.3	4.4	25.4	3.8	14.1	1:1.2	3-4
Blue lupine	14.0	29.5	6.2	36.2	2.9	11.2	1:2.0	3–5
Marsh pea	—	26.4	1.8	48.6	3.2	6.6	1:2.0	3–5
Peas	—	26.6	1.9	53.0	2.7	5.4	1:2.6	4–5
Lentil	13.0	25.2	1.7	53.0	3.4	3.8	1:2.4	3–5
Soybean	10.0	33.2	17.5	30.2	4.7	4.4	1:1.9	3–5
Beans	14.3	25.4	1.5	48.5	3.2	7.1	1:2.3	3–5
Grain crops:								
Maize	13.0	10.4	4.4	68.7	1.3	2.2	1:9	4–6*
Rye	13.4	12.3	2.0	68.4	2.0	1.9	1:7	4–5
Barley	13.0	11.6	2.1	65.8	2.4	3.9	1:8	4–5
Rye waste	13.4	12.3	2.0	68.4	2.0	1.9	1:7	5
Flour and bran:								
Rye meal	12.6	12.5	1.9	68.5	3.0	3.6	1:6	4–6
Barley meal	13.2	11.6	2.3	65.1	2.9	3.0	1:7	4–6
Rye bran	14.0	15.0	2.4	62.0	4.5	3.4	1:4	4–7
Wheat bran	13.2	15.9	4.2	54.6	5.9	10.2	1:4	4–7
Fine sweepings	13.4	13.3	2.0	63.5	3.4	4.4	1:5	—
Oil cakes:								
Sunflower	8.5	35.0	5.4	30.0	8.1	13.0	1:1.2	3–5
Linseed	7.5	30.6	5.7	39.4	7.4	9.4	1:2	4
Mustard	7.0	32.2	7.3	29.9	8.5	15.3	1:1.6	4
Colza (cabbage seed)	12.3	30.9	7.5	27.8	9.0	12.5	1:1.4	4–9
Safflower	11.5	53.8	6.8	15.1	6.2	6.8	—	8
Falbeflax	12.0	32.9	11.1	—	7.9	11.4	—	6
Cotton	10.4	36.9	2.0	33.5	7.3	9.9	—	6
Hemp	9.01	30.98	1.32	42.1	14.18	1.6	—	5
Soybean	7.4	40.0	9.3	32.3	5.8	5.2	—	5
Coriander	7.8	13.1	22.6	21.2	6.1	29.2	—	8
Castor	11.3	46.2	0.9	12.58	7.3	21.72	—	8
Groundnut	8.2	42.5	5.1	40.0	4.2	—	—	—
Rapeseed	8.3	35.6	11.6	38.3	6.2	—	—	—
Tung**	11.2	21.8	0.9	58.0	8.1	—	—	—
Perilla	10.5	34.4	2.6	44.9	7.6	—	—	—
Perilla meal	9.6	38.1	0.5	20.8	10.8 (Raw)	20.2	—	6
Other plant products:								
Boiled potato	—	2.1	0.2	20.7	0.9	1.1	1:17.2	20
Horse chestnut	49.8	4.3	1.5	40.9	1.6	2 5	—	7
Seeds of weed grasses from grain and vegetable crops	—	—	—	—	—	—	—	6

Feed	Water	Raw protein	Fat	Carbo-hydrates	Ash	Cellulose	Protein ratio	Food coefficient
Dough wastes from bread factories	—	—	—	—	—	—	—	6
Animal products:								
Fish meal	10.7	59.2	6.4	—	23.4	—	1:0.2	1.5–2.0
Blood meal	9.0	83.9	2.5	—	4.2	—	1:0.008	1.5–2.0
Meat meal	10.8	72.3	13.2	—	3.8	—	1:0.1	1.5–2.0
Fresh horse meat	—	—	—	—	—	—	—	3
Trash fish and viscera (fresh)	—	—	—	—	—	—	—	5
Spleen and other slaughterhouse waste	—	—	—	—	—	—	—	4
Frogs and fresh tadpoles	—	—	—	—	—	—	—	5
Larvae of chiro-nomids	87.2–88.3	6.6–8.2	0.5–1.4	2.4–3.1***	1.0–1.5	—	—	—
Larvae of insects and beetles	—	—	—	—	—	—	—	5
Molluscan meat	9.5	39.63	4.08	33.8	12.5	—	1:0.2	2.0–2.5
Daphnia	90.7	5.4	0.6	4.1***	1.7	—	—	—
Silkworm cocoons	10.9	57.7	20.2	—	7.4	—	1:0.8	2

*According to the data of Hungarian authors, 2.3.

**In using tung cake in the form of meal in quantities of more than 20 to 45%, the growth rate of carp is reduced. Maize cake is currently being used for feeding carp in foraging ponds.

***Nonnitrogen soluble matter.

Nota bene:

1. The food coefficient given in the Table is that of two-year carp. For younger fish it should be decreased by 25% and for older carp increased by 25%.

2. The food coefficient of soybean according to A.I. Kuzema and V.A. Movchan is 1.7.

following data (%): water 7.1, protein 61.5, fat 3.9, carbohydrates 9.0, ash 3.8 and chitin 2.1.

Feed Mixtures and Supplements. Carp should be fed a mixture of nutritional substances constituting a balanced diet since feeding with one type of food (as a rule, incomplete) leads to metabolic disorder and loss of weight. Feeding carp, for example, a feed rich in carbohydrates causes a more significant loss in protein than even total starvation. A correctly composed mixture is well eaten, reducing the food coefficient, and thus more economical.

Sufficient information regarding the nutritional value of various feeds has yet to be accumulated. A complex assessment of the nutritional value of feed and feed mixtures mainly involves the quality and quantity of proteins (of special significance), the mineral matter, and vitamin content. The daily requirement of carp for proteins may be around 14 g. The "protein ratio" has a special significance for the evaluation of feeds (see above).

In the preparation of a feed mixture products of animal origin are usually combined with products of plant origin, which have a wider protein ratio. To augment the nutritional value of feed mixtures it is appropriate to add pastes from aquatic (duckweed, pond weed, foxtail, water thyme, arrowhead, and young branches of reed, sedge, and cattail) and terrestrial plants, together with the leaves of cabbage, carrot tops, beet root, etc. To avoid deterioration in the quality of the paste (a significant loss in carotene content) it should be prepared on the day of its admixture with the feed.

Several duckweed species have a high food value and thus constitute valuable components in feed mixtures. Common duckweed contains on a dry weight basis 25.75% raw protein, 4.65% fat, 27.24% nonnitrogen soluble matter, 24.57% cellulose, and 17.79% ash. Greater duckweed contains the following percentages of these components respectively—20.94, 2.66, 35.18, and 14.62; and ivy-leaved duckweed—30.40, 2.70, 24.00, 20.80, and 22.10. Duckweeds also contain such trace elements as iodine, cobalt, copper, sulfur, phosphorus, magnesium, and calcium. E'rman (1966) reports 21 amino acids in the aforementioned duckweed species collected from the channels of the fish farm "Yakot" of the Moscow district; the quantity of individual amino acids differed significantly in the three species.

The quantity of plant material added to feeds depends on the conditions of the natural food base in the ponds, quality of the artificial feeds, density of stocking, and techniques followed in farm management, especially in feeding the fish. When 30% plant material was added to a feed mixture for two-year fish (on the basis of raw weight or at the rate of 7% dry weight), the consumption of food per unit growth rate was reduced by 11.5%, an excellent result. Eventually aquatic plants were substituted for 7% of the concentrate feed, with no deterioration in its quality, nor in the quality of the fingerlings, yearlings, and two-year fish compared to those raised on concentrate feed alone. Expenditures on concentrate feed were reduced 11.5% and total fish productivity, through the addition of both aquatic and terrestrial plants to the feed mixture, increased almost 2.5 times (fish farm "Gorodok" of the L'vov district).

In addition to higher aquatic and terrestrial plants, algae in dried and powdered form or raw are also added to feed mixtures. Tests on the addition to artificial feeds of 10% marine phyllophorous algae, known to contain a high percentage of proteins and trace elements and to have a good binding property, have shown that the expenditure on feed per unit growth rate reduces and the assimilation of proteins increases. Among the lower protococcus algae (*Chlorella* deserves special attention as a component of feed mixtures. It is rich in proteins, fats, and carbohydrates. The proteins in *Chlorella* contain the most valuable amino acids: arginine, histidine, isoleucine, lysine, methionine, tryptophan, and others. Aspartic and glutamic acids, glycine, serine, alanine, asparagine, glutamine, and β-alanine have also been found. Of the vitamins, the following occur: A, B_1, B_2, B_6, B_{12}, C, D, and K. Nicotinic acid, pantothenic acid, folic acid, and biotin are also present. Because of its rich composition of amino acids and vitamins *Chlorella* is utilized as an ingredient in feed mixtures for fish to widen or narrow the protein ratio. According to experience abroad, 75 to 100 tons of dried or 4,500 to 6,000 tons of fresh *Chlorella* can be harvested from a single hectare. Only a few species of algae have been adopted for cultivation on a commercial scale, primarily protococcus algae from genera *Chlorella, Scenedesmus, Micractinium,* and *Ankistrodesmus.*

An important component of feed mixtures for carp can be hay meal or cone meal prepared from spruce and pine. Hay meal enriches the feed with proteins, vitamins, and mineral substances. According to data compiled by the Institute of Animal Husbandry and Veterinary Medicine of Latvian SSR, 1 kg of conifer meal contains (mg): calcium—12.25, iron—320, phosphorus—103, manganese—101.6, zinc—4.93, copper—2.41, cobalt—0.56, and carotene—77. Cone meal additionally contains almost all the vitamins of the B group, vitamin K, and provitamin D. The quantity of carotene in 1 kg of conifer meal varies from 77 to 200 mg.

In Ukrainian SSR pine needles, with a moisture content of 40.1%, contain 4.56% protein, 12.59% fats and resins, 32.9% carbohydrates, and 150 to 250 mg% vitamin C; fresh spruce needles contain carotene, vitamins E, B, and K, provitamin D, and of the trace elements—cobalt, manganese, etc.

The optimum norm for conifer meal is 5 to 10% of the total quantity of feed. The fat content in the body of carp increases as a result of such an addition and the growth rate is enhanced. Conifer meal should be stored in glass containers in a dry, shaded place since sunlight decomposes the carotene present in it. It can be stored for 6 to 12 months.

In the fish farms of Hungary poultry droppings are added to feed mixtures for carp. In a dry state the droppings contain 93.82% total dry matter, 67.83% organic matter, and 34.05% raw protein. Poultry manure is mixed with the concentrate and binding component in the following ratio: concentrate 62.5%, fresh poultry manure 25%, and clay 12.5%. The average water content of such a mixture is 32%. Carp feed very well on such a mixture. Poultry manure can replace the concentrate at the rate of 3.5 to 4.0 quintals of droppings to one quintal of concentrate.

For the enrichment of feed mixtures with proteins and vitamins of the B group, hydrolyzed yeast is sometimes added (at the rate of 3% of the total quantity of a given feed), containing 8 to 11% moisture, 45 to 50%

protein, 1.1% fat, 33.7% nonnitrogen soluble matter, and 7.5 to 10.0% ash. The weight of the fish with this feed increases up to 15.3%.

Mollusks, tadpoles, frogs, and small trash fish are valuable components in the ration of carp. They are passed through a mincer and added to the feed mixture with an equal quantity of dry feed. Frogs are better boiled or dried. If dried, they are pulverized and may constitute 30% of the total feed. Such a mixture is eaten well by yearlings and two-year fish, which increases total fish productivity considerably.

Rodina (1966) has analyzed the food value of detritus, which harbors many groups of microorganisms. Different types of bacteria live in detritus, which are rich in amino acids and capable of meeting the nutritional requirements of the aquatic animals feeding on them.

Sapropel can also be used as a mineral supplement in feeds. It is a greenish, olive-green, brown, or gray-colored slime formed from the residue of plant and animal organisms at the bottom of fresh-water resources under an oxygen deficiency, primarily under anaerobic conditions. In the upper layers sapropel looks like a yellow jelly; in the deeper layers it is slightly condensed and called saprocol. Sapropel is rich in macro- and microelements, vitamins, and biostimulants. In addition to considerable carotene, the sediment from fresh-water lakes contains vitamins D, B_1, B_2, and B_{12}, folic acid, etc. It also contains calcium, phosphorus, sulfur, and iron, and such trace elements as iodine, cobalt, copper, manganese, molybdenum, bromine, boron, zinc, nickel, titanium, etc. Microorganisms releasing antibiotics have also been found in sapropel; they suppress the activity of pathogenic microbes. Sapropel is rich in proteins, carbohydrates, lipoids, organic acids, etc.

Sapropel is thus a very valuable food item containing medicinal properties and has already been tried in farm animals. Feeding with sapropel increases productivity in cows, pigs, ducks, and poultry. In pond fisheries sapropel, as a supplement in carp feed, has been tested in Latvian SSR (Vimba, 1964). With an addition up to 20%, the unit weight of carp increased more than 46%, pond productivity by 51.9%, and the survival of fingerlings by 32%, compared to ponds in which sapropel was not used. It is better to use dried, powdered sapropel. One should remember that sapropel varies in qualitative composition in different fresh-water resources; hence its chemical properties should be assessed before use.

It is readily apparent that the items available for feeding carp are highly variable. At collective and state fish farms a reliable feed base should be created as is done in other fields of animal husbandry (sowing of fodder crops, utilization of local resources, etc.). The Scientific-Technical Council of the Ministry of Agriculture of the USSR has recommended the following feed mixtures for collective and state fish farms (Table 29).

In the German Democratic Republic where carp are cultured in ponds

Table 29. Feed mixtures recommended for collective and state fish farms

Components of mixture	Two- and three-year carp, %	Yearlings and brood fish, %	Remarks
Oil cakes and meals (not less than 2 types in equal proportions): sunflower, cotton, soybean, rapeseed, and hemp	40	40	These recommendations also include addition of cobalt chloride and Terramycin. This has already been discussed. Fortifying the feed with vitamin B_{12} is appropriate at the rate of 12 mg cyanocobalamin per ton of feed meant for fingerlings, two-, and three-year brood fish.
Oil cakes and meals: mustard, wintercress, groundnut, sesame, flax, perilla, fallflex, and castor	10	9	
Legumes: lupine, marsh pea, pea, lentil, vetch, and fodder beans	10	15	
Grain crops: rye, wheat, and barley	24	20	
Brans: wheat or yellow maize	6	4	
Edible yeast	4	4	
Fish meal	3	5	
Grass meal after artificial drying, or conifer meal	2	2	
Powdered chalk or travertine	1	1	
Total	100	100	

and enclosures with a high degree of stocking, granulated feed prepared on the basis of the formula given in Table 30 is used.

Tests done in the German Democratic Republic on the food value of dry formula feeds for carp have shown that better results are obtained with feeds containing more than 30% proteins (Steffens, 1969). The chemical composition of the formula feed is as follows: dry matter 87.5 to 91.3%, raw protein 23.6 to 41.9%, fat 2.2 to 4.7%, and ash 3.9 to 12.9%.

Carp are fed five to six times a day at the rate of 1.4% of their body weight. Temperature should be maintained in the range of 16.8 to 23.2°C.

Table 30. Feed mixture used in the German Democratic Republic in carp rearing under high density of stocking

Components of mixture	% of ration	Components of mixture	IU/1 kg of feed	Quantity in mg/kg of feed
Fish meal	25	Vitamin A	6,000	—
Meat meal	7	Vitamin D_3	2,000	—
Dry skimmed milk	8	Vitamin B_1	—	10
Edible yeast for animal feed	10	Vitamin B_2	—	20
Soybean meal	10	Vitamin B_6	—	10
Groundnut meal	5	Vitamin C	—	20
Barley flour	10	Pantothenic acid	—	20
Oat meal	10	Nicotinamide	—	15
Wheat flour	16	Choline chloride	—	100
Biologically active substances	1	Vitamin K_3	—	3
		Vitamin B_{12}	—	20/μ mg

The percentages of nutritional constituents (proteins, fats, and carbohydrates) are determined from the following formula:

$$X = \frac{(Zv) + (Z_1v_1) + (Z_2v_2)}{100},$$

where X = percentage composition of given group of nutritional substances in feed mixture;

v, v_1, v_2 = percentages in a given group of nutritional substances in different feeds used in preparing mixture;

Z, Z_1, Z_2 = percentages of a given kind of feed in mixture.

If a particular mixture contains more than four to five components, the values in the numerator for Z_3 and Z_4 and v_3, v_4, etc. are added accordingly.

The ingredients included in the feed mixture may have different food coefficients. The food coefficient of the mixture is calculated by the following formula:

$$Z = \frac{100}{(v{:}Z) + (v_1{:}Z_1) + (v_2{:}Z_2)},$$

where X = food coefficient of mixture;

v, v_1, v_2 = percentages of components in feed mixture;

Z, Z_1, Z_2 = food coefficients of components.

In the case of increase in the number of components in the mixture, v_3, v_4, and Z_3, Z_4, etc. are added in the denominator.

In the course of practical utilization of the aforementioned data and recommendations, problems relating to the food value of the ratio and increase in fish productivity are simultaneously resolved.

The taste qualities imparted to the fish flesh must be kept in mind; taste is affected considerably by the surrounding environment and the main supplementary components of the feed. Carp smacks of maize when fed intensively with this grain and the flesh is lighter in color. A pleasant smack occurs in carp fed on barley. Taste should therefore be taken into consideration in composing feed mixtures.

For a better storage of artificial feeds and greater convenience in their delivery it is important to have proper storage facilities.

Special attention should be paid to the periods of storage of dried artificial feeds. The time should not exceed six months. Feeding with overstored dry formula feeds causes an enlargement of the liver in fish and leads to production losses.

Food Value of Feeds and their Digestibility. Food, after ingestion, digestion, and assimilation in fish, is used for cell and tissue formation and as a source of energy for performing vital activities. The proteins, fats, carbohydrates, mineral salts, vitamins, and water which enter the organism with natural food and artificial feeds, participate in a large number of anabolic and catabolic processes, which together constitute the metabolic process. Metabolism is a process which influences all the functions of the organism and is closely associated with many external and internal factors. Proper metabolism depends on a rational program of fish feeding, an important aspect of intensification of fish culture.

The nutritional value of a feed is assessed in terms of its containing those properties requisite for fish and depends correspondingly on the quantity and quality of the ingredients of the feed. The nutritional value of a feed is not the same for different fish species, varying according to specific biological peculiarities, differences in productivity, physiological condition, intake of natural foods, and conditions of maintenance. Hence the nutritional value of a given feed, used in a given set of concrete conditions, cannot be expressed by a definite index.

To estimate the nutritional value of a feed mixture for carp cultured in ponds, Schaperclaus took the nutritional value of 100 kg of barley as a standard value of 100, to which he related the nutritional values of the components in the feed mixture (Table 31).

The digestibility of feed, i.e., the capacity of its ingredients to convert into soluble substances under the influence of the digestive enzymes secreted by the alimentary canal and associated glands, is very important. The productive action of feed is calculated on the basis of the digestibility of the nutritional substances in the ingredients.

Shcherbina and Surina (1970) have estimated that in carp, in distinction from those fish and other vertebrates which have a stomach, the breakdown of biopolymers of food and the absorption of their hydrolyzed products takes place all along the digestive tract. In the initial stage of digestion, during the

Table 31. Nutritional values (in %) of certain feed items used for feeding carp (from Schäperclaus, German Democratic Republic)

Type of food	Dry matter	Ash	Protein	Fat	Nonnitrogen substances	Cellulose	Standard (barley)
Barley	88.0	2.3	8.9	1.7	69.6	5.5	100
White lupine	90.0	4.2	34.2	8.2	34.9	8.5	100
Maize	88.0	1.7	10.6	4.4	68.9	2.4	110
Oat	88.0	2.7	10.7	5.1	60.4	9.1	100
Wheat	88.0	1.8	12.3	1.9	70.0	2.0	100–110
Vetch	86.7	3.2	26.0	1.7	49.8	6.0	97
Pea	86.8	3.0	22.7	1.9	53.2	6.0	100
Soybean	90.0	4.7	33.2	17.5	30.2	4.4	120
Sunflower	89.0	8.9	31.5	1.8	24.2	22.6	20–40
Potato	24.0	1.2	1.9	0.1	21.5	0.7	27–42
Rye flour	88.0	1.8	8.7	1.6	73.1	2.8	100
Wheat bran	85.2	6.0	14.9	4.7	50.7	8.9	50
Rice husk	90.0	10.3	12.5	14.3	44.5	8.4	65
Meat meal	91.5	12.5	66.0	10.5	2.5	—	125
Fish meal	88.4	18.0	61.8	8.6	—	—	62

first 0.5 to 1.0 hours, almost all the easily digestible carbohydrates in the feed taken by fish are hydrolyzed and absorbed. The most active absorption of the hydrolyzed products of raw proteins takes place in the first three sections of the intestine. The active process of absorption of the lipid fraction and phosphorus also takes place here.

According to the findings of Chepik (1966) the digestive capacity of the intestinal enzymes of carp changes in the course of the year; the intensity of digestion of proteins, fats, and carbohydrates therefore varies in different periods. In winter it is notably lower compared to the autumn period (protein hydrolysis drops by 47.5 to 65.7%, carbohydrates 58.8 to 67.1%, and fats 37.0 to 65.5%). It increases in March and peaks in April–May (on the average over the spring period the increase in proteins is 56.3 to 155.2%, carbohydrates 133.5 to 218.0%, and fats 25.0 to 200.0%.) The digestive capacity over summer differs compared to spring because the dynamics of different enzymes vary. The digestion of proteins increases by 36.0 to 53.0%, fats 20.0 to 52.8%, and carbohydrates 60.7 to 113.0%. In this case the activity of proteolytic, amylolytic, and lipolytic enzymes decreases with a reduction in temperature, but the rate of decrease is not uniform. Not only temperature, but the conditions of external environment and the physiological conditions of the fish exert an influence on digestive activity. The digestive enzymes are more efficacious in carp yearlings at temperatures of 15 to 20°C and, compared to two-year fish, the digestive process intensifies with regard to proteins by 11.7 to 32.0%, carbohydrates 11.7 to 16.5%, and fats 20.0 to 33.3%. Differences have been noted in the intensity of protein, carbohydrate, and fat digestion in the yearlings of scale and mirror carp, as well as in hybrids of carp and Amur sazan. Maximum values have been recorded in hybrids, intermediate in scale carp, and minimum values in mirror carp.

The digestion of fats in yearlings, two-year carp, and hybrids of carp and Amur sazan throughout the year is less than that of proteins and carbohydrates. In December and January under temperatures of 2°C fat digestion is totally absent, and occurs at a low rate at 5°C.

The digestibility of food is determined on the basis of differences between the substances supplied in the feed and the substances released in the excreta. In animal husbandry the percentage ratio of the quantity of digested components of feed to their total quantity in the feed is known as the coefficient of digestibility. The digestibility of feed depends on the fish species, age, and individual peculiarities, as well as quantity and composition of the feed—the protein, carbohydrate, vitamin, cellulose, and other nutritional substances present. The preparation, taste, and quality of feed are all important aspects.

Shcherbina and Sorvachev (1967, 1969), studied the digestibility of substances present in sunflower and cotton oilseed meal and established with the help of inert materials (chromium dioxide)[7] that in the three anterior sections of the intestine up to 80% of the amino acids available to carp are absorbed; the nonnitrogen substances of these meals are less amenable to the action of the digestive enzymes in carp. Hence in preparing a feed mixture containing oilseed meal, materials which are rich in carbohydrates (cereals) should be included. The most deficient amino acid in oilseed meals is lysine. Its absorption in the anterior section of the intestine from sunflower meal is 22.6 times less and from cotton meal 37.6 times less than that required for meeting the protein needs of fish. If oilseed meal is to be used in fish feed, therefore, it is recommended that synthetic lysine be added and methods developed to prevent nutritional losses.

Data are presented in Tables 32 and 33, compiled by Shcherbina, Mochul'skaya, and E'rman (1971), and by Shcherbina (1969), on the digestibility of the following substances and their value in terms of amino acids: groundnut meal, pea, and barley *per se,* and a feed mixture of groundnut meal (40%), linseed oil cake (10%), barley (22%), pea (10%), protein-vitamin concentrate

[7]The formula for calculating digestibility developed by Shcherbina and Sorvachev is as follows:

$$P = \frac{\text{Percentage of nutritional substances in experiment}}{\text{Percentage of nutritional substances in feed}} \times \frac{\text{Percentage of } Cr_2O_3 \text{ in feed}}{\text{Percentage of } Cr_2O_3 \text{ in experiment}} \times 100$$

where P = digestibility of nutritional substances as percentage of quantity in feed.

Table 32. Digestibility of nutritional substances

Components of feed and parameters	Groundnut meal	Pea	Barley	Mixture
Dry matter	62.4*/624	44.3/443	57.7/577	50.4/504
Raw protein	85.3/310	77.5/201	81.4/104	89.3/.271
Carbohydrates	64.7/97	44.9/153	73.9/406	61.6/180
Lignin and cellulose	43.7/148	6.6/11	0.6/1.0	13.0/28
Calcium	0/0	0/0	19.1/0.4	0/0
Magnesium	36.3/1.6	59.5/1.7	46.0/1.2	0/0
Phosphorus	38.5/2.1	41.0/1.7	44.7/2.3	39.1/2.2
Ratio of nitrogen and nonnitrogen components of feed	1 : 1	1 : 1.2	1, 4, 5	1 : 0.9

*Numerators = percentage of nutritional substances taken in feed; denominators = quantity in g/kg of feed.

(4%), macaroni by-product (5%), fish meal (4%), grass meal (3%), and chalk (2%).

Iančarik (1964) has shown that in addition to endogenous factors (bile, pancreas, and intestinal mucus), extracts from major food components (daphnia, tubificids, earthworms, and larvae and pupae of mosquitoes) also participate in the digestive process. It has been demonstrated that 60 to 80% of natural food is digested and assimilated by fish, and in some cases 90% and more. Absorption of calcium in the anterior and middle parts of the intestine takes place in the first hours of digestion. Two-year fish and under-yearlings of carp assimilate 40% of the calcium content in feed.

With regard to proteins, not only their composition but also their digestibility should be taken into account. Nering (1965) established for three-year

Table 33. Digestibility (%) of raw proteins and individual amino acids available in feeds (g/kg) (from Shcherbina)

Amino acids	Type of feed							
	Barley		Pea		Groundnut meal		Feed mixture	
	g	%	g	%	g	%	g	%
Raw proteins	104.8	81.3	202.4	78.5	309.6	85.3	265.8	88.0
Lysine*	4.2	90.1	15.2	90.0	13.5	81.4	15.4	93.5
Histidine*	3.5	90.2	6.7	88.6	12.0	87.9	12.9	95.6
Arginine	7.2	93.4	18.3	88.7	38.4	94.2	43.3	97.3
Aspartic acid	14.7	92.1	26.6	79.4	33.4	80.0	16.7	80.2
Serine	9.0	93.0	12.7	79.6	18.6	81.3	11.7	75.2
Glycine	6.1	80.8	11.7	79.9	21.9	79.8	13.5	84.5
Glutamic acid	35.0	95.6	35.6	84.7	51.2	95.4	41.0	91.9
Threonine*	6.5	80.2	11.4	82.3	14.6	86.0	16.6	89.3
α-alanine	5.4	85.9	10.9	87.0	17.1	88.0	15.2	90.2
Tyrosine	3.3	89.7	4.5	74.2	10.5	89.7	5.8	92.0
Valine*	3.7	89.7	6.8	81.0	6.0	83.8	7.9	90.9
Methionine*								
Phenylalanine	3.8	84.9	7.2	82.1	15.7	93.3	17.8	95.5
Leucine*	8.4	90.8	15.4	87.0	19.2	91.3	23.6	93.9
Isoleucine*								

*Essential amino acids.

carp the following coefficient of digestibility[8] of raw proteins for different kinds of feed items (in %):

Maize	66.4	Pea	78.6
Soybean	80.7	Oat	64.3
Spring wheat	82.7	Yellow nonalkaloid lupine	85.4
Spring barley	64.2	Winter rye	62.5

Maize, wheat, and rye contain large quantities of leucine and isoleucine, which are deficient in natural foods; natural foods are rich in lysine, which is deficient in cereals. The raw proteins of barley, however, contain that quantity corresponding to the needs of two-year carp for amino acids.

[8]Coefficient of digestibility (KP) calculated by the equation:

$$KP = \frac{a-v}{a} \times 100,$$

where a = quantity of nutritional substances (in g) as quantity of chromium oxide in feed (g);

v = quantity of nutritional substances (in g) as quantity of chromium oxide in excreta (g).

Lysine is present in the least amount and, compared to it, the other amino acids excessive (Shcherbina, 1969). Feeding carp with barley is very economical.

Digestion and assimilation of feed by carp depend on ration. Overfeeding leads to a considerable reduction in digestion and assimilation, readily apparent from the data presented in Table 34.

Table 34. Efficiency of moderate and excessive feeding of carp with lupine (from Eleonskii, 1946)

Degree of feeding	Digestion and assimilation, %		
	Protein	Fat	Carbohydrate
Moderate	35.2	5.4	16.8
Excessive	18.9	3.4	8.8

In feeds containing a large fraction of ballastic or bolus-forming substances of plant as well as animal origin, benthos and plankton are digested to a poorer extent under the effect of excessive feeding compared to moderate feeding. For example, in excessive feeding with zooplankton in experiments the quantity of nitrogen excreted per unit of food intake was three times greater than under moderate feeding. In excessive feeding with benthos the undigested nitrogen excreted was 1.5 times more, and with feed of plant origin 2.5 times greater.

Concept of Food Coefficient and Cost of Feeding. The food coefficient is a number indicating the quantity of feed (expressed in kg) required by a fish to effect a growth rate of 1 kg. If, for example, the food coefficient of any feed is equal to 2, this means that to achieve an increase in body weight of 1 kg the fish must consume 2 kg of feed.

The food coefficient should not be confused with the feed cost, represented by the ratio between food intake and fish growth expressed in kg. The magnitude of the food coefficient is associated with the nutritional value of the feed, which depends not only on its preparation, quality, and mineral content, but on the following factors: fish-rearing conditions in the ponds; extent of pond preparations; conditions of the external environment; age and health of the fish; methods of feeding, maintenance, and care of the fish; and a correct estimation of ration.

The interrelated factors which affect the magnitude of the food coefficient most are as follows:

1. Fish-rearing conditions and level of pond preparation. In nondrainable ponds artificial feeding is not suitable since the food coefficient in such ponds is comparatively high, as feed wastage increases (a large part of the food remains unutilized).

2. Preparation of the feed. Feed should be prepared in such a way that the fish can consume it with a minimum loss of energy. Hence cereal by-products are either ground or soaked in water for six to eight hours. Larger items (pea, beans, maize, etc.) and oilcakes are fragmented or coarsely ground and mixed with water. Potato and other tuber crops are boiled. Castor-oil cake, which contains ricin, is steamed to improve its taste and eliminate bitterness (although fish will consume it even without prior steaming). Bitter lupine should also be steamed. Overheating the feed and increasing the salt quantity in it deteriorate its food value: a salt content greater than 2.5% of the feed causes inflammation of the intestine and leads to mortality in carp. Highly salted feed is soaked in fresh water beforehand, even though in this process it loses not only the salt but also valuable nutritional substances. If the feed is well prepared, the food coefficient is reduced two to three times.

3. Rotten and fungal products should not be given to fish since they may cause intestinal disease, poisoning, and mortality.

4. The lower the value of the food coefficient, the higher the rate of natural fish productivity.

5. Factors such as water temperature, oxygen content, and other hydrochemical parameters, in particular pH, as well as the oxidizability and salt composition affect the food coefficient. Carp have the best food consumption in a temperature range of 20 to 29°C. At 4 to 8°C they just barely take food, at 13 to 14°C feeding is still rather poor, and at 30°C ingestion and digestion of food slow down. The food coefficient changes accordingly. It increases with a reduction in temperature; at 14 to 15°C it doubles compared to the value at optimum temperature. The value of the food coefficient also doubles when the oxygen content in the water drops from the average value of 3 to 6 mg/l to a deficient level (0.2 to 0.5 mg/l).

6. Feeding carp of different ages with the same feed under similar conditions does not yield the same food coefficient. For example, for yearlings fed lupine the food coefficient is 2.5, for two-year fish 4.5, for three-year 5.8, and for four-year 6.3. Diseased fish as well as those obtained from poor quality brood fish digest food poorly, as a result of which the food coefficient increases.

7. The more perfect the techniques of managing a fish farm, the more economical the consumption of food and, consequently, the lower the value of the food coefficient.

8. Containers in which feeds are prepared should be clean; they should be washed and dried every day. Wooden containers should be washed with lime water once a week.

9. With a deficiency in mineral substances in the feed food assimilation is poor and the food coefficient increases. It is recommended that up to 5% calcium phosphate be added to the feed; the addition of up to 3% chalk to groundnut oilcake reduces the food coefficient from 6 to 4.

10. A correct estimation of ration is very important. That part of the feed remaining undigested in the case of overfeeding increases the food coefficient (see Table 34).

The food coefficient of the feed used for the carp is calculated on the basis of increase in growth (weight) due only to the feed given. For this the weight of the fish at the time it is stocked and the growth rate due to natural foods are deducted from the total weight of the fish when netted from the pond. The growth rate due to natural foods is calculated on the basis of the fish productivity for a given locality (see Chapter II); for ponds filled with water for the first time, data from the project documents are used.

Role of Natural Foods in the Feeding of Carp. Monotypic and unbalanced feeds given to fish in ponds—oilseed cakes, feed mixtures, etc.—do not correspond to the nature of the feeding habits of carp and, obviously, cannot be considered complete food. Since carp are capable of digesting plant food well, this fact is exploited in the practice of artificial feeding. Any deficiency in one or the other nutritional substance in the feed, carp supplement from the physiologically complete natural food available in the pond in the form of plankton and benthos. But if the stocking density is high, the stocks of natural food may be so depleted that the fish requirements are not satisfactorily met, resulting in poor carp growth and various diseases. Thus, the efficiency of feeding, especially in the case of an unbalanced feed or one lacking the most important nutritional susbstances, and density of stocking are related to a great extent to the conditions of natural food resources—their quantity and quality. Hence in determining the density of stocking and artificial feeding of carp one must consider the state of the natural food base in the ponds. In modern methods of feeding natural foods constitute 30% and artificial 70% of the feed given. The chemical composition and caloric value of some of the invertebrates serving as food items for fish are presented in Table 35.

On the basis of experiments conducted in the Max Planck Institute, it can be seen that carp feeds on extraneous feed only if it has a composition corresponding to its needs. In that case carp does not need natural food.

The relations between an increase in quantity of natural foods of carp and different levels of stocking density are shown in Table 36.

The effectiveness of feeding depends not only on the conditions of the natural food base of the pond, the significance of which can never be overvalued, but also on the factors of external environment.

Considering the importance of natural foods in the carp ration, manuring of fish ponds is done (see below) to multiply the most important constituents, in terms of food value, of zooplankton and benthos. Daphnids, chironomids, and oligochaetes (*Enchytraeus albidus*) are the primary objects. To further enrich the ponds with food organisms ultraviolet lights are sometimes installed, which attract aquatic insects in large quantities. In the case of a

Table 35. Chemical composition of some invertebrates

Food organisms	Live weight, mg	Moisture	Total quantity of proteins	Pure proteins	Chitin	Fat	Carbohy-drates	Mineral matter	Ratio of nitrogen to nonnitrogen substances	Caloric value, g (g/cal)
Chironomus plumosus	21.72	88.28	6.63	—	—	0.51	3.08	1.50	1:0.6	549
Phryganea grandis	584.60	70.42	13.57	12.78	0.79	7.08	8.17	0.75	1:1.8	1,752
Limnophilus rhombicus	216.10	78.71	11.31	10.28	1.03	1.58	6.71	1.70	1:0.9	810
Agrion sp.	30.06	83.80	11.07	—	—	1.35	2.14	1.63	1:0.5	853
Ephemera vulgata	47.65	82.06	11.37	6.56	4.75	2.92	2.80	0.92	1:0.8	975
Perla cephalotes	305.60	83.44	12.33	—	—	1.17	1.07	2.00	1:0.3	855
Gammarus pulex	24.00	78.44	11.32	8.93	2.39	1.28	2.69	6.28	1:0.5	845
Asellus aquaticus	27.17	80.23	10.15	—	—	0.88	1.69	7.02	1:0.4	744
Daphnia pulex	0.34	90.67	5.42	1.47	1.47	0.61	4.07	1.70	1:1	371
Sphaerium sp.	69.70	75.75	3.06	—	—	0.26	2.89	18.04	1:1.1	314
Dreissencia polup	310.90	56.02	3.41	—	—	0.24	1.62	38.71	1:0.06	281
Bithynia tentaculata	139.90	67.19	5.04	—	—	0.29	3.28	24.21	1:0.08	445
Physa fontinalis	88.20	82.19	7.92	—	—	0.73	7.92	1.23	1:1.2	837

Table 36. Quantities of benthos at different stocking densities of carp (g/m^2)

Months	N	5N	10N
May	—	18.64	6.89
June	31.37	38.20	24.91
July	39.85	14.40	12.94
August	18.75	12.60	2.25
September	35.00	4.66	2.67

deficiency in aquatic plants, wreaths or bundles of slightly dried reeds, cattail, bullrush, sedge, water plantain, and other plants are placed near the source of light.

Luminescent lamps are the most convenient, but common lamps with a bright heating filament can also be used (up to 300 to 500 W).

The special rearing of live foods, as well as attracting aquatic insects by light, can significantly increase the food base of rearing and foraging ponds, which makes it possible to economize on artificial feed.

INCREASING THE DENSITY OF FISH STOCKING. The food resource can be increased through the introduction of artificial feed and the enrichment of natural foods, which makes it possible to increase the total fish productivity.

Total fish productivity due to natural foods should be distinguished from that due to artificial feeds. In the case of additional feeding carp utilize not only the natural foods but the supplementary. Some of the growth rate obtained from the unit pond area over one vegetative growth period is directly attributable to extraneous feeding.

An increase in density of stocking leads to a more complete utilization of the natural foods, the multiplication of which is stimulated by the unutilized part of the feed (see below) and by the excreta serving as manure. Consequently the natural productivity of the pond also increases. An increase in the stocking density by two, three, five and more times compared to the normal level is correspondingly designated 2N, 3N, 5N, etc. Under normal density of stocking carp are usually not artificially fed, since it is believed that with a correct determination of natural fish productivity the food resources in the pond will promote a definite level of unit weight.

At the present level of fish-farm management techniques, stocking densities of 5N to 10N can be recommended with no significant reduction in the unit weight of two-year carp (500 g). Data from the central belt of the European part of the USSR (Moscow district) follow: at 2N stocking density the unit weight did not reduce, at 4N it reduced by 4%, at 6N by 8%, at 10N by 20%, and at 15N by 40.0 to 43.5%. The increasing reduction in unit weight with higher levels of stocking density is explained firstly by the fact that the quantity of desirable quality food decreases in the most intensive growth period of carp (carp feed actively toward the middle of the vegetative growth period, i.e., in the second half of July–August); secondly, a deterioration in the hydrochemical regime occurs. The oxygen content decreases under the influence of toxic products of metabolism, which increase with the growth of fish (as a result of the absence of water flow), and oxidative processes are activated as a result of the churning of the silt by the fish in a thorough search for food.

At 15N stocking in foraging ponds of standard stocking material the growth of carp slows down even toward June, at 10N it becomes discernible in July, and at 5N lags only slightly behind (4.5 to 10.0%) growth at N level of stocking. It is highly probable that the quantity of fat in the yearlings at the time of stocking and the ratio between fat and proteins also influence the growth rate. For example, yearlings of carp containing 4% fat during stocking (on raw weight basis) with a protein-fat ratio of 5:1 had a better growth at N, 5N, 10N, and 15N levels of stocking than yearlings containing 1.6% fat (on raw weight basis) and a protein-fat ratio of 1:12. With much higher densities of stocking, growth slows down correspondingly earlier.

To neutralize or at least moderate the harmful effect of factors affecting fish growth, especially at significant levels of stocking densities, the following measures should be implemented when artificial feed is supplied:

1. Rearing ponds must be thoroughly prepared before fish stocking.

2. The hydrochemical (mainly gas) regime of the ponds should be improved by aeration of the water and increasing the flow. This is especially necessary on windless days and during the peak of the vegetative growth period; otherwise a deficit of oxygen occurs which slows down the growth rate and can lead to fish mortality. According to the data of Havelka and Volf (1966) the quantity of feed given should not exceed 35 q/ha, since a higher rate of feeding not only increases the food coefficient but also reduces the oxygen content.

3. Feeding should preferably be done with feed mixtures.

4. Live foods should be multiplied to supplement any deficiency in natural foods.

5. Only healthy material obtained from good, fast-growing brood fish should be used as stock.

6. In the second half of summer fish weighing about 500 g should be caught and sold, and those which have a good external appearance and growth rate kept for breeding work.

With an increase in stocking density the total fish productivity increases.

The above-listed levels of stocking density and indexes of output of produce are not proven limits. Under optimum conditions of environment and excellent biotechniques of carp rearing the density of stocking in a given pond area can be significantly increased. That quantity of fish per unit of pond area which would be analogous to the number of agricultural animals under conditions of stall maintenance is a feasible goal.

TECHNIQUE OF CARP FEEDING. The efficiency of feeding fish depends primarily on the preparation of the ponds, type of feed, and quality of stocking material. Proper techniques of feeding should not be ignored. Feeding fish in neglected ponds overgrown with emergent coarse plants, silted or cluttered ponds with poor quality water and abundant small trash fish, and ponds with spring water (not more than 17°C temperature), does not give the desired results even with a completely healthy and high quality stock. The conditions of external environment in such ponds are unfavorable for the growth and development of carp, especially under high density of stocking. Feeding should be done in well-prepared ponds and better still in drainable ones.

Feed is normally ground into flour,[9] then mixed and given to fish in the form of a paste. However, a mixture prepared in this way when added to water loses up to 30% of its nutritional substances in the first hour. Feed prepared in the form of pellets has a number of advantages over paste: the leaching of nutritional substances is reduced, the ponds are less polluted with the feed residue, and hence the maintenance of a favorable oxygen regime is better ensured. Both the weight of the fish and the efficiency of food utilization increase in this process, the food coefficient decreases by 25 to 30%, the food value of the fish improves (more fat is accumulated), and the food resources are utilized more economically.

Studies on the water-resistant property of individual items included in the feed of pond fish (Logvinova and Lysenko, 1968) reveal that the most waterproof pellets are those prepared from sunflower and cotton meals and from sunflower oilseed cake. They retain their form well for more than 24 hours and serve as binding material in the "moist" method of pellet formation. Pellets made of barley flour, linseed cake, milk serum, and BVK* have a relatively low water stability (from 40 min to 4 hr). Pea, linseed cake, and sunflower meal are ground together finely since in this case the water resistance is higher. For all other ingredients the level of grinding need not be so fine.

Pisciculturists in Rumania use granulated feed of the following composition: 2.6 to 3.2% water, 30.1 to 31.4% proteins, 6.1 to 6.6% fat, 54.7 to 59.9% nonnitrogen soluble substances, and 5.2 to 6.0% ash. This feed has the consistency and taste of natural food, gives off a strong odor in water which attracts carp and invertebrates, is eaten well by fish, and does not disintegrate in water for 24 hours. The use of this granulated feed has

[9]In foreign practice cereals are not pulverized but used in their natural form after softening them in water.

*Protein-vitamin concentrate—General Editor.

improved the biological composition of fish meal, reduced food consumption (food coefficient 1.7), and increased fish productivity to 1.5 ton/ha.

Compared to pasty feeds, pelletized formula feed has proved better. Experiments carried out at the fish farm "Luban" (Belorussian SSR, Minsk district) have shown that formula feed prepared in the form of pellets is used 16 to 28% more effectively by two-year carp in foraging ponds than the same food provided in the form of paste (Grib and Lyakhnovich, 1968). Formula feed in the form of pellets yielded a saving of 0.95 to 1.50 kg of food per kg of increase in fish productivity. The pellets in this case were prepared in the form of cubes with rounded corners, 90 mm×45 mm×20 mm in size. In water they lose their nutritional substances to a lesser extent than dispersed feeds. Formula feed pressed into cylindrical pellets 4 mm in diameter and 6 to 8 mm long fed to two-year carp in foraging ponds was 7 to 11% more effective than the same feed prepared in the form of a paste. The economy of such pellets was 0.4 to 0.7 kg of feed per kg of increase in fish productivity.

Care must be taken with regard to water temperature and hydrochemical regime, especially the oxygen content, when feeding fish. At optimum temperatures the daily ration amounts to about 3 to 4% of the weight of the carp, and at the end of the vegetative growth period, when the temperature is lower, 1.5 to 2.0%. Even a small change in temperature (just a fraction of a degree) has an influence on the intensity of feeding of fish. The maximum daily ration eaten by carp is consumed within a temperature range of 23 to 29°C. With a reduction in temperature to 18 to 22°C, food consumption reduces 1.5 times; with a further reduction in temperature to 15 to 17°C, it reduces 2.5 to 4.0 times. The daily ration also shrinks with a reduction in the quantity of oxygen dissolved in water (see above), due not only to its utilization for respiration by aquatic organisms, but also for oxidation of the unutilized part of the feed and the solid and liquid excrements of the fish. Shpet and Feldman (1962) found that with the oxidation of the solid and liquid excrements from fish stocked at the rate of 20 q/ha, the daily oxygen consumption was more than 3.5 mg/l. The quantity of excrements, according to their data, constituted approximately 20% of the weight of a given quantity of feed. In a pond area of 0.5 hectare, with a stocking density of 12N, and a daily application of 200 kg feed, the oxygen spent on the unutilized part of the feed was 1.5 to 2.5 mg/l per day. This consumption would be higher still with an accumulation of released substances.

The following norms must be observed in the course of feeding carp under dense stocking: the quantity of feed correctly calculated and prepared on the basis of daily requirements for different months of the vegetative growth period and number of feed pans; the quantity of stocking material determined and transferred to the farm (if not produced at the farm itself); the type and number of feed pans selected, having planned their maintenance over the vegetative growth period; the hours of feeding stipulated; the gas regime systematically observed; laborers assigned for pond maintenance and feed preparation; the growth, condition, and rate of food consumption regularly checked in the fish; and the periods of control catches determined. The results of control catches are recorded in a special register (Table 37).

The quantity of feed required is calculated by the formula:

$$K = P \times G \times a(N-1),$$

where K = total quantity of feed required under certain set of conditions (quantity to be determined in kg);
P = natural productivity of pond (kg/ha);
G = area of pond (ha);
N = index representing density of stocking;
a = food coefficient of a given feed.

If, for example, the area of a foraging pond is 100 ha, its natural fish productivity 200 kg/ha, food coefficient of the feed 3, and stocking density 5N, the total quantity of food required would be:

$$K = 200 \times 100 \times 3\ (5-1) = 240{,}000 \text{ kg}.$$

Taking into consideration the monthly growth rate of two-year fish in the Moscow district reared on natural foods (see Table 9), the distribution of artificial feed over different months of the vegetative growth period would be as follows:

May	6%	(14,400)
June	28%	(67,200)
July	29%*	(69,600)
August	35%*	(84,000)
September	2%	(4,800)
	100%	240,000 kg

Thus, under the conditions in the Moscow district 92% of the total food required is consumed in just three months (June, July, August). In the southern regions of the USSR, where the period of vegetative growth is more suppressed and includes April and October as well as June, July, and August, the distribution of feed differs and corresponds to the approximate growth rate of carp given in Table 9.

Feed prepared per the calculations given above is distributed in accordance with need each month from special storehouses located near the ponds (Figure 86).

*The index for growth rate in July and August can be interchanged in certain areas.

Table 37. Register for recording data on control catches of a foraging pond and food consumption by carp*

Carp breed	L, cm	l, cm	H, cm	C, cm	Q, g	Average daily temperature, °C	Food consumption			Remarks on food intake
							Type of food or feed mixture	Weight of dry matter, kg		
								per feeding pan	for entire pond	

*Data for *L*, *l*, *H*, *C*, and *K*, are essential in research work but can be omitted under commercial production conditions; data on weight (*Q*) are essential for commercial fish cultivation.

Figure 86. Storehouse for feed and its loading in motor boat.

Carp should be fed early in the morning (6:00 to 7:00 a.m.), when the requirement for food is greatest, and under favorable temperature conditions a second time between 4:00 to 6:00 p.m. In the conditions of enclosure-rearing of carp the number of feeds per day are increased (see below). The food supply is adjusted according to the time of its most effective utilization under definite conditions. The results of experiments on the comparative effectiveness of one, two, and three feedings of carp, carried out at the fish farm "Osenka" (Moscow district) with a stock of 10,000 yearlings per hectare of foraging pond, are given in Table 38. It can be seen that dividing the daily ration is accompanied by a lower quantity of food supply and an increase in pond productivity.

Fish feed with different intensity at different times of the day. Feeding of carp is far more rational if the daily rhythm of feeding in different zones and even different months and years is studied. Feeding rhythm is influenced by temperature, light, and weather conditions. The absence of a stomach in carp makes the ingestion of a large quantity of food impossible during a single feed. Hence hours of active feeding alternate with almost total abstention, depending on sex, age, condition of the gonads, etc.

The monthly quantum of feed is divided by the number of days in a month. With a sharp reduction in water temperature, when the carp

Table 38. Comparative efficiency of repeated carp feeding

Parameters	Single feed at 8:00 a.m.	Two feeds, at 8:00 a.m. and 4:00 p.m.	Three feeds, at 8:00 a.m., 12:00 noon, and 4:00 p.m.	Remarks
Average weight in catch, g	497.3	527.7	544.6	Feed mixture made according to formula for pellet feeds developed by the All-Union Research Institute of Pond Fisheries
Fish productivity, q/ha	48.4	52.8	54.5	Rate of food supply, 18 to 25 g per fish per day
Food coefficient	4.7	4.0	3.6	

utilize and assimilate food poorly, the daily norm is reduced or no food provided, and the ration subsequently increased under optimum temperatures. The water temperature should be measured three times a day (7:00 a.m., 1:00 p.m., and 7:00 p.m.) and, depending on the temperature level, the daily ration and frequency of feeding regulated.

Food must be given strictly at the hours determined by the fish farm authorities. Carp adjust to feeding hours and congregate unbidden at the feeding places. The daily ration is predetermined over small intervals of time (ten days). This makes it possible to give better consideration to the changing conditions of external environment (temperature, gas regime, etc.), as well as the quantity and quality of the feed and density of stocking (Table 39). In planning the ration use is made of the data on growth rate collected over several years, together with the total growth rate in the current year, and an assessment of the quality and quantity of food. If, for example, the daily ration is being calculated for the last decade of July, and in previous years the growth rate over this period was 45 g, i.e., about one-tenth of the total projected growth over the vegetative period (470 g), then the amount of feed given in previous years in this particular decade is given in the current year also. In other words, if of the quantity of food calculated for the month of july (69,600 kg), 6,960 kg were given in the last decade, then the daily ration for the current year would be 232 kg [*sic*].

The Scientific-Technical Council of the Ministry of Agriculture of the USSR has accepted the following tentative growth rate of under-yearlings, two-year, and three-year carp for different zones of the country under dense stocking and artificial feeding (Table 40).

Table 39. Register for recording fish feeds

Pond No.——————Area——————ha. Category of pond——————Stocking density——————No. of fish stocked——————Type of food provided. Number of feeding points——————Unit growth rate according to plan toward autumn——————

Dates	April			May			June			July			August			September			Total
	Water temperature	Hours of day	Quantity of food, kg	Water temperature	Hours of day	Quantity of food, kg	Water temperature	Hours of day	Quantity of food, kg	Water temperature	Hours of day	Quantity of food, kg	Water temperature	Hours of day	Quantity of food, kg	Water temperature	Hours of day	Quantity of food, kg	

Table 40. Approximate daily growth rate of under-yearlings, two-year, and three-year carp under dense stocking and artificial feeding (g)

Months	Decades	Northern Siberia and northwestern zone			Central, central chernozem, and southwestern zone			Southern regions		
		Under-yearlings	2-year fish	3-year fish	Under-yearlings	2-year fish	3-year fish	Under-yearlings	2-year fish	3-year fish
April	3	—	—	—	—	—	—	—	—	—
May	1	—	—	—	—	—	2	—	—	3
	2	—	—	2	—	—	3	—	1	4
	3	—	1	3	—	2	4	—	2	5
June	1	—	3	4	—	3	6	—	4	8
	2	—	3	7	0.1	4	8	0.1	5	10
	3	0.1	4	9	0.1	5	12	0.2	6	14
July	1	0.2	5	10	0.2	5	12	0.3	6	14
	2	0.3	5	10	0.3	5	14	0.4	6	16
	3	0.4	6	12	0.4	6	14	0.5	6	16
August	1	0.5	6	12	0.5	6	14	0.5	6	16
	2	0.5	6	12	0.5	6	14	0.4	6	16
	3	0.4	4	8	0.4	5	10	0.4	6	14
September	1	0.3	2	4	0.2	2	8	0.3	3	10
	2	0.2	—	2	0.1	1	4	0.2	2	5
	3	0.1	—	—	0.1	—	—	0.1	1	3
October	1	—	—	—	0.1	—	—	0.1	—	—
	2	—	—	—	—	—	—	—	—	—
Total growth over season		30	450	1,250	30	500	1,500	35	600	1,550

Feed is given at special feeding places 2 m² in size. The silt at these places is thoroughly cleaned, compacted, and applied with lime at the rate of 250 to 300 g of lime per 1 m². Lime application is essential for the disinfection of the pond soil; it also facilitates compacting the ground in somewhat silted and plant-infested areas and in places with a considerable overgrowth of aquatic plants. Feeding places are selected and prepared in autumn after draining the ponds and drying the bottom. Rectangular or square feed pans or trays 1 to 2 m² in size and 5 to 8 cm in height are placed at the feeding points (Figure 87). In the process of growth the fish inhabit deeper areas of the pond and so the pans must be transferred to the places where fish concentrations are maximum (i.e., the food must follow the fish). A bottom

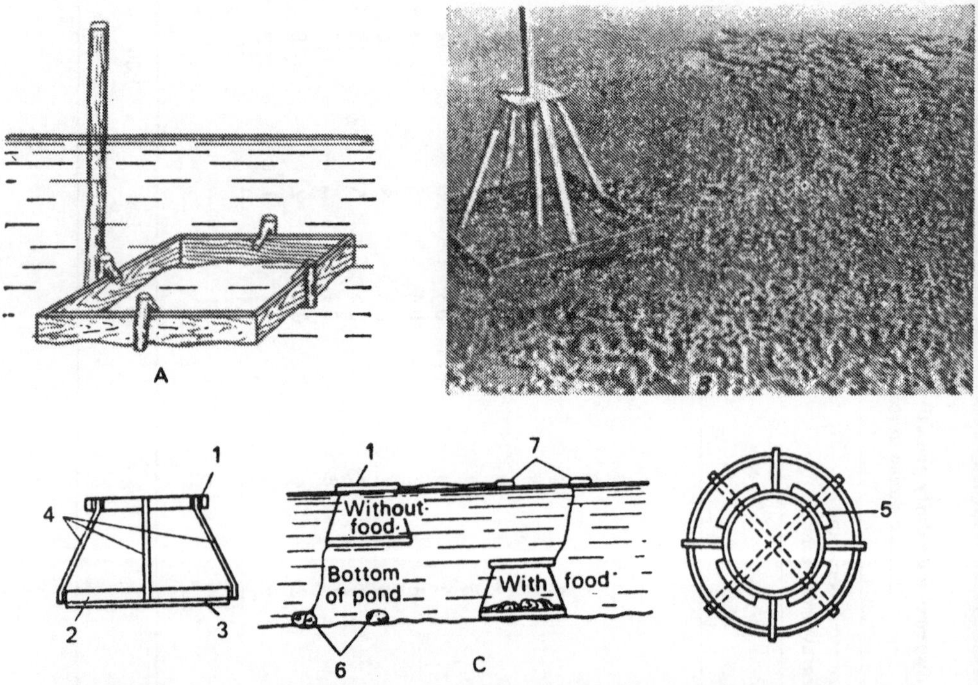

Figure 87. Feeding devices for carp.
A—Bottom feed tray; *B*—Feed-table device and bottom of pond disturbed by carp in search of food at 10 times stocking density; *C*—Self-floating metal feed table.
1—upper frame (d=400 mm); *2*—lower frame (d=700 mm); *3*—bottom made of galvanized iron; *4*—four supporting rods 400 mm long made of 5 mm wire; *5*—four floats made of cork or foam which provide lifting strength to tray released in water (0.2 to 0.3 kg); *6*—ballast sinker (anchor) weighing 3 kg; *7*—floats made of dry wood or cork.

feed-table device (see Figure 87) is very convenient as it can be submerged in the initial period of feeding (up to August) at 0.6 to 0.8 m, and later, with an increase in the weight and size of carp and a reduction in water temperature, shifted still deeper—to a depth 1.3 m. When the food is brought to the fish, they expend less energy in searching for it. The advantage of the bottom feed-table device, compared to feed pans, is not only the fact that it can be kept at the required depth, but also that it can be transferred anywhere else when the original site is no longer convenient. Changing the placement of the feed trays is difficult once the pond has been filled with water. Metal floating feed-table devices are also used (see Figure 87).

The location of the bottom feed trays is marked out by pegs. The feed-table devices, the panels of which are either raised or lowered along pegs, are marked by the portion of the peg raised above the water surface.

Before placing the feed-tables the congregations of fish in the pond are noted. The tables are positioned at the places of fish congregation, preferably in a single line and spaced 3 m apart. In large ponds the tables may be arranged in two to three lines with not less than 3 m between them. This ensures delivery of the feed from boats moving between the lines of tables. Tables inconveniently placed are not visited by the fish. Such tables should be transferred immediately as carp quickly become accustomed to feeding places. Frequent transfers of the feeding tables compel the fish to search for food, which could lead to a loss in feeding days and quality of feed and a decrease in fish weight. The transference of feeding trays over small distances in slightly deeper waters does not cause such harmful consequences.

Boats with an outboard engine are convenient for distributing feed in ponds because the food can be deposited without stopping the boat. More sophisticated self-propelled feed distributors are also used.[10]

The number of feeding devices depends on the feed load for each age group. One feed pan should be allowed for every 3,000 to 5,000 under-yearlings, every 250 to 400 two-year fish, every 200 three-year fish, and every 20 brood fish. The quantity of food delivered in one feed tray is determined by dividing the daily quantity of food by the number of feeding times per day and the number of feed trays. Smaller quantities of food are initially provided during the feeding period, the full norm given in the middle of the month, and the rate increased in the second half of the month to ensure that the planned (calculated) quantity of food is delivered in full. At the end of the feeding period, when a reduction in water temperature occurs, the daily norms of food supply are lowered.

Before the supply of each food ration, the feeding table is raised and the consumption of the previous ration checked; fresh feed is added accordingly. If all of the feed has been consumed the daily ration should probably be

[10]The problem related to mechanization of production processes in fisheries, especially the mechanization of food preparation and distribution, is covered in agricultural universities in a separate course entitled "Mechanization and Electrification of Animal Husbandry."

increased and, conversely, if some remains, the daily ration decreased. Feed unconsumed in the previous delivery can be left for one more day, after which it must be collected and replaced with fresh feed.

Carp gradually becomes accustomed to a particular feed. If one type is suddenly replaced by another, the fish stop feeding. Changes in feed should be done gradually by mixing increased quantities of the new feed in the old one until the fish have adjusted to the new feed.

Feeding places eventually become polluted with carp excreta and decomposed residual feed or food particles dispersed in the water which settle to the bottom. Feed allowed to accumulate decomposes in summer under high temperatures; its stench repels carp and the fish stop visiting the feed pans. In addition, blue-green algae develop in the polluted areas which deplete the oxygen in the water in the same way that decomposing food and excreta do. Hence feeding places should be systematically cleaned by collecting feed residue and excreta in dip nets.

The requirements of the stocking material, when artificial feeding is undertaken, is calculated by the following formula:

$$A = \frac{P \times G + \frac{K}{a} \times 100}{(V - v) \times r},$$

where A, P, G, 100, V, v, and r are the same parameters used in the formula for normal stocking as given in Chapter IV: K = total quantity of feed; a = food coefficient.

The magnitude of fish productivity (P) is taken as the average for the two previous years at N level of stocking.

On the basis of the foregoing calculation for food requirement and taking the unit weight of carp stocking as 25 g, the final unit weight of marketable fish as 500 g, and output as 90%, one gets:

$$A = \frac{200 \times 100 \; \frac{240{,}000}{3} \times 100}{(500 - 25) \times 90} = 203{,}158 \text{ fish.}^{11}$$

The rearing of fish under dense stocking and feeding are controlled in the same manner as described in Chapter IV. With a view to achieving the best results from feeding, this work is assigned to trained workers held responsible for its efficacy.

The output of fish is calculated as a percentage—number of fish caught of total number of fish stocked.

[11]In the case of combined rearing of two- and three-year carp, stocking of yearlings and two-year fish in the foraging ponds is done at the ratio of 5 : 1. A higher rate of stocking of three-year carp is irrational.

Feeding of Brood Fish and Replenishing Stock. The poorer the feeding conditions of brood fish, the more inferior their progeny: lower viability, slower growth, lower resistance to unfavorable conditions of external environment, and higher susceptibility to disease. Needless to say, proper feeding of brood fish is essential to fish productivity.

In most cases carp brood fish are maintained on natural foods because many researchers believe that artificial feeds cause a deposition of fat in the gonads, which leads to a reduction in breeding capacity. However, data on the feeding of brood fish with cereal wastes from the time of their transfer from wintering ponds to spawning ponds and before plantation for foraging, reveal that the fish are tougher after winter emaciation and consequently spawning and susceptibility to disease are reduced. The feeding of brood fish prior to spawning has a special significance for the northern regions where the prolonged winter causes a considerable depletion in carp, reducing their reproductive capacity. In State fish farms feeding carp brood fish with plant and animal food has proved successful.

The feeding of brood fish and replenishing stock is carried out in the same way as described above. The feed mixture should be ground into a fine flour or bran, soaked, mixed thoroughly, and then given to the fish. The feed mixture recommended by the Ministry of Agriculture of the USSR for carp brood fish has been presented above (see page 229).

The daily norm of feeding brood fish and replenishing stock is 2.5 to 3.0% of the weight of the carp depending on feed consumption. Brood fish should be fed moderately.

Detailed studies on the feeding of carp brood fish with different feeds under dense stocking (different degrees of density) and under changed conditions of life need to be undertaken. Questions pertaining to the effect of feeding on replenishing stocks and brood fish, the changes caused in the internal organs of the fish (by various quantities and qualities of feed), and how these are reflected in the reproductive capacity and progeny beg for resolution. Any study on the feeding problem of carp should commence with the embryonic period and cover the various developmental stages from juvenile to adult.

In practice, brood fish are fed without distinction of sex. Classical works on indigenous animal husbandry (M.F. Ivanov, P.N. Kuleshov, M.I. Pridorogin, and others) point out that paternity exerts a significant influence on progeny.

In one experiment with rabbits and sheep males were fed concentrated feed enriched with proteins and vitamins in the form of fresh blood, yeast, and carotene. The females and a control group of males were given their usual food. It was found that in the experimental group the motility of the spermatozoa doubled, fertility increased, and the litter was larger and more viable. Losses were reduced by 50%. The experimental males were given the special feed only for a period of two months. This affirms the high sensitivity of sex cells to external influences, a fact mentioned even by Charles Darwin in his book, *Cross-Pollination and Self-Pollination.*

Most probably, a better carp progeny could be obtained from differential feeding of males and females than from indiscriminate feeding. Differences in the metabolism of male and female animals cannot be denied. This problem needs to be studied in pond fisheries at an early date.

Experience has shown that feeding of brood fish during the prespawning period is especially important under dense stocking. The calculated norms of growth rate for females are considered to be 20% higher than those of males in the postspawning period, after stocking of brood fish in summer stock ponds. This is associated with the greater weight loss of females during the spawning period.

In the course of feeding brood fish and the replenishing stock the density of stocking should not exceed 3N, so that the natural food in their ration is not less than 30%. Increasing the density of stocking reduces the per capita consumption of natural foods, curtails fat deposition, and results in a progeny with reduced viability.

Strel'nikov (1966) experimentally established that under-yearlings and older age groups of carp fed with feeds primarily of animal origin (50 to 90%) during the fourth stage of maturation produced 7% more large-sized eggs than carp fed feeds of plant origin (up to 90%). Moreover, the output of fry increased by 34%, spawning took place 10 to 15 days earlier, the progeny was characterized by better feeding activity, using the food base more thoroughly, and hence showed greater growth and better survival.

Feeding of Carp Fry. According to the recommendations of the Ministry of Agriculture of the USSR and the All-Union Lenin Academy of Agricultural Sciences, the stocking of fish in rearing ponds in the northern, eastern Siberian, and northwestern zones should not exceed 30,000 to 40,000 fish per hectare; in the central, central chernozem, southwestern, and southeastern zones, and in southwestern Siberia—40,000 to 50,000 fish per hectare; and in the southern regions, Trans-Caucasus, and Central Asia—80,000 to 100,000 fish per hectare.

In the course of feeding of fingerlings it is essential that the feeds used have a normal protein ratio (up to 1 : 2), which is closer to the protein ratio of natural food (from 1.05 to 2.50). The feed given to fingerlings should be of high quality, prepared more carefully, and fractionated (by grinding) up to a level easily ingested by young fish. Even with such careful preparation the consumption of feed initially proves difficult for very small fish. Hence before stocking fry in rearing ponds the natural food base in them should be increased by manuring the ponds to ensure that those food organisms requisite for a given stage of fish development will be abundant. Only after the fry have grown well, attaining a body weight of 3 to 4 g, can they effectively utilize artificial feeds. Feeds given prior to such weight are usually ineffective.

The feed mixture detailed in Table 29 is recommended for fingerlings. Other mixtures are also used, however. According to British authors (1967), for young fish from hatching up to fry of 15 days the best mixture is one prepared from notonectids, prawns, and cowpea in the ratio of 5 : 3 : 2.

In the USSR fucus crumbs[12] are sometimes used to supplement the ration. These crumbs are added at the rate of 5% by weight of dry feed mixture to the feed of fingerlings. The mixture consists of 60% sunflower cake, 30% formula feed of poultry, 4% hydrolyzed yeast, and 6% fish meal. In a rearing pond stocked with 60,000 fish/ha 17 days old a feed mixture containing 5% fucus meal was eaten well. Toward the end of July the guts were filled up to 60% with this feed, and toward the end of the vegetative growth period up to 90%. Losses of fingerlings given this feed were only 9.6% over the winter period, versus 20% among the control group.

Better results were recorded when a phosphatide concentrate prepared from the by-products obtained in the process of purification of sunflower oil was added to the feed. Fingerlings given a ration containing 20% phosphatide concentrate (in relation to weight of dry feeds) had a unit weight of 43.8 g toward autumn, versus 28.9 g in the control group. The output from the wintering ponds was 97% versus 87% from the control ponds.

Feed prepared in the form of rather small pellets is utilized best by fingerlings.

In providing artificial feed to fish it should be kept in mind that the proportion of natural foods in rearing ponds should be maintained at a higher level than that in foraging ponds where two-year fish are kept. The higher the density of stocking and the lower the total quantity of natural foods in the fish ration, the more complete the artificial feed must be in terms of nutritional substances.

Japanese pisciculturists, in a 30-day experiment with carp fingerlings, found that their weight increased proportionate to the increase in protein content from 0.4 to 54.0% in the feed provided. Maximum assimilation occurred at a protein level of 38%.

Regular feeding of fingerlings is continued in autumn until transplantation to wintering ponds. Feed mixtures rich in digestible carbohydrates (potato, cereal waste, etc.), as well as proteinic food should be given in this period. Carbohydrates do not compensate for a deficiency in proteins and without the latter cause excessive fattening. On introducing feed trays (Figure 87) the shallower (0.6 to 0.8 m) areas of the pond should be selected since the fry

[12]Fucus (*Fucus vesiculosus*) is a brown algae. It is dried and crumbled in the White Sea region and then added to the feed of farm animals. Fucus is a carbohydrate feed rich in mineral substances.

initially congregate in such places. In the second half of the year, when the fish have grown and their natural foods become more diverse, the feeding trays should be transferred to deeper areas.

Distribution of the feed over different months of the vegetative growth period for fingerlings under conditions of the southern regions of the USSR should be as follows: June—5%, July—25%, August—40%, September—25%, and October—5%.

The number of feeds per day plays an important role, correlated with the hours of the most intensive feeding under the given conditions. Experience has shown (Moldavian SSR) that in the case of three distributions of feed per day the weight of fingerlings increased by 34% compared to a single feed. The normal daily ration is divided by the number of feeds given each day. Fingerlings are also given feed in autumn (after their transplantation to wintering ponds) until the water temperature remains above 4°C, and in spring until they are transplanted to foraging ponds (at a temperature of 6°C). Carbohydrates should predominate in the ration.

The following norms of daily ration are recommended.

1. After transplantation to wintering ponds at a temperature of 10°C, the daily ration should be 3% of the body weight; with a reduction in temperature it is gradually reduced to 1%. If the fingerlings consume the food, feeding is continued even after ice formation.

2. In spring, before the transfer of yearlings to foraging ponds, when the water temperature is higher than 6°C, the norm is initially 1% of body weight, and subsequently gradually increased to 3% of the total quantity of food consumed on the previous day.

In wintering ponds under-yearlings are fed from feeding tables installed before fish transplantation at the rate of one table for 10,000 fish. No interval between feeding in the rearing and wintering ponds is permitted. The intake of feed is strictly controlled so that its residue does not pollute the water in the wintering pond.

* * *

The theoretical basis and technique of feeding fish species other than carp and trout (see Chapter XV) cultured in warm-water fish farms have not been worked out. Data on feeding practices of one fish species cannot be applied to another species. Methods of feeding each type of fish cultured in ponds must be studied separately. According to existing information tench digests plant food less well than carp. This fish should be fed with more finely ground food, better boiled and steamed. Tench is mostly reared in carp farms as an additional fish. Its feeding in monoculture is the same as that of carp. Afternoon is a better period for feeding tench.

Feeding Carp in Cage (Enclosure) Culture. The founder of the method of culturing carp in enclosures is the Japanese pisciculturist Tanaka who, in 1954, raised more than 8.5 tons of marketable fish in two fish enclosures, each 62 m^2.

Culturing fish in enclosures and cages involves rearing them in a limited volume of water at a high density of stocking and feeding (200 to 300 fish/m^2 instead of 0.25 to 0.30 fish/m^2 in ordinary ponds).

The first experiments in the USSR on rearing carp in tanks were conducted by N. Prokulevich (1940). He used cement tanks (Voronezh city) with an area of 300 m^2 and depth of 0.9 m, and stocked 250 yearlings with an average weight of 18 g in each. From May 15 to September 20 the fish were given a feed mixture containing 76% cotton seed cake, 18% dried molluscan meat, and 6% rye bran; 150 kg carp were collected with an average unit weight of 600 g. The water in the tanks was totally replaced every 12 to 15 days. Plankton and benthos were negligible and hence did not play a significant role in fish feed.

The problem of rearing a number of fish species in floating fish enclosures installed in reservoirs was resolved by Mikheev, Meisner, and Mikheev (1970). They not only designed floating enclosures (hatchery, nursery, rearing, foraging, and hibernation type), but organized a system whereby the supply of natural foods to fish is improved by attracting planktonic organisms, aquatic insects, etc. by light.

In the case of carp in floating enclosures it has been demonstrated that with a stocking density of 200 to 500 fish/m^2 and the relatively low summer water temperature in the usual reservoirs located in the central belt of the USSR, rearing of fingerlings and two-year fish primarily on plant food is not very profitable. Carp grow better on feed prepared from animal items (*Dreissena,* fresh-water mollusks, chironomids, trash fish) and plant food. Fingerlings stocked at 100 to 300 fish/m^2 reach a weight toward autumn of 25 to 30 g and two-year fish almost 300 g. Culturing carp in fish enclosures gives better results in the conditions of the reservoirs located in the south.

V.P. Mikheev and P.V. Mikheev have constructed a special automatic feed distributor for releasing food into floating enclosures (see Figure 88), which works electrically on AC current. It consists of a feed distributor and master equipment. Several feed distributors can be installed above the reservoir and controlled by the master equipment located in a house on the bank.

Trial runs with the feed distributor in carp cultured in floating enclosures and fed pellets 5 mm in diameter and 10 to 30 mm long gave encouraging results. The feed is released in the enclosures from the bunker of the feed distributor every minute for two seconds over a period of five to six hours per day.

Experiments carried out with carp, trout, sander, and whitefish have shown that they quickly adjust to such a system of food supply; conditional

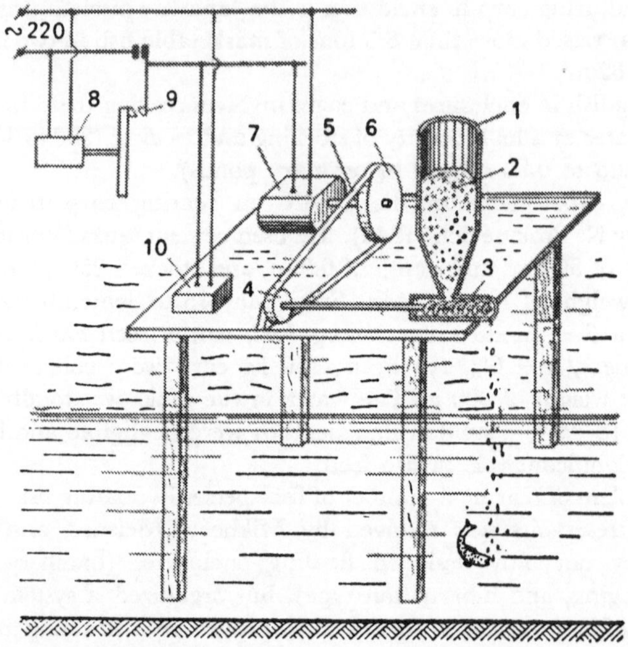

Figure 88. Automatic feed-distributor system designed by V.P. Mikheev and P.V. Mikheev. *1*—bunker covered with lid; *2*—feed; *3*—vermiform mechanism releasing feed in water; *4* and *6*—pulleys or belt sheaves; *5*—V-shaped belt; *7*—electric motor; *8*—electric engine of AOLB type (rated input 15 kw, 220 V, speed 2 rpm); *9*—contact breaker located on shaft of electric engine for regulating duration of switching on/off the electric motor of feed distributor; *10*—electric bell.

reflexes develop in the fish toward sound and light signals emitted during the process of feeding. Initially carp take feed from the bottom of the enclosure, but subsequently also remove it from the water column. At night the fish concentrate in the illuminated space and feed on the copepods attracted by the light.

Since fish react well to sound also, the authors recommend the use of automatic food distributors in combination with a sound signal, preferably an electric bell which rings simultaneous with the food delivery.

If electricity is available at the fish farm, food distributors can also be used in foraging ponds; this system makes it possible to economize on feeds, reduce labor costs, and ensures a better organization of fish feeding.

In floating enclosures carp spawns every year on artificial nests made of capron fiber, the percentage of fertilization is high, and the fry raised from the eggs exhibit no deviation from normal fry. The stock of brood fish raised under enclosure conditions produced normal progeny.

Rearing Lake Chudskoe and Lake Ladoga whitefish and sander in floating enclosures is also quite feasible, but for a large-scale culture of sander methods for their feeding need to be developed. Other objects for enclosure cultures could be hybrids of beluga and sterlet, phytophagous fishes, and others. Pike have proved unsuitable for this type of culture: they change over to a predatory mode of life very early, display considerable cannibalism (resulting in losses of up to 80 to 90% of juvenile fish and 53% of two-year fish), and grow very slowly.

Research work on methods for organizing enclosure culture has also been done in the Donets District by L.V. Gribanov, who provides a very detailed description of the conditions and methods for rearing two-year carp in net enclosures. He reared not only carp but also grass and silver carp in a canal in an enclosure of 81 m^2 and in ponds with an enclosure area of 700 m^2.

In Ukrainian SSR carp culture in tanks was initiated in 1957. It has been underway since 1959 in Georgian SSR (O.G. Buchuladze and A.N. Korneev), the Moscow district (A.N. Korneev and L.A. Korneeva), Belorussian SSR (V.I. Belyaev), and in Kazakh and Moldavian SSR.

Similar studies are being made abroad. In Japan enclosure culture has become widespread but one must recognize its peculiarities and limitations due to the small size of the culture areas.

The area of fish enclosures varies from 15 to 20 to 50 to 60 m^2, and they are mostly found in farmsteads. If there is sufficient flowing water and feed available in the enclosures, the yield of marketable fish can be higher than 200 kg/m^2, versus that from fish tanks with stationary water of about 0.5 kg/m^2. A sufficiently strong flow of water facilitates its enrichment with oxygen, making a high stocking density possible. The enclosures must be cleaned regularly, removing dirt, food residue, and excreta. Water temperatures below 20°C and above 32°C are unfavorable for culturing fish in these conditions.

The best shape of tank is rectangular or triangular in terms of water circulation; this depends however on the tank area. Such tanks are usually 1.2 to 2.0 m deep, their sides covered with small pebbles, and the cracks filled with concrete.

In the period of most intensive growth the maximum quantity of water is supplied by opening the sluice; special arrangements for collecting dirt and debris must be made in the water discharge canal. The tank is covered on top with a net to trap falling leaves. On the average, in a tank area of 50 to 100 m^2 the water is changed 10 to 15 times per day.

Carp are released in the tanks when they have reached a weight of 70 to 150 g (at the end of March–beginning of April) at a water temperature of 10 to 15°C. Only fish of similar weight are released in each tank.

Granulated feed prepared according to a special formula is recommended, but feed mixtures prepared at the farms may be used, depending on the availability of one or the other component.

Feeding commences when the water temperature reaches 15°C (and sometimes even lower). The granulated feed is added with the pupae of mulberry silkworm in the following ratios, depending on the water temperature (Table 41).

Table 41

Composition of feed, %	Water temperature, °C			
	Below 12	12–17	17–21	Above 22
Granulated feed	100	80	70	60
Pupae of mulberry silkworm	0	20	30	40

Feeding terminates when the carp start darting about after ceasing active food consumption.

The principal of commercial carp rearing in Japan includes their culture in small tanks of 50 to 100 m^2 area, with warm-water, a high water flow, and intensive feeding with high-calory feeds. The temperature is maintained at 25 to 32°C and water exchange takes place every 3 to 4 min. The stocking material (yearlings) weighs 50 to 90 g and the density of stocking is 90 to 130 fish/m^3. The fish are cultured from April to November and the unit weight after harvest is 600 to 1,200 g with a productivity of 100 kg/m^3.

Methods of carp culture vary in European countries.

In the German Democratic Republic the following arrangements are made for culturing carp: (a) fish enclosure in hot spring waters; (b) thermal ponds with cooled water; (c) ponds or tanks with cold water; (d) heating arrangements; and (e) heating troughs with cold water.

Carp are cultured in heated waters in the usual complex of pond fisheries and reared up to autumn. With a reduction in water temperature they are transferred to thermal ponds and fed entirely on a feed mixture. In late autumn and during winter the temperature is maintained at a level which ensures fish growth and toward the end of winter their weight usually reaches the level achieved in a two-year rotation under pond conditions. Thus, in this system the period of raising marketable carp is significantly reduced (from 2.5 years to 1.5 years), which makes it possible to widen the rearing of stocking material in the given area without additional capital investment, making the farm more profitable. Under-yearlings are transplanted in the second half of autumn, before a sharp reduction in water temperature occurs; with the onset of spring if the need arises for carp of higher body weight, they are transferred to the ponds.

The first thermal unit in the German Democratic Republic was constructed in the Kotbus District in 1967 and consists of six rectangular iron concrete tanks or raceways 10 m × 1 m × 1 m, arranged in a covered space.

The water exchange rate is very high, with almost total replacement every 10 min. Even at such a high water exchange rate, given a high stocking density, a considerable reduction in oxygen content occurs at the outlet end of the raceways. The water temperature is 25.4°C. One year 2,456, 4,174, and 7,795 carp were released in three tanks with an area of 10 m^3 each. The increase in weight of carp in all the tanks over a six-month period was 296.5 kg/m^3. The carp were fed pellets of trout feed which contained a high proportion of fish meal, skimmed powdered milk, edible yeast, and biologically active substances. Feeding was done six times a day at a norm of 4 to 5% of the fish weight. The consumption of feed per unit of growth was 1.66 to 2.92.

Another experimental farm, in the Leipzig region, consists of 10 raceways of the open type made of reinforced iron concrete: dimensions 8.0 m × 1.5 m and volume 4 m^3. Water circulation is of the closed type, water consumption 200 m^3/hr, and water replaced 10 times in one hour. This farm is well mechanized, with automatic feeding pans operated by remote control ensuring six feeds per day. The store for reserve feeds is in the form of a silo with a capacity of five tons. It is loaded with the help of a pneumatic conveyor and unloaded by opening the damper (slide or baffle plate) in the lower part of the tower. Distant thermometers and other mechanisms make it possible to control and regulate the water temperatures in the raceways. An accident-signalling system has been installed. The arrangements for preparing feeds are very efficient.

The annual yield from this farm is 250 kg/m^3 with a food coefficient of granulated feed of 1.7 to 2.0.

In the Frankfurt-on-Oder district very good results have been obtained by culturing fish in cages with a volume of 6.5 m^3. Stocking carp with an average weight of 68 g at a density of 160 and 320 fish/m^3 has yielded a harvest of 66 and 120 kg/m^3 respectively, with a food coefficient of 2.2 (granulated trout feed) and a unit weight of 429 and 346 g.

In the Federal Republic of Germany, at the Max Planck Institute in Hamburg, carp are raised in polyethylene aquaria 60 cm × 40 cm × 20 cm with a capacity of 40 l, equipped with discharge and overflow openings. Water exchange is of the closed type; water from the container passes through a number of sedimentation tanks and undergoes biological purification thereafter. Water is pumped into the cleaning apparatus by means of a pump fitted with an electric heater, which thermostatically maintains the temperature around 24°C. The water returns to the aquarium through a water conduction system in which a standard air compressor and diffusers have been installed for aeration.

Aquaria used for mass-scale carp rearing are constructed of glass reinforced plastic. The water in these aquaria is replaced every 15 min. At a constant temperature of 23°C the carp reach 800 g in body weight in 12 months and 1,200 g in 17 months. Larger carp, with an initial weight of 130 g, attain a body weight of 2,600 g in 12 months and 4,000 g in 17 months.

Such experiments prove that carp can be raised in very small volumes of water. The main conditions are: biological purification of water, temperature maintenance, and proper feeding with nutritive food. Depending on the size of the fish food is given at the rate of 3 to 5 g for each 100 g of live weight six times a day at an interval of 2 hr. Mixtures used for carp rearing at high densities are also suitable (see page 230). If at the beginning of the experiment (November) the average weight of carp fingerlings was 10 g, 5 months later (April) it had increased to 60.8 g, and 7 months later to 80.5 g. Carp under pond culture with warm-water and in aquaria with a constant temperature of 24°C fed only on artificial feeds, according to both Soviet (A.N. Korneev and L.A. Korneeva, 1968) and foreign studies, mature in two years, i.e., twice more rapidly than in common ponds.

The use of warm-water for rearing stocking material during winter using fry obtained in August and a little later is also of practical interest.

Rearing carp in fish enclosures has recently been undertaken by the All-Union Research Institute of Pond Fisheries under the leadership of A.N. Korneev. Thermal waters, primarily those from electric power stations are being used. The area of cooling reservoirs of hydroelectric stations and thermal electric power stations presently covers about 70,000 hectares, which is expected to increase in the near future. Experience has demonstrated that carp can be reared in enclosures in warm industrial waters on a carefully balanced feed mixture only at the optimum temperatures of 23 to 33°C. Growth is retarded at 22°C and almost totally ceases at 20°C.

Organization of the production processes necessary for the management of fish culture in enclosures perforce includes problems associated with the production of spawners as well as individual age groups of carp raised under these conditions. As revealed by studies at the All-Union Research Institute of Pond Fisheries (Moscow district), female fish under the conditions of cooling waters even in the beginning of April had reached the fourth stage maturation and after a pituitary injection ovulated in 8 to 10 hr. This was the result of temperatures suitable for spawning, obtained in the cooling reservoirs even in the first decade of April. Hence spawning in enclosures can occur 1.0 to 1.5 months earlier than is usually possible in the ponds of this zone.

The brood fish in this experiment were raised in fish enclosures under artificial feeding in the warm-waters of the cooling reservoirs of hydroelectric power stations. It is therefore possible to obtain viable progeny from brood fish at an earlier age than heretofore expected.

The commercial collection of eggs and their fertilization and incubation in the case of enclosure culture would be analogous to the procedures already described in Chapter IV.

Rearing younger stages in enclosures in the warm-waters of hydro electric power stations requires the development of methods for working with prolarvae, larvae, and early fry, and an investigation of the nature of their feeding under such conditions.

Experiments on carp culture in fish enclosures in the USSR have shown that in these conditions considerable attention must be given to the preparation of complete feed mixtures for different age groups. Feeding of fry in fish enclosures in the cooling reservoirs of hydroelectric power stations is already being done and positive results have been obtained when the feeds for each age group were correctly balanced. Of the known carp types, fingerlings of scale and mirror carps have been reared more successfully.

According to the data of A.N. Korneev, L.A. Korneeva, and L.N. Titareva (1968), even at the age of 10 days fry consume food well from feeding trays, when the ration consists of blood meal, pupae of mulberry silkworm, egg powder, powdered milk, hydrolyzed yeast, groundnut cake, and a number of vitamin supplements. With fry growth, the authors decreased the quantity of animal feed and increased the feed of plant origin. A small quantity of daphnids caught from nearby waters were also given to the fry during the first month of their development.

Stocking larvae at more than 2,000/m^3, according to the foregoing authors, increased losses and caused a cessation of growth.

In the winter rearing of fingerlings in fish enclosures in the warm-waters of hydroelectric power stations favorable results were obtained in the experiments of the All-Union Research Institute of Pond Fisheries (Moscow District) with the following feed composition (%): formula feed 40, sunflower cake 20, soybean meal 10, linseed cake 10, fish meal 10, phosphatides 5, and yeast 5. Dokukina (1969) has recommended the practice of feeding in accordance with food consumption (2% of body weight) at a temperature of 7 to 10°C,[13] as well as the enrichment of the feed mixture with all the essential nutritive substances, including vitamins.

Stocking during hibernation is permissible up to 1,000 fish/m^2. With an initial weight of 22 g at the commencement of hibernation and feeding in accordance with food consumption, the weight of experimental fish toward the end of hibernation increased to 36.8 g (by 65%) with an output of 97% of the stocked material toward spring. With moderate feeding (1.5% of body weight) weight increased by 33%, and with feeding at the normal level (0.5% of body weight) increased 11%. The output from the latter two feeding programs was 96%.

It is known that the results of further raising depend on the conditions

[13]According to K.N. Dokukina, A.N. Korneev, and L.A. Korneeva, who conducted experiments on hibernation in fish enclosures in cooling reservoirs, winter feeding of fingerlings is more effective at water temperatures of 8 to 12°C.

of fish maintenance during the winter period. In the present case, with further enclosure rearing during the summer period, the group of fish fed according to their consumption rate during winter, achieved a greater weight compared to the other groups.

Thus, under the conditions of maintenance of fingerlings in warm industrial waters and the necessary feeding of fish during spring, the weight of fry does not decrease as happens in the usual wintering ponds; instead it increases significantly and the quality of the stocking material toward spring improves. Here, too, the waste during hibernation decreases notably.

The quality of the ration, similar to rearing replenishing stock and brood fish, is also important in raising two-year fish in fish enclosures.

The best utilization of food by carp in fish enclosures has been recorded for fish weighing from 20 to 700 g.

According to the preliminary recommendations of the Ministry of Fisheries of the USSR and the All-Union Research Institute of Pond Fisheries (1968), the following parameters have been accepted for raising marketable carp in net enclosures in warm-waters:

Ratio of enclosure area to cooling reservoir, 1:1,000.

Density of stocking of carp yearlings, 100 to 200 fish/m^2.

Average weight of stocking material, not below 30 g.

Average weight of marketable carp, 400 to 450 g.

Food coefficient 5.0 to 5.5.

Loss of two-year fish, 15 to 20%.

Frequency of feeding of fish in enclosures 6 to 12 times daily.

Use of substandard stocking material is prohibited.

Norms of growth of two-year carp in enclosures in warm-waters in the central belt: May—1%, June—13%, July—32%, August—38%, and September—16%.

Cage fish culture is also being organized in the USSR. For example, using the waters released from the Berezovsk hydroelectric power station (Brest district), a cage carp farm has been established with 200 metal net cages, 100 pontoons, 37 wooden sections for fixing the fish cages, and other essential equipment (Figure 89).

Cages of three sizes are used: (1) 4 m$\times$2 m$\times$1.5 m; (2) 3.8 m$\times$1.8 m $\times$1.5 m; (3) 3.6 m$\times$1.6 m$\times$1.5 m.

The water for the cooling reservoir of the Berezovsk hydroelectric power station is provided by Lake Beloe, which covers an area of 492 hectares. The cages for rearing fish have been installed in the discharge canal, which is 2.5 km long, 24 m wide, with a rate of flow of 50 cm/sec (Figure 90). The average monthly water temperature in this canal fluctuates from 23.7 to 20.5°C in May, 29.0 to 25.0°C in June, and 19.7 to 17.0°C in October.

Fish cages containing carp yearlings with an average weight of about 30 g were released from May 8 to 25. Upon introduction the fish remained

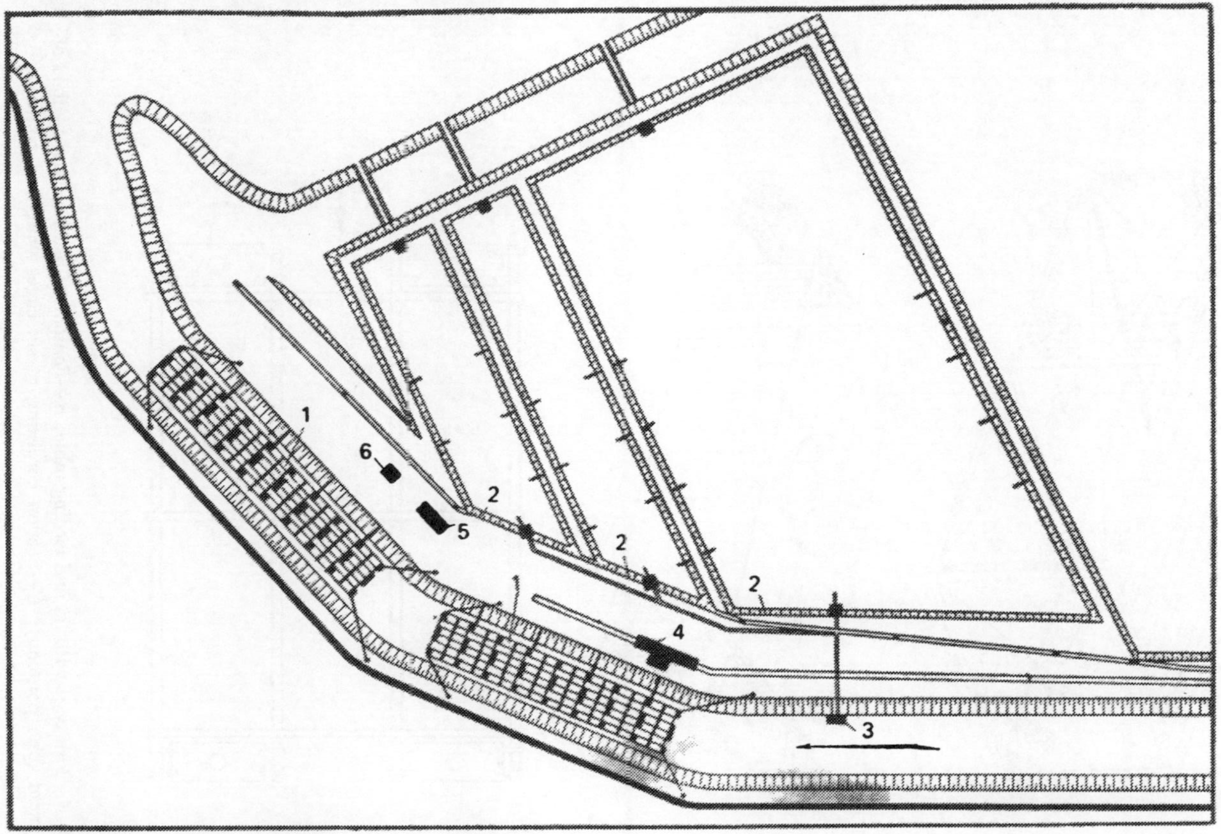

Figure 89. Berezovsk cage-culture farm in Belorussian SSR (diagrammatic).

1—netted fish cages; *2*—earthen pits; *3*—pumping station; *4*—bunker for feeds; *5*—wagon house; *6*—administrative center.

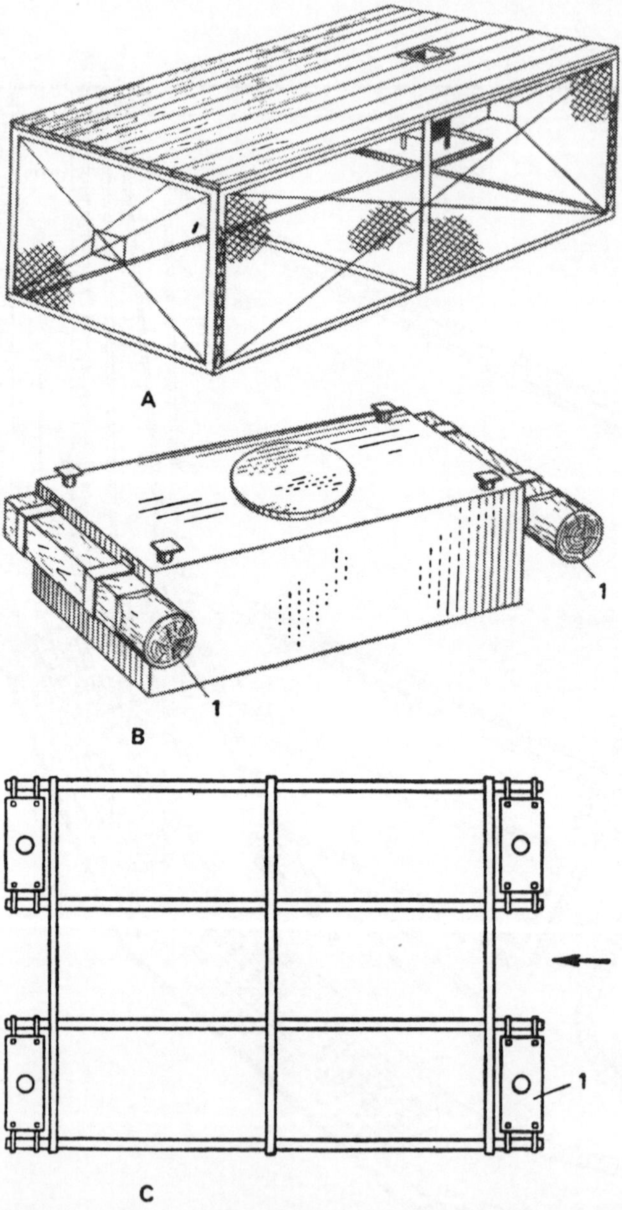

Figure 90. *A*—Fish cage with lid and feeding table; *B*—Pontoon of fish cage and method of its attachment (one pontoon); *C*—Section consisting of six cages with four pontoons.

in the surface layer of the cages and continually nibbled at the sides. Within about two hours they became highly agitated, so much so that they tried to jump out of the cages and consequently injured themselves against the sides and lid of the box. Such injuries under the conditions of high temperatures could lead to saprolegniosis. Restless behavior continued for almost two days, after which the fish gradually calmed down and the majority settled to the bottom of the cage. Only the weak and injured fish remained in the upper part of the cage, together with those infected with fungus. These fish accounted for the significant loss; nonetheless the loss over the entire vegetative growth period (May to October) did not exceed 23.7% (16.5% in May).

The ration consisted of feed pellets weighing 100 to 200 g each. Feeding was done six times a day, commencing at 6:00 a.m., with intervals of three hours up to 9:00 p.m. The feed, distributed on feeding tables, was consumed within 10 to 15 min.

V.P. Lyakhnovich and E.P. Leonenko (BSSR, 1970) conducted an experiment in carp culture in cages in a canal with running water, rich in zooplankton. In spite of a high stocking density (212 fish/m^2) and repeated feeding with specially prepared feed pellets, the gut contents of two-year carp contained more than 30% zooplankton. Selectivity was observed for different species of zooplankton, with preference given to daphnids and cyclopoid copepods. After feedings, selectivity for cyclopoid copepods decreased while that for daphnids increased.

The total weight of zooplankton in the gut contents of carp changed significantly over different months before the supply of feed (25 to 70% of the ration) and 30 minutes after feeding (36.5 to 68.0% of the ration).

The zooplankton population in the upstream and downstream of the cages was three times higher than at the source of the canal.

The consumption of feed over different months was as follows: (%): May—2.8, June—14.8, July—26.8, August—24.8, September—23.1, and October—7.7.

The area of the fish cages was 611.6 m^2 and the average stocking norm 214 fish/m^2 (total 13,700 carp yearlings with an average unit weight of 29.6 g); 347.6 quintals of marketable fish with an average weight of 413 g, ranging in unit weight from 275 to 1,200 g were obtained, and 81% of the total number of fish collected had a weight greater than 400 g (Figure 91).

Further improvements in the biotechniques of raising marketable fish would better the production indexes and reduce fish mortality.

In view of the absence of natural food availability in cages the development of a complete ration is of paramount importance. In netted cages without a forced flow of water and with a density of stocking of 400 fish/m^2, 160 kg fish/m^2 have been obtained. If the yield of fish is taken as 100 kg/m^2, this figure is commendable since the per hectare yield is more than 1,000 tons versus the maximum currently obtained from pond cultures of 20 to 30 and, rarely, 40 to 50 q/ha. It has been demonstrated that food

Figure 91. Selection of larger specimens from fish cages for early marketing (August).

consumption is reduced to half with repeated daily feedings compared to a single feeding, and that the best food consumption occurs with six feeds at a daily norm of 4 to 5% of the fish weight.

To develop cage fisheries in warm-waters, especially in the northern and central regions of the USSR, the following tasks must be done: organization of breeding work for specific carp stocks; rearing of stocking material with a unit weight not below 50 to 80 g; producing early fry, culturing live food items for the same; examining the possibilities for raising marketable underyearlings; developing methods for winter maintenance of fish in cages, establishing frequency of feeding, and standardization of daily rations; determining the optimum conditions in thermal waters during the course of fish culture; and deciding what type of fish cage would be most suitable as well as their optimum arrangement.

Important features here include exclusion of small trash fish from the cages and control of plants, etc. which block the cage netting.

In addition to carp, other warm-water fish are also being raised in fish cages in cooling reservoirs of hydroelectric power stations, in particular silver carp and its hybrids and hybrids of beluga and sterlet.

Cage culture of carp appears quite promising: several problems require further study, however.

With the development of cage culture of fish, pisciculture could become an industrial enterprise.

Utilization of Thermal Waters for Intensive Carp Culture in Ponds. Warm-waters obtained from deep wells as a result of deep boring have a special place in fish culture. In western Siberia (Omsk and Novosibirsk districts) even now a fairly large number of wells exist. The daily water discharge from the 122 wells in the Omsk district is about 100,000 m^3. The water temperature at the place of release is 29 to 30°C and even a little higher. Carp can therefore be raised throughout the year, even during the severe Siberian winters.

Studies on the use of thermal waters for carp culture were initiated at the Omsk Agricultural Institute by A.V. Fedyushin and A.S. Zybin in 1965, and in the Novosibirsk district in 1967 (Novosibirsk Agricultural Institute, G.M. Krivoshchekov, M.V. Timofeeva, and others).

In an experiment conducted in the Omsk district carp yearlings maintained in an unfrozen pond during the three winter months of 1965 (November, 1965—January, 1966), attained an average weight of 354 g, with some specimens weighing almost 500 g. In 1967 to 1968 spawning of carp was effected and fingerlings of a unit weight of 25 g raised; the average weight of marketable two-year fish reached more than 500 g. In 1970, at the experimental fish farm of the Omsk Agricultural Institute constructed on thermal waters, 70 quintals of marketable fish were obtained and 480,000 fingerlings raised. The farm has now created its own replenishing stock of brood fish.

These encouraging results led to the establishment of the first experimental fish farm in the USSR, using thermal waters, at Demonstration Farm No. 2 of the Omsk Agricultural Institute. A deep tubewell was sunk with a discharge of 1,000 m^3/day and a temperature at the outlet of the tubewell of 32°C. The planned capacity of the farm is 1,000 quintals of marketable fish and one million fish for stocking purposes.

Satisfactory results have also been obtained in the Novosibirsk district. Despite unfavorable conditions in the summer of 1967 and with a low level of feeding, two-year carp weighed almost 430 g by September 5. Thus, for this district in western Siberia (and possibly for other districts in this zone) a totally new opportunity has appeared for organizing and developing pisciculture.

An important peculiarity of ponds with thermal water is that higher temperatures can be maintained in them in winter; since the water doesn't freeze better natural aeration can be achieved. Hence feeding can be done in winter avoiding much of the waste occurring in wintering ponds.

It should be kept in mind that thermal waters can be used for round-the-year carp culture in fish cages.

References

Alekseev, N.K. 1962. Uvelichenie kormovoi bazy ryb privlecheniem nasekomykh k prudam (Increase in the Food Base by Attracting Insects to Ponds). Izd. AzNIPRKh, Baku.

Ananichev, A.V. 1959. Pishchevarital'nye fermenty ryb i sezonnaya izmenchivost' ikh aktivnosti (Digestive enzymes of fish and seasonal changes in their activity). *Biokhimiya,* vol. 24, no. 6.

Aronovich, T.M. 1969. Sovremennoe sostoyanie rybovodstva v teplykh vodakh za rubezhom (The present state of warm-water fisheries abroad). Sbornik *Rybovodstvo na Teplykh Vodakh SSSR i za Rubezhom.* Moscow.

Belyaev, V.I. 1969. Karp v sadkakh (Carp Culture in Cages). Minsk.

Bodrova, N.V. and B.V. Krayukhin. 1962. Nekotorye voprosy fiziologii pishchevareniya u ryb (Some Problems in the Physiology of Digestion in Fish). Kiev.

Bogatova, I.B., A.N. Korneev, L.A. Korneeva and L.N. Titareva. 1969. Vyrashchivanie v prudakh molodi karpa ot rannego neresta (Pond-raising of carp fry from early spawning). Sbornik *Rybovodstvo na Teplykh Vodakh SSSR i za Rubezhom.*

Bogoyavlenskaya, M.P. and G.S. Karzinkin. 1956. Nekotorye dannye po izucheniyu kaltseivogo obmena pri pomoshchi radioaktivnogo izotopa Ca^{45} (Some data on calcium metabolism using radioactive isotope Ca^{45}). *Trudy Soveshchaniya po Fiziologii Ryb.* Moscow.

Bottesch, A. 1958. Variatili cantitative si calitative ale grăsimii crapului de cultura in timpul unui an, *Bul. Instit. de Cereeta ri pi Scicole,* vol. 17, no. 3.

Chepik, L.G. 1966. Deyatel'nykh pishchevaritel'nykh fermentov karpa i faktory, vliyayushchie na nee (Active digestive enzymes of carp and factors influencing it). Avtoref. Kand. Diss., Leningrad.

Chitino, P. 1968. Piscicoltura e ittiopatologia, *Riv. Zootechn,* vol. 41, no. 9.

Csigo, F. 1967. Tenyé szpontyok gazdasàgos téli takarmányozása, *Halaszat,* vol. 13, no. 5.

Deeva, T.A. 1961. Ispol'zovanie zelenoi vodnoi rastitel'nosti v kormovykh ratsionakh karpa (Use of Green Aquatic Vegetation in Carp Feed). Izd. VDNKh, Moscow.

Dokukina, K.N. 1969. Rezul'taty opytov po normirovaniyu kormleniya segoletkov karpa vo vremya zimobki v sadkakh na teplykh vodakh (Results of experiments on standardization of feeding carp under-yearlings at the time of hibernation in warm-water fish ponds). Sbornik *Rybovodstov v Teplykh Vodakh SSSR i za Rubezhom.* Moscow.

Dokukina, K.N., A.N. Korneev and L.A. Korneeva. 1969. Rezul'taty opytov zimovke karpov v sadkakh na vodoemakh—okhloditelyakh teplovykh e'lektrostantsii (Results of experiments on hibernation of carp in fish cages placed in water cooling tanks of thermal power stations). Sbornik *po Prudovomu Rybovodstvu.* Izd. VNIRO.

Dyman, V.K. 1958. Teoreticheskie osnovy kormleniya sel'skokhozyaistvennykh zhivotnykh (Theoretical basis of feeding farm animals). *Bull. N.-tekhn. Inform. Mosk. Rybovodno-meliorativnoi Opytnoi Stantsii,* no. 2.

E'rman, E.Z. 1966. K voprosu o kormovoi tsennosti ryaski prudovykh ryb

(On the problem of feed value of duckweed for pond fish). *Trudy VNIIPRKh.* Moscow, vol. 14.

E'rman, E.Z. 1969. Ob azottoberegayushchem e'ffekte uglevodov v karpa (On the nitrogen-preserving effect of carbohydrates in carp). *Vopr. Ikhtiol.,* 9, 4, 5.

E'rman E.Z. 1969. Monosakhar v iskustvenykh kormakh (Monosaccharides in artificial feeds). *Rybovodstvo i Rybolovstvo,* no. 5.

Erokhina, L.V. 1960. Povyshenie e'ffektivnosti kormleniya karpa putem sovershenstvovaniya metoda prigotovleniya kormov (Increasing the efficiency of carp feeding by improving the method of preparing feeds). Sbornik *Rybn. Khoz-vo.* Izd. VNIIPRKh, Moscow.

Fedyushin, A.V., A.S. Zybin and L.D. Zemlyanitsyna. 1969. Opyt zimnego vyrashchivaniya karpov v geotermal'noi vode (iz glubokikh skvazhin Omskoi oblasti) [Experience in winter raising of carp in thermal waters (from deep wells in the Omsk district)]. *Trudy Omsk. S.-kh. In-ta,* vol. 71.

Fedyushin, A.V., A.S. Zybin, L.D. Zemlyanitsyna and K.S. Rastegaeva. 1969. Razvedenie karpov v geotermal'noi vode iz glubokoi skvazhiny (Carp rearing in thermal waters from deep wells). *Trudy Omsk. S.-kh. In-ta,* vol. 77.

Gerasimova, T.D. 1967. Ekonomit' belkovy korm (Economizing on protein feeds). *Rybovodstvo i Rybolovstvo,* no. 4.

Grib, V.K. and V.P. Lyakhnovich. 1968. Sravnitel'naya e'ffektivnost' kormleniya dvukhletkov karpov briketirovannym, granulirovannym i testoobraznym kormom (Relative efficiency of feeding two-year carp with pelletised, granulated, and pasty feeds). Sbornik *Prudovoe Rybn. Khoz-vo i Razdevenie Ugrya.* Riza, vol. 1, pt. 2.

Gribanov, L.V., A.N. Korneev et al. 1967. Nekotorye voprosy biotekhniki i kormleniya karpa pri sadkhovom vyrashchivanii an termal'nykh vodakh (Some problems in the biotechnique and feeding of carp in the course of their culture in cages in thermal waters). *Trudy VNIIPRKh.* Moscow, vol. 15.

Gridnev, A.T. 1969. O vyrashchivanii touarnogo karpa v prudakh s protochnoi vodoi v Yaponii (Raising marketable carp in fresh-water ponds in Japan). *Mirovoe Rybolovstvo,* no. 2.

Gruch, W. 1955. Vorläufige mitteilung über fütterungsversuche mit antibiotika an karpfen, *Archiv f. Fischereiwissenschaft,* vol. 6.

Havelka, I. and F. Volf. 1966. Sledováni zdravotniho stavu ryb ve zhutšěnych obsàdkàch, *Živočišna Vŷroba,* vol. 11, no. 9.

Hirasi, A., J. Masuda, T. Mimura, T. Saito and A. Komo. 1968. Requirement of young carp for vitamin A. *Bull. Japan. Soc. Scient. Fish.,* vol. 34, no. 10.

Iančarik, A. 1964. Die verdanung der hauptnährstoffe beim karpfen, *Z. Fischerei,* vol. 12.

Ianeček, V. 1966. Pruzkum efextivnosti zhuštěných obsádekv chovuu kapra a jejich vliv na rybniční prostředi, *Živočišna Výroba,* vol. 11, no. 9.

Jerásek, J. 1967. Odchow kapriho pludku přisvýšených obsádkach. *Acta Vniv. Agricult.,* vol. 15, no. 4.

Karzinkin, G.S. 1962. Ispol'zovanie radioaktivnykh izotopov v rybnom khozyaistve (Use of Radioactive Isotopes in Fisheries). Pishchepromizdat, Moscow.

Karzinkin, G.S. 1967. Sbornik Obmen veshchestv i biokhimiya ryb (Metabolism and Biochemistry of Fish). Nauka, Moscow.

Karzinkin, G.S. 1971. Metodicheskie osnovy kormleniya karpov (Methodical Basis of Carp Feeding). Izd. VASKhNIL, Moscow.

Korneev, A.N. 1969. Opyt sadkovogo vyrashchivaniya v subtermal'nykh vodoemakh (Experiments on cage culture of carp in subthermal waters). Sbornik *Rybovodstvo v Teplykh Vodakh SSSR i za Rubezhom.* Moscow.

Korneev, A.N. 1969. E'ffektivnost' granulirovannykh kormov pri vyrashchivanii tovarnogo karpa v sadkakh na subtermal'nykh vodoemakh (Efficiency of granulated feeds during culture of marketable carp in fish cages in subthermal waters). Sbornik *Rybovodstvo na Teplykh Vodakh SSSR i za Rubezhom.* Moscow.

Korneeva, L.A., L.N. Titareva and A.N. Korneev. 1969. Rost segoletkov karpa razlichnykh genotipov pri vyrashchivanii v sadkakh na teplykh vodakh (Growth of under-yearlings of different genotypes during their culture in fish cages in warm-waters). Sbornik *Rybovodstvo na Teplykh Vodakh SSSR i za Rubezhom.* Moscow.

Krayukhin, B.V. 1959. Vliyanie odnoobraznogo korma na pishchevaritel'nuyu sistemu i obmen u karpa (Influence of same type of food on the digestive system and metabolism in carp). *Zool. Zhurn.,* vol. 38.

Krayukhin, B.V. 1963. Fiziologiya pishchevareniya presnovodnykh kostistykh ryb (Physiology of Digestion in Fresh-water Teleosts). Izd-vo Akad. Nauk SSSR, Moscow.

Krivoshchekov, G.M. and M.V. Timofeeva. 1968. K voprosu ispol'zovaniya termal'nykh vod dlya tselei rybovodstva (On the problem of utilization of thermal waters for the purpose of fish culture). *Puti Uvelicheniya Produktov Zhivotnovodstva Sibiri.* Novosibirsk.

Kudryashova, Yu.V. 1961. Vliyanie razlichnykh kormov na rost i razvitie karpa (Effect of different feeds on the growth and development of carp). *Dokl. TSKhA,* no. 61.

Kudryashova, Yu.V. 1961. Vliyanie uslovii vyrashchivaniya segoletkov karpa na ikh zimovku (Effect of raising conditions of carp under-yearlings on their hibernation). *Dokl. TSKhA,* no. 69.

Lakshmannan, M.A.V., D.S. Murty, K.K. Pillai and S.C. Banerjee. 1967. On a new artificial feed for carp fry, *FAO Fish. Repts.,* vol. 3, no 4.

Lavrovskaya, N. 1966. Vitaminnye vodorosli (Vitamin-rich algae). *Rybovodstvo i Rybolovstvo,* no. 4.

Lider, V. 1965. Das eiweiss in der nahrung der karpfen, *Dtsch. Fisch. Ztg.,* vol. 12, no. 1.

Lobacheva, L.L. 1959. Izmenenie v vode pitatel'noi tsennosti iskusstvenno vnosimykh kormov, upotreblyaemykh dlya kormleniya ruby (Changes in the nutritional value of artificial feeds added to water for fish feeding). *Rybnoe Khoz-vo,* no. 5.

Logvinova, V. and E. Lysenko. 1968. Vodostoikie granuly (Water-stable granules). *Rybovodstvo i Rybolovstvo,* no. 1.

Lovern, D.A. 1953. Khimiya zhirov i zhirovoi obmen ryb (The chemistry of fats and fat metabolism in fish). Sbornik *Biokhimiya Ryb.* II, Moscow.

Malikova, K.M. 1958. Biokhimicheskii sostov kormovykh bespozvonochnykh (Biochemical composition of invertebrates as food). *Trudy Soveshchaniya po Fiziologii Ryb.* Moscow.

Martyshev, F.G. 1954. Biotekhnika prudovogo rybovodstva (Biotechnique of Pond Fisheries). Sovetskaya Nauka, Moscow.

Martyshev, F.G. 1968. Bol'she ostorozhnosti (Be more careful!). *Rybovodstvo i Rybolovstvo,* no. 6.

Maslova, N.I. 1968. Nakoplenie zhira v karpov pri raznykh usloviyakh vyrashchivaniya (Fat accumulation in carp under different culture condition). *Izv. TSKhA,* no. 3.

Maslova, N.I. 1968. Vliyanie razlichnykh uslovii vyrashchivaniya na vodno-solevoi obmen karpov (Effect of different culture conditions on the water-salt exchange in carp). *Izv. TSKhA,* no. 6.

Maslova, N.I. 1968. Nekotorye pokazateli fosforno-kaltsievogo obmena tela dvukhletkov karpa, vyrashchennykh v prudakh na torfyanykh kar'erakh nizinnogo tipa bolot (Some indices of the phosphorus-calcium metabolism in the body of two-year carp raised in ponds in peat quarries of lowland (bogs). *Dokl. TSKhA,* no. 141.

Melikova, L.A. 1969. Soderzhanie aminokislot—triptofana i oksiprolina v belkakh nekotorykh ryb Volgo-Kaspiiskogo basseina (The quantity of amino acids—tryptophan and oxyproline—in the proteins of some fish of the Volga-Caspian basin). *Rybn. Khoz-vo,* no. 5.

Mikheev, P.V. and E.V. Meisner. 1961. Privlechenie zhivogo kozma ryb s pomoshch'yu sveta (Attracting live food for fish with the help of light). *Trudy VNIIPRKh.* Moscow, vol. 10.

Mikheev, P.V. and E.V. Meisner. 1962. Vyrashchivanie ryby pri stoilovom soderzhanii v setnykh plavuchikh sadkakh (Fish raising in floating net enclosures). *Voprosy Prudovogo Rybovodstva. Trudy VNIIPRKh.* Moscow, vol. 11.

Mikheev, P.V., E.V. Meisner and V.P. Mikheev. 1965. O kormlenii i povedenii ryb v sadkakh (Feeding and behavior of fish in fish cages). *Rybovodstvo i Rybolovstvo,* no. 3.

Mikheev, P.V., E.V. Meisner and V.P. Mikheev. 1967. Sadkovoe rybovodnoe khozyaistvo na vodokhranilishchakh (Fish Culture in Fish Enclosures in Reservoirs). Pishchevaya Promyshlennost', Moscow.

Mitterstiller, J. 1966. Abraknemück resrelges potlasa, *Halasrat,* vol. 12, no. 3.

Müller, W. 1956. Über die perspektiwen des antibiotika—verfütterung in der teichwirtschaft, *Deutsche Fischerei Zeitung,* vol. 3, no. 4.

Mukhina, R.I. 1964. Znachenie aminokislotnogo sostava kormov dlya karpa (Significance of amino acid composition of feeds for carp). Sbornik *N.-tekhn. Inform. VNIRO,* no. 9.

Mukhina, R.I. and V.I. Koroleva. 1964. Aminokislotnyi sostav segoletkov karpa (Amino acid composition of under-yearlings of carp). Sbornik *N. tekhn. Inform. VNIRO,* no. 7.

Nering, D. 1965. Die ausnutzung verschiedener getreidearten und hülsenfrüchte durch karpfen, *Z. Fischerei,* vol. 13, no. 4.

Nikulesku, M. 1966. Ispol'zovanie granulirovannykh kombinirovannykh kormov dlya karpa v rybovodnykh predpriyatiyakh Rumynii (Use of granulated formula feeds for carp in the fish-rearing farms of Rumania). *Materialy VII Sessii Smeshannoi Komissii po Primeneniyu Soglasheniya o Rybovodstve v Vodakh Dunaya.* Kiev.

Polyakov, G.D. 1958. Istoshchenia kak odna iz prichin gibeli segoletkov karpa zimoi (Emaciation as one reason for the mortality of under-yearlings in winter). *Trudy Soveshchaniya po Fiziologii Ryb.* Akad. Nauk SSSR, no. 8.

Popov, I.S. 1965. Aminokislotnyi sostav kormov (Amino Acid Composition of feeds). *Izd. TSKhA,* Moscow.

Prosyanyi, V.S. and B.A. Nikolyuk. 1965. Opyt basseinovogo otkorma ryb v Donrybokaombimate (Experiments on feeding of fish in tanks at the Don fish farm). *Rybnoe Khoz-vo,* no. 8.

Rodina, A.G. 1966. Kormovoe znachenie i stroenie detrita (The feed value and composition of detritus). *Biologicheskie Resursy Vodoemov, Puti ikh Rekonstruktsii i Ispol'zovaniya.* Izd-vo Akad. Nauk SSSR, Moscow.

Sabotash, V.M. 1970. Dinamika soderzhaniya tsinka v karpe na rannikh e'tapakh ego razvitiya (Dynamics of zinc content in carp in the early stages of its development). *Gidrobiol. Zhurn.,* vol. 6, no. 3.

Schäperklaus, W. 1965. Versuchte mit verfütterung von spurenellementen an karpfen, *Dtsch. Fisch. Ztg.,* vol. 12, no. 6.

Sergeev, Yu.S. 1967. Metodika otsenki intensivnosti obmena ryb, ikh ratsionov i koe'ffitsienta ispol'zovaniya imi pishchi (A method of evaluation of the intensity of metabolism in fish, their rations, and coefficient of food utilization). Sbornik *N.-tekhn. Inform. VNIRO,* no. 1.

Shabalina, A.A. 1968. K voprosu o primenenii khloristogo kobal'ta v prudovom rybovodstve (On the problem of using cobalt chloride in pond fisheries). Sbornik *Rybokhozyaistvennoe Izuchenie Vnutrennikh Vodoemov.* Leningrad.

Shcherbina, M.A. 1962. Soderzhanie aminokislot v syrom proteine prudovykh ryb i nekotorykh kormov (Amino acid content of proteins from pond fish and some feeds). *Voprosy Prudovogo Rybovodstva. Trudy VNIIPRKh.* Moscow, vol. 11.

Shcherbina, M.A. 1969. Dostupnost' aminokislot iskusstvennogo korma prudovym rybam (Availability of amino acids from artificial feed to pond fish). *Vopr. Ikhtiol.,* 9, 6, 59.

Shcherbina, M.A. and K.F. Sorvachev. 1967. Rezorbtsiya aminokislot is kusstvennykh kormov v protsesse prodvizheniya pishchi po kishechniku karpov (Resorption of amino acids from artificial feeds in the process of movement of food along the intestine of carp). Sbornik *Obmen Veshchestv i Biokhimiya Ryb.* Nauka, Moscow.

Shcherbina, M.A. and K.F. Sorachev. 1969. Nekotorye dannye o vsasyvanii, aminokislot v pishchevaritel'nom trakte dvukhletnikh karpov (Some data on absorption of amino acids in the digestive tract of two-year carp). Sbornik *Vopr. Prudovogo Rybovodstva.* Moscow, vol. 16.

Shcherbina, M.A. and O.P. Surina. 1970. Osobennosti vsasyvaniya osnovnykh pitatel'nykh veshchestv v pishchevaritel'nom trakte karpov (Peculiarities of absorption of major nutritional substances in the digestive tract of carp). *II Vses. Biokhimicheskii S''ezd. Tezisy Sektsionnykh Soobshchenii.* Moscow.

Shcherbina, M.A., V.F. Mochul'skaya and E.L. E'rman. 1970. Izuchenie perevarimosti pitatel'nykh veshchestv iskusstvennykh kormov prudovymi rybami. Soobshch. 1. Perevarimost' pitatel'nykh veshchestv arakhisovogo shrota, gorokha, yachmenya, kormovoi smesi dvukhletnim karpam (Studies on the digestibility of nutritional substances of artificial feeds by pond fish. 1. The digestibility of the nutrients in groundnut meal, pea, and barley feed mixture in two-year carp). *Vopr. Ikhtiol.,* 10, 5, 64.

Shekhanova, I.A. 1959. Izuchenie fosfornogo obmena u molodi karpovykh i osetrovykh ryb s primeneiem radioaktivnogo fosfora (Study of phosphorous metabolism in juvenile carp with the use of radioactive phosphorus). *Trudy VNIRO,* Moscow.

Shpet, G.I. 1965. Izuchenie aminokislotnogo sostava kormov i trebovaniya k nemu v karpovodstve (Study of amino acid composition of feeds and its requirements in carp culture). Sbornik *N.-tekhn. Inform. VNIRO,* no. 4.

Shpet, G.I. and M.B. Feldman. 1962. Vliyanie iskusstvennykh kormov i produktov vydeleniya ryb na kislorodnyi rezhim prudov (Effect of artificial feeds and fish excreta on the oxygen regime of ponds). *Trudy Soveshchaniya Ikhtiologicheskoi Komissii Akad. Nauk SSSR.* Moscow, no. 14.

Steffens, W. 1969. Karpfenfutterungsversuche mit trockenfutter unterschiedlichen eiweßgehalt, *Z. Fisch.*, vol. 17.

Steffens, W. 1969. Der vitaminbedart des karpfens, *Dtsch. Fisch. Ztg.*, vol. 16, no. 5.

Strel'nikov, S.I. 1966. E'kologiya vyrashchivaniya stolovykh segoletkov karpa v prudakh Kubani (Ecology of raising carp under-yearlings for table purposes in the ponds of Kuban). *Trudy Azovskogo In-ta Rybn. Khoz-va,* no. 8.

Sukhoverkhov, F.M. and F. Dikushnikova. 1968. V zimovali—zhiznestoikikh segoletkov (Taking viable under-yearlings to the wintering ponds). *Rybovodstvo i Rybolovstvo,* no. 5.

Szumies, J. 1966. Vdtzial pokarmu naturalnego przy zywie niu karpi, *Acta Hydrobiol.*, vol. 8, Suppl. 1.

Titareva, L.N., A.N. Korneev and L.A. Korneeva. 1969. Rannee poluchenie mododi zavodskim metodom s ispol'zovaniem teplykh promyshlennykh vod (Early production of carp fry by the commercial method with the use of warm industrial waters). Sbornik *po Prudovomu Rybovodstvu.* VNIIPRKh, Moscow.

Tsirkova, M.N. 1967. Usovershenstvovanie iskusstvennykh kormov dlya ryb (Perfecting artificial feeds for fish). Sbornik *N.-tekhn. Inform. VNIRO,* no. 4.

Tsypina, A. 1969. Opyt vyrashchivaniya karpa v termal'nykh vodakh Berezovskoi GRES (Experiments on carp raising in the thermal waters of the Berezovsk Electric Power Station). Sbornik *Rybovodstvo na Teplykh Vodakh SSSR i za Rubezhom.* Moscow.

Vimba, B.Ya. 1964. Sapropeli—vitaminno-mineralnaya podkormka (Sapropel as vitamin-mineral nourishment). *Vestnik S.-kh. Nauki,* no. 12.

Vladimirov, V.I. 1969. Zavisimost' e'mbrional'nogo razvitiya i zhiznestoikosti karpa ot mikroe'lementa tsinka (Relationship between embryonic development and viability of carp and zinc). *Voprosy Ikhtiol.,* vol. 9, no. 5.

Wolny, P. 1959. Winiki zynickia karpi sorgien, *Gosp. Rybna,* no. 11.

Chapter XII

MANURING OF FISH PONDS

The primary purpose of manuring ponds is to create conditions which facilitate an increase in the stocks of good quality natural foods, thereby increasing fish production. An increase in fish production is associated with density of stocking.

Theoretical problems of a complex nature must be solved before manuring of fish ponds is undertaken. In ponds the complex process of conversion of assimilable chemical biogenic substances in the presence of carbon dioxide, water, light, and heat, contributes to the organic matter of green plants, plankton, and benthos, which are consumed by fish. Academician S.A. Zernov wrote: "Ultimately the growth of fish is the result of this system: bottom⇌water and aquatic⇌pond organisms." The complexity of the problem of manuring of ponds is contained in that statement. The definite chemical interaction between the water and the bottom of the pond depends on a number of factors: temperature, photosynthesis, nature of the soil, etc. This interrelationship is distributed as a result of fertilizer application, since the quantitative interactions between the organisms and its life factors lead to qualitative differences in the medium to which fish invariably respond. The processes taking place in ponds under the influence of manures cannot be examined in isolation or as static factors, therefore, but rather as dynamic interrelated phenomena.

Compared to agriculture, where manures are applied directly to the soil, through which plants obtain their nutrients, in pond fisheries the effect of manures is brought about through a longer chain of conversion: soil→water column→fertilization→bacteria→aquatic plants→zooplankton and zoobenthos→fish. The initial links in this chain are bacteria and algae. The manures applied also have an influence on the extent of development of the bacteria found in the soil as well as in the water. The presence of bacteria in ponds depends on their condition, zonal peculiarities, and methods of pond management. Significant fluctuations in bacteria occur in ponds in relation to the application of chemical and organic manures; the total population of microorganisms in ponds considerably increases.

The average bacterioplankton count in manured carp ponds in the European part of the USSR, according to Akimov (1966), is 3.18 million/ml.

A gradual increase in the number of bacteria occurs from June to August and a decline in September.

Bacteria play an important role as a direct source of nutrition for zooplankton and zoobenthos in the conversion processes. As revealed by Rodina (1966), detritus is a complete food for cladocerans, larvae of *Chironomus plumosus,* mollusks, and oligochaetes. It contains a complex of rich bacterial flora which serve as a source of proteinic food, vitamins, and biostimulants for animals.

Chemical fertilization (in particular, phosphorus and calcium), while promoting an increase in the quantity of food in the above-mentioned chain, can also be used by fish directly through their skin, scales, and gills (see below).

Manures used in pond fisheries are divided into chemical fertilizers (phosphate, nitrogen, calcium, and potash), organic manures (dung, liquid manure prepared from dung, green manure, etc.) and organic-chemical manures.

Chemical fertilizers contain a known quantity of biogenic elements; hence their quantity and effect on productivity can be determined fairly precisely. The composition of organic manures is highly variable and complex, and depends on the animal species cultured, the nature of its feeding, etc. The excreta, urine, and droppings present in dung (compost), for example, contain complex organic substances which have some effect on the biochemical and physical processes, complicating the direct action of these manures. The role of dung in soil is especially important; it enriches the soil with humus and improves its structure and buffer properties, and the effectiveness of chemical fertilizers. Manures applied to fish ponds also act indirectly by creating conditions for the mass-scale development of microorganisms—an important link in the food chain.

The effectiveness of manuring depends on many factors: water temperature, pH, oxygen (and in general gas) regime, soil of the pond (structure and silt deposition), and technical state of the pond.

Water temperature influences the effect of manures quite differently and hence the activity of aquatic organisms (flora and fauna), including microorganisms, each species of which develops under a definite temperature range. For most aquatic microorganisms, under the conditions of the central belt of the USSR, the optimum temperature is 22 to 25°C, and for the southern regions 25 to 30°C. Any deviation up or down in temperature retards or even stops the development of microorganisms; however, they can withstand low temperatures better, while an increase in temperature causes mortality. Therefore, the periods of application of organic manures are determined in accordance with temperature conditions.

The effectiveness of manuring with a favorable reaction of the medium (neutral or somewhat basic) and good oxygen regime. The optimum pH range for the development of nitrifying bacteria is 6.5 to 8.0, although these bacteria can also develop at a pH level ranging from 4.0 to 13.0. Different species of *Azotobacter* are more demanding and develop only in the pH range of 6.4 to 8.8. A sufficient level of water with a good oxygen content and the required aeration of soil not only facilitate normal activity in fish, but also create conditions for the development of beneficial bacteria—nitrogen-fixing (*Azotobacter*) and nitrifying (*Nitrosomonas* and *Nitrobacter*). Some aerobic bacteria utilize not only the free oxygen of the air, but also the bound oxygen of chemical substances; they also influence the oxygen content in the water.

Quite a large quantity of oxygen is consumed by *Azotobacter*; this is noticeable with a strong reduction in the oxygen content of water and with an oversaturation of water with oxygen.

When aerobic bacteria develop, the oxygen consumption may increase to such an extent that depletion occurs and conditions develop which encourage harmful anaerobic processes: decomposition of assimilated nitrogen compounds and degradation of carbohydrates from the organic matter of the soil (methane fermentation).

The effect of fertilizers on the soil is caused not so much by the initial chemical composition and soil structure as by the secondary formation of systematically settling silt. This silt differs from the initial soil by a high absorption capacity, which is significantly greater than the capacity of the best cultivated soils and more so of sandy soils. Chemical fertilizers do not produce a similar effect on the population of bacteria in different soils. Bacteria develop more vigorously on soddy-loamy soils than on sandy and heavy clayey soils (however, clayey soils with remnants of roots are rich in bacteria). Bacteria occur in greater numbers in pond soil than in the water. The depth of penetration of bacteria into the soil depends on the nature of the pond and conditions of manure application. In unmanured ponds in Latvian SSR and Krasnodar *Azotobacter* was found at a depth of 30 cm, while in manured ponds it did not always penetrate to such depths. This shows the significance of the conditions under which manures are applied.

Chemical fertilizers applied to pond soil are initially absorbed by the soil and then gradually released. They increase the fish productivity of not only the current season but of subsequent seasons. However, in the first year the action of manuring is more effective. On heavy soils absorption is more intense and the effect as a result of manuring greater.

The practical application of manures should be based on a prior determination of the need for requisite nutrients in a pond; this requirement may change even within the course of a single vegetative growth period. These changes, similar to the need for biogenic elements, should be systematically assessed to ensure a rational utilization of manures.

In an experiment two-year carp were reared under dense stocking and manure applied in accordance with biological needs; its consumption, compared to the recommended norms of application of nitrogen fertilizers (niter), was reduced by 3.2 to 13.3 times (Lavrovskaya, 1967). Thus, before the application of manures it is essential to have data on the requirements of a given pond for biogenic elements and changes in the course of vegetative growth.

A thorough preparation of the ponds helps to create favorable conditions in the external environment for fish and food organisms. The use of manures is inadvisable in unprepared ponds. Ponds intensively used for feeding ducks should be manured. Ponds with stagnant or mildly flowing waters may be effectively manured; in those with a strong current the manure is washed away with the water and does not yield the desired impact. The application of manure is also senseless in ponds which have heavy percolating soils, for example, sandy soils, since the applied substances are leached through the dikes and the bottom into deeper unproductive soil layers. It is better to apply manures to ponds located on nonseeping grounds. Ponds put to use for the first time should not be manured since their natural fish productivity has to be known in order to calculate the efficiency of manures applied later.

CHEMICAL FERTILIZERS

Phosphate Fertilizers. Phosphorus is necessary for skeleton formation, muscular and neural activity, and as a constituent of blood plasma. About one-third of the fish skeleton consists of phosphorus, in the form of calcium phosphate $Ca_3(PO_4)_2$ (tricalcium phosphate). Phosphorus also enters into the composition of complex proteins, many fatty substances, and carbohydrates. Phosphorus is essential for phytoplankton and bacterioplankton for the formation of their cells; its deficiency has an adverse effect on their development. A deficiency in phosphorus, especially in the initial stage of postembryonic development, likewise has an adverse effect on fish growth, which cannot be compensated with the application of phosphorus in subsequent stages of fish culture.

Under certain conditions phosphorus fertilizers stimulate the development of nitrogen-fixing bacteria (*Azotobacter* and *Clostridium pasteurianum*). However, in ponds with acidic peat soils, or deficient in certain salts or essential trace elements, the application of phosphate fertilizers does not create the conditions requisite for *Azotobacter*, which are very sensitive to the environment. These fertilizers stimulate the growth of *Azotobacter* only if other conditions are favorable (oxygen regime, pH, composition of mineral salts); growth is retarded with a deficiency of phosphates and organic substances. Some phosphate fertilizers stimulate nitrification poorly (oxidation of ammonia into nitrous and nitric oxides). In ponds, nitrification leads to the formation of the nitrates required as nutrients by phytoplankton, and mobilizes phosphates through the action of the acids formed by bacteria on the phosphatic salts (which are less soluble), resulting in the improved growth of aquatic plants. The conditions for the growth of nitrifying bacteria under different kinds of manuring and their distribution in ponds with different soil-climatic and hydrochemical properties have not been studied well to date.

In Belorussian SSR phosphate fertilizers used alone did not exert an influence on the plankton in the ponds. At fish farms in Germany the application of 4 quintals of 16.8% superphosphate per hectare increased fish productivity by 32% (1 kg of carp for every one kg of P_2O_5 applied to the ponds). A slightly greater yield was obtained when calcium was combined with the application of phosphate fertilizer.

The most common phosphate fertilizer applied to fish ponds is the simple superphosphate obtained by treating natural phosphate (apatite) with sulfuric acid. Its value as a fertilizer is determined by the quantity of total P_2O_5 present (the total of water soluble and nitrate soluble P_2O_5) which, on the average, varies from 15 to 20%. Precipitate and phosphorite powder which are not soluble in water, but only in mild acids, are used as fertilizers.

Ponds are applied with phosphorite powder when they have acidic soils where the tricalcium phosphate is changed into dicalcium or monocalcium phosphate. The precipitate is better applied to ponds with an acidic medium.

In addition to simple superphosphate, more concentrated double superphosphate is also used which contains 38 to 50% P_2O_5, mainly in the form of water-soluble monocalcium phosphate. Thomas slag can also serve as a phosphate fertilizer in ponds if it contains not less than 14% P_2O_5. Thomas slag, similar to the precipitate and phosphorite powder, is more useful in ponds with an acidic medium and in ponds located on light soils. Besides phosphoric acid, Thomas slag contains calcium oxide (48%), manganese oxide (13%), silicon oxide, ferric oxide, and sulfuric oxide.

Phosphate fertilizers applied to ponds are absorbed by the soil and often converted into insoluble compounds. Microorganisms change these phosphorus compounds into assmilable forms. This produces the beneficial effect of phosphate fertilizers; their efficiency in the second year of application depends on the conditions of the aquatic medium and the pond soils.

The approximate norm of application of phosphate fertilizers varies from 15 to 20 kg P_2O_5 per hectare. Sometimes beneficial results are obtained by higher doses (30 to 35 kg P_2O_5 per hectare). The growth rate of carp in ponds treated with phosphoric acid is approximately 0.44 to 1.22 kg of fish weight per kg of phosphoric acid added.

According to Müller (1969) in the German Democratic Republic phosphate fertilizers are usually applied at the rate of 35 kg P_2O_5 per hectare of effective pond area. A correct use of phosphate fertilizers increases the nitrogen content of phytoplankton, since it facilitates the fixation of water-soluble nitrogen by nitrogen-fixing bacteria and some blue-green algae.

As revealed by Soviet and foreign experience, the best effect is obtained by combining phosphate fertilizers with others, notably nitrogen and potash fertilizers. Nitrogen and phosphate fertilizers are used more often because their combined application enhances the effect of each.

The following norms of phosphate (superphosphate) treatment of carp ponds abroad have been reported: in France—17 to 150 kg per hectare, Holland—300 to 400 kg per hectare, Yugoslavia—400 kg per hectare, and central Europe—300 to 400 kg per hectare.

NITROGEN FERTILIZERS, as many years of experience have revealed, increase fish productivity in ponds and reduce the quantity of feed required. In nature, nitrogen occurs predominantly in a free state, and the compounds of nitrogen in the form of nitrates, nitrites, and ammonium salts are used in ponds by green plants and microorganisms which, in turn, serve as food for aquatic invertebrates which, in turn, are consumed by fish. Free molecular nitrogen is used by only a few organisms, including nitrogen-fixing bacteria and a few forms of algae. Most microorganisms, algae, and higher plants need fixed chemical compounds of nitrogen. Ponds can be supplemented with these important biogenic compounds either through the conversion of free molecular nitrogen into fixed nitrogen or through their application in the form of fertilizers.

However, the application of nitrogen fertilizers alone suppresses the activity of nitrogen-fixing bacteria. According to A.G. Rodina ammonium salts inhibit the growth of *Azotobacter* and nitrogen in the form of nitrate, if the medium contains nitrates, leads to the stoppage of nitrogen fixation by *Azotobacter*. The harmful effects of some nitrogen fertilizers may not occur if the phytoplankton development is fairly vigorous under the influence of applied fertilizers. In this case the quantity of chemical compounds of nitrogen is reduced at such a rapid rate that they cannot exert a suppressing effect on nitrogen-fixing bacteria.

The distribution of *Azotobacter* and anaerobic nitrogen-fixing bacteria (*Clostridium pasteurianum*) in ponds located in different climatic zones is far from similar; it depends on soil, nature of pond, and manures applied. This aspect of pond fish culture has been poorly studied. *Azotobacter* may be totally absent in the soil and water of ponds. Favorable conditions are essential for its growth, in particular the organic matter determining the extent of redox potential and pH. *Azotobacter* not only speeds up nitrogen-fixing, but serves as a source of proteins and vitamins for different food organisms (plankton and benthos). It has been demonstrated (Rodina, 1969) that it lives in the water column as well as in the soil and on lower and higher aquatic plants. Conditions are favorable for *Azotobacter* in heated ponds with a good oxygen regime, extremely slow water current, and rich in nutrients. *Azotobacter* also needs calcium ions; it develops poorly in water deficient in calcium (soft waters). Moreover, it also needs potassium, magnesium, iron, and sulfur salts, and among the trace elements—boron and molybdenum. The pH should be in the range of 6.5 to 8.0. An acidic medium is not favorable and at pH 6.0 most *Azotobacter* species are totally suppressed.

Among the anaerobic bacteria, the nitrogen-fixing species is the sporulating microbe *Clostridium pasteurianum,* but it is far less prolific than *Azotobacter. C. pasteurianum* also grows in ponds with a good oxygen regime, in the water as well as in the soil, in symbiosis with aerobic organisms.

Experience has shown that the application of inorganic nitrogen to ponds in the form of niter and ammonium sulfate yields favorable results, especially in ponds where the aftermath of phosphate fertilizers is apparent. The best effect of nitrogen fertilizers is obtained in combination with phosphate and potassium fertilizers. Experiments conducted for three years with ammonium sulfate during the rearing of three-year carp in foraging ponds showed that at the rate of 960 kg/ha (200 kg/ha of pure nitrogen), the yield of fish increased by 137 kg/ha, while with a supplementary feeding with rye the yield increased to 174 kg/ha. The food coefficient decreased by 23% (2.3) (Schäperclaus. 1968).

Ammonium nitrate (35% nitrogen), ammonium sulfate (20.5 to 21.0% nitrogen), ammonium water (25% nitrogen), and carbamide (synthetic urea) are commonly used as nitrogen fertilizers.

Liquid nitrogen manuring is much cheaper than solid fertilizers. The rate of application of nitrogen fertilizers is calculated on the basis of maintenance of a concentration of 2 mg pure nitrogen per liter of water.

A combined application of nitrogen and phosphate fertilizers has caused cases of fish poisoning, attributed mainly to the release of free ammonia (NH_3). The formation of free ammonia is facilitated by the active water reaction of pH 7.5 to 11.0. At pH 8.0 mixtures containing ammonia should no longer be applied (Kempinska, 1968). Neither should ammonium fertilizers be applied when aquatic plants are in full bloom or the water temperatures high. To avoid the lethal effects of ammonia of fish during hot summer days at a pH level above 8.5, the quantity of nitrogen fertilizers should be reduced or they should be applied before sunrise or during rains. Ammonium fertilizers are more toxic to carp fry than nitrate fertilizers (Müller, 1969).

Calcium Fertilizers. Calcium is an essential element for all members of aquatic flora and fauna. It is used in the formation of the skeleton in fauna, tissue structure in fish, and influences chemical and physical processes in the pond water and soil by facilitating an improvement in environmental conditions. Calcium causes a precipitation of colloidal humus, reducing its absorption capacity and thereby releasing earlier absorbed nutrients which enrich the water. Under the influence of calcium the process of bacterial activity changes. As a fertilizer calcium is added to ponds rich in organic matter and ponds located in heavy clayey soils, on acidic highland bogs where a sharp deficit of calcium is observed. In ponds poor in organic matter calcium fertilizers should be combined with organic manures.

At high concentrations of calcium and a deficiency in nutrients (for example, phosphorus or potash) lime has an adverse effect; it sterilizes the water. The positive values of calcium have been discussed in Chapter X. As a fertilizer calcium acts in different ways; it is difficult to isolate what percentage is ameliorative and how much serves as a fertilizer.

In the case of a calcium deficiency in ponds peat mosses (*Sphagnum* species), horsetail (*Equisetum*), sedge (*Carex acuta*), and other aquatic plants develop; the presence of water thyme (*Elodea canadensis*) or stonewort (*Chara flagilis*) indicates that the quantity of calcium is sufficient.

Calcium carbonate ($CaCO_3$) and unslaked (burnt) lime (CaO) are used as fertilizers for ponds. The rate of calcium fertilizers depends on the zone, type of soil, and other factors (condition of the pond, quality of water, its acidity, etc.). Usually, with a systematic annual application of calcium fertilizers to ponds, 2 to 4 quintals of lime/ha are recommended; nevertheless the conditions of the pond and other factors must be considered. For improving the quality of water in ponds located on highland bogs, 50 quintals/ha of unslaked lime are applied as an ameliorative measure.

The norms of lime application to ponds as a fertilizer are given in Table 42.

Table 42. Total yearly quantity of lime applied to ponds as a fertilizer (from Il'in, 1955)

Characteristics of pond	Norms of lime application, g
Ponds located in marshy flood land, with acidic soil and water (pH of water 5 to 6, of salt extracts from soil less than 4 to 5)	5–8
Same as above, but with marshy catchment area	10–12
Ponds located in podsolic and slightly marshy area (pH 6.0 to 6.5 and salt extracts from soil 5 to 6)	4–5
Heavily silted ponds with high acidity and transparent water during summer	up to 4
Ponds continuously exploited for more than 20 years with poor silt deposits	up to 3
Ponds located on black soils with mild silt deposits	up to 1
Ponds located on chernozem soils without silt deposits	0.3–0.5
New ponds in which grassy layer has not yet decomposed	0.3–0.5
Ponds fed with water rich in calcium	0.3–0.5

Depending on the particular conditions in a pond, lime is applied as a fertilizer at the rate of 0.3 to 50 q/ha. The finer ground the lime, the stronger its action as a fertilizer. With the application of 10 and 20 quintals of lime per hectare, according to the findings of German researchers, fish productivity is increased by 20 and 35% respectively. In other words, fish productivity is raised by 1 kg for every 3 to 6 kg of lime applied.

In addition to lime, excreta acts well as a calcium fertilizer.

The following norms of application of calcium compounds to carp ponds are followed in European countries: in France—200 kg/ha, Holland—600 to 700 kg/ha, and Yugoslavia—500 kg/ha.

POTASSIUM FERTILIZERS. Potassium as a fertilizer is rather controversial, primarily because its effects on fish productivity have not been properly assessed. In Belorussian SSR a large number of experiments with potassium fertilizers yielded negative results due to the fact that usually a sufficient quantity of potassium was already present in the ponds. Potassium is mostly deficient in ponds located on sandy, podsolic, marshy, and moss-covered soils.

The potassium content of different soils is shown in Table 43.

Table 43. Total quantity of potassium present in major soil types in the USSR

Type of soil	Total potassium quantity in layer 20 cm thick	
	%	Tons/ha
Deep chernozem	2.0–2.5	60–75
Serozem	2.1–2.6	63–78
Chestnut	2.0–2.4	60–72
Gray forest	1.9–2.3	57–69
Podsolic loamy	1.7–2.2	51–66
Podsolic sandy-loamy	0.6–0.8	18–24
Red soils	0.7–0.9	21–37
Peaty	0.02–0.05	0.4–0.8

Chernozem, serozem, and chestnut soils contain large quantities of total potassium, while peaty soils contain little. However, a large quantity of potassium in different soil types does not guarantee its higher content in ponds, since in the soil 98 to 99% of potassium is in the form of stable, alumosilicate minerals, mainly feldspar and partly mica and other stable compounds, which under certain conditions do release calcium into the solution but very slowly and in a negligible amount.

So-called exchangeable potassium and water-soluble potassium are present in soils in extremely small quantities. The quantity of the latter in a cultivated layer of different soils is shown in Table 44.

Aquatic plants serve as an index to the potassium content of ponds. For example, water thyme, arrowhead, and water plantain attest to a high degree of potassium, while horsetail attests to a deficiency. Yellowish-brown leaves on these plants also indicate a potassium deficiency.

Table 44. Approximate quantity of water-soluble potassium in cultivated layer of different soils

Soil type	mg/100 g of soil	kg/ha
Podsolic sandy loam	2–3	60–90
Leached chernozem	4–5	120–150
Solonets	6–7	180–210
Solonchak	7–10	210–300

Potassium is applied to those ponds in which its deficit has been affirmed.

Potassium fertilizers help the growth of soft submergent plants and emergent aquatic plants; the latter is not desirable (see Chapter X). At the same time the suppressing effect of an excessive quantity of potassium on emergent aquatic plants has been observed, especially on species adapted to conditions of potassium deficiency (for example, horsetail); this is a favorable situation. Potassium fertilizers facilitate the development of phytoplankton and hence zooplankton. The influence of this fertilizer on the development of bacterioplankton has not been studied. Practical experience has shown that potassium fertilizers should be applied to hatching and rearing ponds which are poor in its content and in which the fry feed exclusively or mainly on zooplankton.

In pond fisheries in the USSR sylvinite or $KCl \cdot NaCl$ (from Solikamsk mines) is used as a potassium fertilizer.

Kainite—$KCl \cdot MgSO_4 \cdot 3H_2O$ (containing 13% pure potassium), potassium chloride—KCl (containing 54.0 to 57.8% pure potassium), and potassium sulfate—K_2SO_4 (containing 42.0 to 52.7% pure potassium) can be used as fertilizers in ponds. Good results have been obtained on acidic marshy soils by using nepheline (a by-product in the extraction of concentrated apatite) which has alkaline properties.

Ordinary domestic ash can be used after moistening as a fertilizer in ponds in which potassium is present in the form of potassium carbonate salt (K_2CO_3), along with alkaline properties. The potassium, phosphorus, and calcium content of ash after burning of various kinds of fuels is shown in Table 45.

The rate of potassium fertilizer applied to carp ponds varies considerably more than the rate of phosphorus fertilizer—from 30 to 100 kg of potassium per hectare. This is related to the differential potassium content in ponds located on different soils. It is best therefore to sample the effect of small doses, starting from the lowest—approximately 20 kg of potash per hectare.

According to reports in the literature, with a successful use of potassium fertilizers the growth rate of carp varies from 0.23 to 1.47 kg/ha of pure potassium application. The effect of potassium fertilization on increasing

Table 45. Potassium, phosphorus, and calcium content in different kinds of ash (%)

Type of ash	Potassium	Phosphorus	Calcium
Large-leaved trees	10.0	3.5	30.0
Coniferous trees	6.0	2.5	35.0
Rye straw	16.2	4.7	8.5
Wheat straw	13.6	6.4	5.9
Buckwheat straw	35.3	2.5	18.5
Sunflower stocks	36.3	2.5	18.5
Lowland peat	1.0	1.2	20.0
Highland peat	0.3	0.5	6.0
Manure briquettes	11.0	5.0	9.0

fish productivity is also reflected in the second year; however, like phosphate fertilizers, improvement depends on the conditions of the aquatic medium and pond soils.

A combination of potassium fertilizers with nitrogen and phosphate fertilizers yields the maximum results.

COMBINED USE OF CHEMICAL FERTILIZERS. Different chemical fertilizers used in combination are more effective. The development of phytoplankton is more profuse in ponds treated with phosphate, nitrogen, and potassium fertilizers. The development of green algae (*Protococcalex*) under the impact of fertilizers is a symptom of higher productivity of carp ponds. Decaying phytoplankton is important in the development of bacteria.

Different combinations of fertilizers are practiced in pond fisheries: phosphorus and potassium, phosphorus and calcium, phosphorus with potash and calcium, nitrogen and potassium, and nitrogen with phosphorus and potassium. Ponds are prepared before fertilizer application by applying calcium to them. The calcium added in each mixture (with the addition of calcium for the purpose of improvement), forms assimilable compounds with phosphorus in the soil, favorably influences the hydrochemical regime of the ponds, and improves soil structure.

A correct proportion of N : P while applying fertilizers to ponds is important in the most effective utilization of the substances. It varies in different zones in the country and depends on the chemical composition of the water and soil, rotation of farm, stocking density, techniques of farm management, etc. Soviet researchers have proposed two major methods of applying nitrogen and phosphate fertilizers: (a) their application in accordance with biological needs (G.G. Vinberg, V.P. Lyakhnovich, I.V. Baranov, and A.V. Frantsev); and (b) increasing the content of biogenic elements to a definite level (M.B. Feldman, V.S. Prosyanyi, and A.V. Sukhovii). Sometimes the periodic application of fertilizers is practiced irrespective of the quantity of biogenic elements present in the water. Experience from combined application of nitrogen and phosphate fertilizers in the USSR as well as abroad has shown that the optimum ratio of nitrogen to phosphorus is in the range of 4:1 to 8:1, i.e., 4 to 8 unit weights of nitrogen for one unit weight of pure phosphorus. Sometimes this ratio is wider. In Belorussian SSR 1.5 to 2.0 kg of superphosphate and 2.0 to 2.5 ammonium nitrate are used in rearing and foraging ponds for each kg of additional growth. If necessary, one kind of nitrogen fertilizer may be replaced by another using the following formula:

$$x = \frac{A}{C} \times 100,$$

where x = required number;
A = nitrogen content in substitute nitrogen fertilizer (e.g., ammonium nitrate);
C = nitrogen content in equivalent replaced nitrogen fetilizer e.g., ammonium sulfate).

The most effective combination for the conditions of Moldavian SSR is superphosphate and ammonium nitrate. The average optimum norms for ponds in Moldavia are 150 to 200 kg/ha of superphosphate and 75 to 100 kg/ha ammonium nitrate for rearing ponds, and 200 to 300 kg/ha superphosphate and 100 to 150 kg/ha ammonium nitrate for foraging ponds. The German Democratic Republic has reported the best development of zooplankton under a phosphorus: nitrogen ratio of 1:10 (Müller, 1969). Considering that part of the nitrogen required by phytoplankton is obtained from water, 35 to 45 kg/ha P_2O_5 and 200 kg/ha of nitrogen are recommended for the total pond area. This constitutes 200 to 250 kg superphosphate and 1,000 kg of ammonium sulfate or 570 kg of ammonium nitrate. In Poland (Paladino, 1968) nitrogen and phosphate fertilizers are mostly used in ponds, but in those located on peaty and sandy-loamy soils potassium fertilizers are also applied. That quantity of chemical fertilizers is applied which is required to obtain the following quantity of the chemical components (kg/ha):

N	P	K
40	20	10
50	25	12
60	30	15
80	40	20
90	45	20

In applying chemical fertilizers one commences with the lowest quantity, increasing the amount in subsequent years. Ponds are first treated with lime and the chemical fertilizers added two to three weeks later.

For northern European conditions the optimum dose for nitrogen compounds (nitrates) is 350 to 450 kg/ha and 300 to 400 kg/ha of superphosphate, and in the south 700 to 1,000 kg/ha and 600 to 800 kg/ha respectively. This dosage ensures a productivity of carp of 300, 600, and 700 kg/ha. In the USA the most widely used nitrogen, phosphate, and potassium fertilizers are applied in the ratio of 8 : 4 : 2 from 8 to 10 times over the period of vegetative growth. Experience has shown that to obtain one kg of growth in fish weight, it is essential to apply 0.41 to 3.0 kg of ammonium nitrate and 0.65 to 2.33 kg of superphosphate.

In the process of preparation of the fertilizer mixture in the form of a concentrated liquid, which dissolves more readily in water, the most appropriate ratio is from 2 to 8 at a pH level close to neutral (7.0 to 7.4).

Experiments carried out by the Laboratory of Hydrochemistry of the All-Union Research Institute of Pond Fisheries (1967) on carp rearing in ponds using ammonia water as a nitrogen fertilizer (containing about 25% nitrogen in the form of ammonia) together with 15% of superphosphate, established a significant increase in fish productivity compared to the average fish productivity of rearing ponds in the fish farms of the central belt. The fertilizer was applied in accordance with the findings of a hydrochemical analysis at the rate of 0.86 q/ha of ammonia water and 3.35 q/ha of superphosphate (0.5 mg/l). The ammonia was diluted with water in a ratio of 1:20 before application.

Peat-ammonia fertilizer is gaining popularity. It consists of peat treated with a water solution of ammonia, which enriches the peat with nitrogen and also causes a mobilization of nitrogen in the peat. Peat decomposition is accelerated, which helps the transition of nitrogen compounds from an almost nonsoluble to a readily soluble form. The application of peat-ammonia water (0.9% N) with the addition of superphosphate at the rate of 15 kg/ton to the pond bottom and peat-ammonia water (0.3% N) with the addition of superphosphate at the rate of 7 kg/ton to the pond water, followed by the addition of peat-ammonia water (0.4% N) with superphosphate at the rate of 6 kg/ton, has given very favorable results. Other alternate doses of ammonia water of peat have also been used. Experience has shown that with the addition of peat-ammonia fertilizer to the ponds the consumption of ammonia in the waters used in making up the mixture is in terms of nitrogen 2 to 3 times, and in terms of superphosphate, 1.5 to 2.0 times, that of fertilizers using ammonium nitrate and superphosphate.

Thus, in determining the rate of different kinds of chemical fertilizers the requirement for one or the other substance must be predetermined.

The rate of chemical fertilizers is calculated in the following manner. If, for example, the natural fish productivity is 175 kg/ha and it is hoped that by fertilizer application it will be increased three times, i.e., to 525 kg/ha, then for each fraction of the fertilizer applied it will be 350 kg/ha (525 less 175 = 350). Say, for example, superphosphate and ammonium nitrate are applied. Taking the consumption of superphosphate at 1.5 kg and ammonium nitrate at 2.5 kg for each kg of increase in fish weight, then a total quantity of 525 kg/ha superphosphate and 875 kg/ha ammonium nitrate will be required.

The approximate rate of phosphate and nitrogen fertilizers used for calculating the total quantity may vary depending on the zonal requirement, characteristics of the ponds, water supply to the ponds, and level of techniques and methods of farm management.

Studies have recently been carried out on the use of trace elements as fertilizers, in particular cobalt chloride in combination with phosphate and nitrogen fertilizers. With the application of a fertilizer complex consisting of 10 kg/ha cobalt, 12 q/ha ammonium nitrate, and 2 q/ha superphosphate, the biomass of plankton and benthos increased markedly, fish productivity almost doubled, and the vitamin B_{12} and cobalt content in the food organisms and fish was enhanced (leading to an improvement in metabolism and an increased rate of growth). Studies have also shown that cobalt applied to the ponds is absorbed and retained by silt deposits for a long time, and is gradually included in the biological cycle. Attention is currently being given to magnesium, which has a great physiological significance for algae. In Latvian SSR experiments have been conducted on the use of such trace elements as boron and molybdenum, applied in the form of ammonium molybdate and boric acid, in conjunction with other chemical fertilizers (superphosphate and ammonium nitrate) in solution form. The maximum effect was obtained with the addition of boron and molybdenum to chemical fertilizers at the rate of 0.07 g/m^3 and 0.01 g/m^3 respectively. This increased the biomass of zooplankton 1.5 to 2.0 times and natural fish productivity by 20%, and reduced the quantity of chemical fertilizers required for application.

Recommendations for the inclusion of trace elements in fish pond melioration can only be made after undertaking an intensified and integrated study of their influence on the medium and its inhabitants, especially fish reared for human consumption.

Techniques of Fertilizer Application to Ponds. Chemical fertilizers are approximately applied to ponds with water to achieve the maximum effect. They are first dissolved in water in any container and then distributed over the pond surface from a boat. Patches overgrown with aquatic plants, especially coarse species, are not fertilized. It is not appropriate to apply chemical fertilizers to the pond bottom since they are absorbed by the bottom quickly and their nutrients fixed for a long time, and higher plants use them for growth.

In the German Democratic Republic fertilizers are spread over water from special boats which have a slit in the bottom. About 50% of the total quantity of fertilizers is applied in spring after the ponds have been filled and the balance added in small doses during the period of vegetative growth.

The application of combined chemical fertilizers is done in a definite sequence. If calcium forms a part of the mixture (especially in the form of unslaked lime), it is applied first and not less than 14 days before the application of any other fertilizer. Other types of fertilizers are applied simultaneously. The frequency of application is significant Large quantities applied in a single dose can suppress the development of bacteria. Smaller portions applied at small intervals effect the development of bacterioplankton. Furthermore, the natural food base develops more uniformly with a consistent concentration of biogenic substances. Hence fertilizers are applied in small portions throughout the period of vegetative growth. The frequency of fertilizer application depends on the type of fertilizer and the characteristics of the ponds. The effect of chemical (mineral) salts on the population of bacterioplankton is short-lived. Bacteria and phytoplankton quickly utilize mineral salts, creating a need for another portion to be applied.

The period and frequency of fertilizer application should be differentiated according to soil type; chemical and physical properties in the soil determine the differential absorption capacity.

Breaking up the total quantity of nitrogen fertilizers is very beneficial; smaller doses can be applied after every ten days and sometimes more often. This is equally applicable to other kinds of chemical fertilizers, including lime, except when the fertilizer is being applied for pond melioration (when the whole quantity of lime is applied in a single dose and added directly to the bottom of a dried pond). If lime is used as a fertilizer, it is added in fractionated doses at small intervals, which has a beneficial effect on the development of nitrogen-fixing and nitrifying bacteria.

It is also appropriate to use fertilizers in a granulated form. A triple granulated fertilizer—Nitrofoska—is manufactured, which consists of ammonium phosphate (or the precipitate), ammonium nitrate, and potassium chloride (or potassium sulfate). Nitrofoska contains 13 to 17% N, 11 to 30% P_2O_5, and 15 to 26.5% K_2O, and the ratio of $N : P_2O_5 : K_2O$ is determined according to the conditions of a given pond.

Fertilizers increase the natural fish productivity of the pond; stocking of fish is therefore determined in accordance with increased productivity. In calculating the required quantity of chemical fertilizers one first considers the expected norms of the active principle and its percentage in the fertilizer to be applied. If, for example, 20% superphosphate is to be applied to a pond at the rate of 20 kg P_2O_5/ha, the quantity of fertilizer required would be:

$$\frac{20 \times 100}{20} = 100 \text{ kg/h.}$$

The consumption of chemical fertilizers should be conscientiously recorded in a register in order to calculate their efficiency at a later date.

1. Republic...District...Region...Collective/State farm........
2. Serial No. (or name of pond)................................
3. Category of pond (rearing, foraging, etc.)...................
4. Soil of the pond...
5. Depth of silt layer in pond..................................
6. Flow rate of water in pond...................................
7. Fertilizers applied with their respective quantities during the last three years and their effect..................................
8. Chemical fertilizers to be applied in the current year and their components (P_2O_5, K_2O, etc.)..............................
9. Quantity of chemical fertilizers applied in current year over entire area and per hectare.....................................
10. Method of application of chemical fertilizers...............
11. Periods of application of chemical fertilizer...............
12. Increase in fish productivity as a result of using chemical fertilizers (indicate precise name of fertilizer) in kg/ha..................
13. Average unit weight of carp during autumn catches..........
14. Remarks and additional information obtained over vegetative growth period..
15. General conclusions...

Chemical fertilizers should be of good quality and must be stored separately in a suitable place. The fertilizers should be dry, friable, and without clods, and must correspond to published standards in nutrient content. If superphosphate is stored together with unslaked lime or limestone it may be intermixed and under the influence of lime convert into the tricalcium insoluble salt, tricalcium phosphate $Ca_3(PO_4)_2$. Iron salts also have a harmful effect on superphosphate. Their combination with superphosphate produces the almost insoluble salt vivianite, $Fe_3(PO_4)_2 \cdot 8H_2O$.

The conditions of storage of different kinds of chemical fertilizers vary and hence the instructions issued with the product should be followed precisely. The quantity of fertilizers utilized is strictly recorded so that information is available from year to year on the efficiency of different chemical fertilizers in the particular conditions of their application.

ORGANIC MANURES

Organic manures used in pond fisheries include dung, liquid compost, compost, and green manure. Artificial feeds of plant origin as well as fish excreta serve as additional manures. Sewage water containing organic matter, after purification, can also be used for manuring fish ponds. In a number of cases organic manures are more effective than chemical fertilizers (for example, in the absence of a silt layer in the pond and in the case of sandy and podsolic soils). They are extremely beneficial in multiplying aquatic fauna. Excessively applied they pollute the ponds with these consequences: reduction in oxygen content, development of such diseases as branchiomycosis (gill rot), the undesirable development of denitrifying processes, and the creation of anaerobic conditions at the bottom of the ponds, leading to the decomposition of cellulose and the formation of methane which is lethal to fish and other aquatic organisms. Hence the norms of application of organic manures in ponds should be carefully assessed.

In contrast to chemical fertilizers, organic manures are more variable with regard to the important biogenic elements (phosphorus, potassium, calcium, nitrogen) and include a percentage of all the nutrients. This makes a determination of their norms of application difficult; an analysis of the material is essential in each instance.

Organic manures which contain all the requisite nutrients are called complete manures.

Dung and Compost Manures. The quality of dung depends not only on the species of animal, but also on the nature of its food, conditions of storage, quantity, type of flooring, and other factors. The composition of fresh dung is given in Table 46.

Dung is a widely available organic manure. Well-decomposed dung of cattle, horse, sheep, pig, and poultry is commonly used. Dung extract is used less often than the dung itself and mostly as a nitrogen-potassium

Table 46. Composition of fresh dung (in %) (from Mamchenkov, 1955)

Components	Dung on straw bed					Dung on peat bed	
	Mixed dung	Horse dung	Cattle dung	Sheep dung	Pig dung	Horse dung	Cattle dung
Water	75.0	71.3	77.3	64.6	72.4	67.0	77.5
Organic matter	21.0	25.4	20.3	31.8	25.0	—	—
Total nitrogen (N)	0.50	0.58	0.45	0.83	0.45	0.80	0.60
Proteinic nitrogen	0.31	0.35	0.28	—	—	0.48	0.38
Ammoniacal nitrogen	0.15	0.19	0.14	—	0.20	0.28	0.18
Phosphorus (P_2O_5)	0.25	0.28	0.23	0.23	0.19	0.25	0.22
Potassium (K_2O)	0.60	0.63	0.50	0.67	0.60	0.53	0.48
Calcium (CaO)	0.35	0.21	0.40	0.33	0.18	0.44	0.45
Magnesium (MgO)	0.15	0.14	0.11	0.18	0.09	—	—
Sulfuric acid (SO_3^{2-})	0.10	0.07	0.06	0.15	0.08	—	—
Chlorine (Cl^-)	—	0.04	0.10	0.17	0.17	—	—
Silicic acid	—	1.77	0.85	1.47	1.08	—	—
Iron and aluminum sesquioxides (R_2O_3)	—	0.11	0.05	0.24	0.07	—	—

manure. It is more appropriate to use dung extract in new ponds where the layer of colloidal silt has still not formed. The most effective dungs are those of pig and poultry and so they are more widely used in pond fisheries, especially in Western Europe. The most important condition for the effective use of dung in ponds is the absence of overgrowth of higher aquatic plants; manure stimulates their development and its influence on fish production thereby is minimalized.

The application of dung for manuring ponds in carp farms in Western Europe is based on different norms. In Poland 20 to 30 tons/ha of dung is applied as manure in several portions, depending on the oxygen regime of the pond water. Dung is placed in small heaps with the tops slightly above the water surface. Dung is also applied on a dry bottom by broadcasting it in a uniform layer but only in ponds without a silt layer. Correctly used, dung increases fish production by 100%. In Czechoslovakia 8 to 14 quintals of dung is applied to one hectare of pond area, depending on fish productivity, humus content of the ground, and the type of manure. It is applied directly before flooding the pond and then broadcast over the water every month thereafter. Havlena (1965) established that by using dry pig dung at the rate of 18 to 100 quintals per hectare, fish production increased by 2.5 to 3.0 kg for each quintal of manure. The best application rate is 50 q/ha of dung when applied in the form of heaps distributed throughout stagnant or mildly flowing ponds.

In Saxony fresh pig manure is applied in the form of a thick slop and special pits are dug for its storage near the ponds. The thick slop is poured into a boat and the manure applied from the boat to certain areas in the pond at an interval of one to three weeks. The best results have been obtained with the rate of 3 to 5 tons/ha. In Yugoslavia, to prevent diseases in fish, this dose is supplemented with 100 to 200 kg/ha lime. In Saxony an additional growth of 30 to 40 kg/ha of carp was obtained for each ton of pig manure applied. This practice was first developed by the Fisheries Research Institute in Budapest (Hungary), where dung is applied with the help of a pump fitted to the boat for diluting it with water (1 m^3 of water for 1 quintal of fresh pig manure). Manure prepared in this way is scattered uniformly over the pond surface.

The Budapest method has proved more suitable than the application of manure through heaps or spreading it in a layer on the pond bottom; in the latter case an anaerobic process occurs which produces methane. In broadcasting dry dung over the surface the main bulk decomposes within 24 hours. It is recommended that 2 q/ha of fresh dung be applied every week; in the warmer periods of vegetative growth this rate can be increased to 3 to 4 q/ha so that manuring is finished by the beginning of August. Such a method of manuring prevents the possibility of any drastic reduction in the oxygen content and eliminates its deficit.

Over the large territory of the USSR, where different soil-climatic peculiarities prevail in different zones, the norms of dung application and manuring with compost vary considerably—from only 18 to 20 to as much as 320 quintals per hectare. The norm also depends on the conditions of the ponds and the nature and quality of the dung or compost.

Compost is prepared by the usual methods followed in agriculture and, when fully prepared, is a homogeneous, dark-colored mass which smells like smut. Well-prepared and well-decomposed, compost is not inferior to fresh dung in its action. Better results are obtained by using compost in ponds poor in organic matter. From a sanitary-hygienic point of view compost is preferable to dung.

In China compost is prepared from cow dung and green plants and used to manure nursery ponds to increase the stock of natural foods for rearing fry. Here, special waterproof pits are dug in which layers of green grass are alternated with layers of dung. The compost pile is covered with a layer of unslaked lime (about 70 kg), flooded with liquid dung manure, and covered with earth. The ratio of the components in the compost should be as follows: green plants—4, dung—2, liquid manure—1 part. For every 100 kg of plants 100 liters of water are added so that the decomposition process is accelerated.

Compost prepared in this manner is applied 4 to 5 days after the stocking of young fish at the rate of 75 q/ha. Prior to this, the ponds are thoroughly disinfected with unslaked lime. The compost is first passed through a sieve with a mesh size of 8 mm, in which it is crushed and nondecomposed grass roots and plants removed. The sifted compost is then dissolved in water and the liquid manure spread from a boat in the pond at the rate of 3,000 to 3,500 l/ha on the first day, and 1/4th to 1/5th of this quantity applied on the second day. The remaining liquid manure is subsequently diluted and applied twice a day, in the morning and evening.

The rates of application of liquid manure have not been defined; hence it should be used cautiously. In Ukrainian SSR liquid manure is applied in ponds with a silt layer of 10 to 15 cm at the rate of 10 q/ha. However, the manure is applied not directly to the ponds, but to the stream supplying the pond with water. The liquid manure is poured into the stream in small portions over a 16-hour period. Application commences two months after the fish have been stocked.

In China liquid manure is prepared in concrete pits where dung and aquatic and terrestrial plants are placed together with lime. It is then broadcasted at the rate of 30 to 40 kg/ha.

Duck droppings also constitute a very valuable manure for fish ponds. Over a period of 45 days between 5 to 10 kg of excreta can be collected from a single duck. Frček (1957) found that with keeping ducks in ponds the maximum increase in fish productivity occurred in the first two to three years. Here the lower the original fish productivity of the pond the higher its relative increase.

In collective and state farms which rear coypu, it is more appropriate to use their excreta for pond manuring. The manure[1] of coypu in composition is closer to pig manure. Over a period of 45 days a single coypu produces about 6 kg of excreta or an average of 130 g/day.

In ponds overgrown with higher aquatic plants up to an area of 70% it is recommended that 20 to 22 coypu/ha be released for the summer period. As a result the natural productivity increases by 100% (Wiltowski, 1958). With a density of plantation of 22 animals/ha in the pond about 1.5 kg/ha of pure nitrogen is obtained in addition to the excreta.

Besides the dung of farm animals and poultry, night soil is widely used for manuring ponds in many Asian countries (China, India, Malaya, etc.). In China and other countries, public latrines and piggeries are constructed near fish ponds. In Malaya, Singapore, and southern China domestic animals are often domiciled near fish ponds.

In Chapter XI it was mentioned that part of the feed introduced into ponds is dispersed and dissolved in the water before it can be consumed by the fish. Some quantity also settles to the bottom or remains in the water column. In ponds where fish are given supplementary food the quantity of bacteria increases significantly. Even if only 5% of a given feed is considered dispersed, a large quantity, especially at a higher density of stocking, actually serves as manure. This should be taken into consideration when supplying feed to fish in ponds.

Various methods of pond application of dung and compost, the most widely used organic manures, have been developed. The following are more common: (a) placement of dung in small heaps in the marginal zone along the water line; (b) placement of dung heaps uniformly throughout the pond; and (c) installation of wattle enclosures on dikes where the dung is stored. Whichever method is used, care must be taken that the organic manure does not reduce the oxygen content.

Cow dung and compost should never be applied to those parts of a pond overgrown with aquatic plants.

GREEN MANURES

Since farmyard manure is in constant demand for agriculture, horticulture, and kitchen gardening, and the preparation of compost is costly in

[1]The coypu *Myopotamus* (*Myocaster*) *coypus* is a pond mammal feeding on the roots of plants, young branches of water chestnut, aquatic plants, mollusks, and fresh-water crustaceans. Coypu is reared at fish farms where the ponds tend to become highly overgrown with coarse plants.

terms of labor, this manure will never be widely available for fish farms.

Green manures are economically profitable and more readily available. They include agricultural crops or grasses cultivated on the pond bed and cut aquatic or terrestrial plants placed in the pond.

Combined crops are cultivated more often, especially a vetch-oat mixture which yields a large quantity of green matter. In the case of incomplete manuring (when the plants are topped) the crop is used as cattle feed. A vetch-oat mixture is beneficial also because vetch enriches the soil with nitrogen, while oat utilizes nitrogen extracted from the deeper and less accessible soil layers. The value of vetch is further increased if the ponds are located on soils of low fertility; filled with water for the first time the colloidal silt layer in such ponds is either absent or negligible. Lupine, vetch, clover, seradella, and other nitrogen-accumulating plants of the leguminous family can be grown. Green manuring helps to increase the nitrogen content of the pond. Manuring may be complete or incomplete. In the case of complete manuring the green growth is flooded with water; in the case of incomplete manuring the plants are mowed for cattle feed and the underground residue flooded with water. The best results are obtained from incomplete green manuring.

Complete green manuring results in an exceptionally high quantity of waste compared to incomplete manuring. It deprives the fish farm of stocking material for the next year, reduces fish productivity, and eliminates the profitable use of green cuttings for cattle feed. However, the method of incomplete green manuring is less effective than zonal manuring with plant material; the increment in population of bacterioplankton, distributed in the water column throughout the pond, is not so great and moreover is short-lived.

In 1947 Isakova-Keo proposed a method for zonal manuring of ponds based on the placement of previously sun-dried green matter (including branches of willow, birch, mountain ash, alder, etc.) in the shallow, quiet, and nonflowing parts of a pond. This method found wide application in commercial fisheries. The efficacy of plant material of different species and groups has been studied and conditions worked out in which green manuring is most beneficial.

Higher aquatic plants and terrestrial shore plants, including herbaceous weeds, are suitable for green manuring. The initial link in the food chain—bacteria—develops better when soft meadow grasses are included. Leguminous plants are more effective than others; they grow more rapidly, especially the smaller types, and provide the necessary nutrients. At a water temperature of 25 to 30°C and a little more the population of bacterioplankton develops well; at lower temperatures, say 16 to 19°C, it does not. If emergent coarse plants are used for manuring, bacteria develop slowly even at higher temperatures. The nutrients from the plants of this group are not released into water as quickly as those from soft meadow plants. Soft plants decompose in 10 to 14 days and coarse plants in 20 to 25 days. With soft meadow plants and a favorable temperature the population of bacteria increases rapidly (at the end of one day by 40 to 50 times), although only for a short time; by the fifth to the sixth day a sharp decline in population is evident due to intensive feeding by zooplankton. With coarse aquatic plants these processes proceed more uniformly and without peaks. The biomass of bacteria with the introduction of soft meadow plants into ponds is higher than with emergent coarse aquatic plants. This is explained by the fact that soft aquatic plants serve as food for different members of chironomids in the process of their decomposition. Observations have shown that in those places of a dry pond covered with rotting soft flora the number of chironomid larvae may be five times greater and oligochaetes two times greater than in areas devoid of flora. Plankton also develops well.

The greater the quantity of plants added at one time, the higher the biomass of bacteria. However, large quantities of this manure should not be added to ponds, especially during high temperatures, since they tend to lead to an oxygen deficiency. A reduction in temperature, although it slows down the development of bacterioplankton even when soft meadow plants are used, permits an increment in the quantity of plants added and promotes a situation whereby the biomass of bacterioplankton remains higher for a longer period. Moreover, in such cases the vegetation rarely needs replacement. Decomposing plants which turn brown and start shredding are removed from the ponds and replaced with new ones.

In spite of the fact that coarse aquatic plants are less effective as green manure than soft ones, they should be used for green manuring purposes. Those which grow profusely are periodically trimmed. Experience has shown that coarse plants used as green manure increase the output of fish by 50 to 100% with a reduction in the cost of production by 20 to 30%.

Coarse emergent and soft submerged aquatic plants, when used for the purpose of green manuring (especially pondweed), promote the development of bacteria and green algae, which serve as food for zooplankton and increase its quantity 10 to 16 times. This increases natural fish productivity significantly. The perishing aquatic plants (especially soft plants) cause a vigorous development of plankton and promote the development of benthos as well.

Preference should be given to soft shoots which are easy to cut. In the case of heavy overgrowth the plants are cut for green manuring purposes from areas adjacent to the open parts of the pond, but care is taken that the zooplankton is not suppressed. The following norms for the application of plants as green manure are recommended, using hatching-rearing ponds as an example:

1. Ponds with quite a favorable oxygen content, unpolluted flowing

water, dense sandy bottom, and open to winds should be manured with 3 to 4 tons/ha.

2. Ponds with a low oxygen content (especially in the predawn hours), mildly polluted but not marshy, with a significant mixture of silt at the bottom but devoid of blue-green algae, should be manured with 1 to 2 tons/ha.

Green manuring should not be done in highly polluted ponds.

The selection of the place for green manure is of practical significance, especially the areas for cutting. To increase the quantity of zooplankton cutting is started from the shore; to increase the quantity of benthos cutting commences from the deeper places at the center of the pond. In large ponds the cut plants are left for two or three days on the water surface for sun-drying; this speeds up the decomposition process. They are then collected into raft-like bundles and positioned with the help of pegs (soft plants are kept 1 m apart and coarse plants 2 to 3 m). The longer the stems of the plants, the more sparse the placement of the pegs. The raft-like bundles are arranged along the shore and submerged at 30 to 40 cm; the central part of the pond is kept free to improve the oxygen regime. Benthos develops profusely on the layer of cut plants, enriching the pond bottom. The fish find abundant food in the space between the pond bottom and the raft bundles.

Plants cropped from small ponds are pulled out on the bank and partially dried in the sun to a green-brown color. They are then tufted and arranged in stagnant places along the shore. The bundles are submerged half their length at the rate of 12 kg plant mass for each meter of shore length. To maintain the oxygen regime at the required level the bundles should not be placed in the water supply channels and troughs. Water outlets and spillways should also be kept free; otherwise the nutrients needed for the development of plankton and benthos will be lost.

The oxygen content of the water is measured regularly starting at 5:00 to 8:00 a.m. during the time of green manuring. Samples are taken near the water releasing points from the water surface (at a distance of 100 m from the water outlet), near the bottom, and from those places where the plants are cut.

To increase the development of zooplankton it is recommended that the plants be tufted or bundled, and to increase the stock of zooplankton and benthos shaped into rafts.

Green manuring can also be done in ponds by piling or heaping the plants loosely and periodically turning them over to keep them loose to encourage decomposition. The ponds are cleared in autumn of all green manures and the nondecomposed plants used in compost.

Experience has shown that if with one compost manuring daphnids and diaphanosomids develop more intensively, by combining composts with green manure containing birch oligochaetes multiply. The use of birch branches as green manure stimulates the development of *Bosmina,* while sedge stimulates diaptomids and chironomids. In using one or the other kind of green manure therefore one must assess its efficiency not only from the viewpoint of increase in total fish productivity, but also its influence on the development of the most valuable and essential food organisms at the given moment. Such a directed application of manures augments the productivity of fish ponds.

Lyakhnovich (1966) has recommended for the conditions of Belorussian SSR a special approach to manuring different categories of ponds. Organic manures are best for hatching ponds since their effect is more rapidly reflected in increasing the food base for fry. Chemical fertilizers are used in spawning ponds when it is necessary to improve the conditions of the aquatic plants in which carp lay eggs. Nitrogen-phosphate fertilizers should also be spread over the surface of these ponds once or twice with an interval of one to two days in the case of a deterioration in their oxygen regime. Spreading should be done directly after flooding the ponds with water at the rate of 30 to 40 kg/l.

The application of manures in rearing and foraging ponds has its own peculiarities. Manures in the form of a solution are applied directly in the water 10 to 15 days before releasing the fish. Organic manures are deposited in small heaps in the marginal zones, making sure that the top of the heap is not covered with water.

In the first half of the vegetative growth period manures are added every 7 to 10 days; in the second half every 10 to 15 days; and throughout the season in rearing ponds 6 to 8 times and in foraging ponds 8 to 10 times. The calculated quantity of manure is divided into respective equal portions. At the time of manuring the intensity of development of algae and the transparency of the water are taken into account. If the transparency is reduced to 20 to 30 cm and the water has "blossomed", manuring is done with restraint. Foraging ponds are better treated with organic manures when they are still frozen. The application of manures should terminate 30 to 40 days before autumn catching.

Such a differential approach to the application of manures in different categories of ponds under the conditions of Belorussian SSR, should also be followed in the other zones of the country.

ORGANO-CHEMICAL MANURES

A rational combination of chemical and organic manures, especially with green manure, is more effective. In the USSR experiments have been done using different combinations and quantities of dung manure, aquatic plants, plant meal, superphosphate, ammonium nitrate, kainite, lime, and others. A greater increase in fish productivity was recorded for such combinations than for organic manures or chemical fertilizers used alone.

Mukhina and Urazova (1967) conducted experiments in the rearing ponds of the Moscow district using different doses of nitrogen (1 mg/l and 5 mg/l), phosphorus (1 mg/l), and liquid manure with a stocking density of 50,000 fry per hectare. The authors found that the physiological conditions of fingerlings under combined manuring with chemical fertilizers and organic manures was better and that the fish were better prepared for hibernation. The output after hibernation was 15 to 20% higher than in the control ponds and reached 90 to 95%. The consumption of reserve matter in yearlings raised under high doses of nitrogen, phosphorus, and composts was less than in yearlings raised in ponds where only chemical fertilizers were used. In the later case a greater reduction in fat, protein, and caloric values and an increase in moisture content and mineral matter were observed.

Experiments carried out in the rearing ponds of Moscow district by Bakhtina (1967) also showed that versus the use of chemical fertilizers alone (nitrogen, phosphate, and potassium), these in combination with organic manures (dung-silt compost) yielded better results. The dung-silt compost was applied over the pond bottom before flooding and stimulated the development of phytoplankton, zooplankton, and benthos, especially in the first half of the vegetative growth period. Better development of the food base was observed in ponds where small quantities of chemical fertilizers and organic manures were used in the season (1.0 to 1.5 q/ha ammonium nitrate, 0.8 to 1.0 q/ha superphosphate, 4 q/ha lime, and 10 q/ha dung-silt compost).

An experiment was conducted in the Astrakhan reserve forests by the North Caspian Directorate of Fisheries and Fishing Industry using a special preparation of organo-chemical manures. A mixture of plant mass and cow dung was kept in air in a special container and regularly sprinkled with solutions of ammonium nitrate and superphosphate. Such an organo-chemical mixture did not affect the oxygen regime of the water in the ponds and contained a large mass of bacteria, which serve as food for zooplankton. With this mixture the quantity of zooplankton increased 380 times versus 53 times with the usual recommended organo-chemical manures.

Studies done by Abramovich et al. (1965) in Ukrainian SSR demonstrate significant differences in the composition of bottom deposits and the water of ponds located in different physico-geographic zones. Hence the use of one and the same manure in all zones is not appropriate. In the ponds of Poles'e where the bottom deposits show an acidic reaction, the water contains a large quantity of organic matter, is poor in free potassium, and poorly mineralized, measures directed toward mineralization of the organic substances and mobilization of biogenic elements are recommended. Estivation, agro-reclamation works, and lime application serve this purpose; potassium and calcium manures are applied over the water and nitrogen and phosphate fertilizers likewise recommended. For ponds located in Carpathia, which are characterized by organic matter deficiency and poor water mineralization, organic manures and calcium fertilizers are essential. They should be applied to the water as well as over the bottom and chemical fertilizers (in particular phosphate) added in small portions throughout the summer period. Broadcasting nitrogen and phosphate fertilizers over the water is recommended for ponds in the forest-steppes where the water is characterized by a higher mineralization and alkalinity, is rich in biogenic elements, and the bottom deposits have a neutral reaction.

Similar studies in other zones of the country would make it possible to use chemical fertilizers as well as organic manures more effectively.

The better utilization of chemical fertilizers and organic manures would be expedited by a well-organized agro-chemical service program, whereby the necessary analyses of the water and soil of ponds could be ensured. This would make it possible to implement the rule: Each pond should be given a manure tailored to its needs.

References

Abramovich, L., Ya. Syara, R. Arenkova and S. Bogonis. 1965. Intensifikatsionnye meropriyatiya i potrebnosti vodoemov (Intensification measures and needs of fish ponds). *Rybovodstvo i Rybolovstvo*, no. 1.

Akimov, V.A. 1966. Obshchaya chislennost' mikroorganizmov v vode rybovodnykh prudov pri intensivnom udobrenii i kormlenii ryby (Total population of microorganisms in the water of fish ponds under intensive manuring and feeding of fish). *Trudy VNIIPRKh.* Moscow, vol. 14.

Bakhtina, V.I. 1967. Vliyanie mineral'nykh i organicheskikh udobrenni na razvitie estestvennoi kormovoi bazy v vyrostnykh prudakh (Effect of chemical fertilizers and organic manures on the development of a natural food base in rearing ponds). *Trudy VNIIPRKh.* Moscow, vol. 15.

Bakhtina, V.I. and A.I. Batenko. 1961. Vliyanie nebol'shikh doz mineral'nykh udobrenii na razvitie estestvennoi kormovoi bazy i ryboproduktivnost' prudov (Effect of small quantities of chemical fertilizers on the development of a natural food base and on fish productivity in ponds). *Trudy VNIIPRKh.* Moscow, vol. 10.

Batenko, A.I. 1958. Vliyanie srokov vneseniya superfosfata na soderzhanie forfora v vode i pochve prudov (Effect of time of superphosphate application on the phosphorous content in pond water and soil). *Trudy VNIIPRKh.* Moscow, vol. 9.

Batenko, A.I. and N.P. Ushakov. 1967. Ammiachnaya voda v rybovodstve (Ammonia water in fish culture). *Rybovodstvo i Rybolovstvo*, no. 6.

Chittino, P. 1968. Piscicoltura e ittiopatalogia, *Riv. Zootech.*, vol. 41, no. 9.

Frček, M. 1957. Vliv chovu kachen na prirozenu vyrobnost rybnikü, *Cescosl. Rybařstvi*, vol. 6.

Havlena, H. 1965. Spusoba vysnam použiti hnoiv z velkovykrnem vepru s hlediska rybaraskeho. *Sb. ¨eskosl. Acad. Zemed. Ved. Zivocisna Virobocn.*

Isakova-Keo, M.M. 1954. Opyty po udobreniyu zonal'nym metodom zimoval'nykh i forelevykh prudov (Experiments on manuring of wintering trout ponds by the zonal method). *Trudy Problemnykh i Tematicheskikh Soveshchanii Zool. In-ta Akad. Nauk SSSR,* no. 2.

Karzinkin, G.S. and S.I. Kuznetsov. 1956. Ispol'zovanie zhestkoi rastitel'-nosti v rybovodnykh khozyaistvakh del'ty Volgi v kachestve zelenogo udobreniya (Use of hard vegetation in fish farms of the Volga delta as a green manure). *Trudy VNIRO.* Moscow, vol. 32.

Kempinska, H. 1968. Wptuw nawosom amonomych na ryby, *Gospod Rybna,* nos. 3–5.

Lavrovskaya, N.F. 1967. Vitaminy i pigmenty vodoroslei iz rybovodnykh prudov (Vitamins and pigments from the algae in fish ponds). Avtoref. Kand. Diss., Moscow.

Lyakhnovich, V.P. 1962. Organicheskie udobreniya prudov (Organic manures for ponds). Sbornik *Vopr. Rybnogo Khoz-va Belorussii.* Minsk.

Lyakhnovich, V.P. 1966. Opredelenie potrebnosti prudov v udobreniyakh i kontrol' za e'ffektivnostyu ikh deistvii (Determining the requirements of ponds for manures and controlling the efficiency of their action). *Peredovoi Opyt Mekhanizatsii Trudoemkikh Protessov i Vedeniya Rybnogo Khozyaistva v Belorussii.* Minsk.

Mamchenkov, P.S. 1955. Organicheskie udobreniya. Spravochnik agronoma po udobreniyam (Organic Manures. Agronomist's Handbook on Manures). Sel'khozizdat, Moscow.

Mints, A.G. and E.N. Khairulina. 1965. Novyi sposob udobreniya prudov, postroennykh na torfyanykh vyrabotkakh (A new method of manuring ponds in peat quarries). *Trudy VNIIPRKh.* Moscow, vol. 13.

Mukhina, R.I. and V.F. Urazova. 1967. Otsenka fiziologicheskogo sostoyaniya segoletkov karpa, vyrashchennykh pri intensivnom udobrenii prudov mineral'nymi (azotnymi i fosfornymi) i organicheskimi udobreniyami [An evaluation of the physiological condition of carp under-yearlings raised under intensive manuring of ponds with chemical (nitrogen and phosphorous) fertilizers and organic manures]. *Trudy VNIIPRKh.* Moscow, vol. 15.

Müller, W. 1969. Anleitung zur kombinierten stickstoff-phosphat-düngung für die teichwirtschaft, *Dtsch. Fisch. Ztg.,* vol. 16, no. 4.

Okhryamkina, N.P., Grotane et al. 1968. Primenenie mikroe'lementov v udobreniyakh dlya vyrostnykh prudov Latvii (Use of trace elements in manures for rearing ponds in Latvia). Sbornik *Prudovoe Rybn. Khoz-vo i Rasvedenie Ugrya.* Riga, vol. 1, pt. 2.

Paladino, J. 1968. Nawozenie Stawow, *Cosp. Rybna,* 20, 12, 5.

Rodina, A.G. 1958. Mikroorganizmy i povyshenie ryboproduktivnosti prudov (Microorganisms and Increasing Fish Productivity of Ponds). Izd-vo Akad. Nauk, SSSR, Leningrad.

Rodina, A.G. 1958. Azotnye udobreniya rybovodnykh prudov i denitrifikatsiya (Nitrogenous manures for fish ponds and denitrification). *Trudy In-ta Biologii Akad. Nauk LatvSSR,* vol. 7.

Rodina, A.G. 1959. Mikrobiologiya udobryaemykh prudov (Microbiology of manured ponds). *Trudy VI Soveshchaniya po Problemam Biologii Vnutrennikh Vodoemov.* Moscow.

Rodina, A.G. 1959. Dinamika chislennosti azotobaktera pri udobrenii prudov (Dynamics of *Azotobacter* population under manuring of ponds). *Trudy V Nauchnoi Konferentsii po Izucheniyu Vnutrennikh Vodoemov Pribaltiki.* Minsk.

Rodina, A.G. 1966. Kormovoe znachenie i stroenie detrita (Food value and structure of detritus). Sbornik *Biologicheskie Resursy Vodoemov Puti ikh Rekonstruktsii i Ispol'zovaniya.* Nauka, Moscow.

Schäperclaus, W. 1968. Ertol greiche drei jährige stickstoffdüngung in den kauppaer versuchsteichen 1965-1967, *Dtsch. Ztg.*, vol. 15, no. 3.

Steffens, W. 1969. Der vitaminbedart des karpfens, *Deutsch. Fisch. Ztg.*, vol. 16, no. 5.

Vinberg, G.G. and V.P. Lyakhnovich. 1965. Udobrenie prudov (Manuring of Ponds). Pischevaya Promyshlennost', Moscow.

Wiltowski, J. 1958. Obserwacye nad chowem nutrii na stawach, *Biul. Zakl. Biolog. Stawow.* PAN, vol. 7, Krakow.

Zonova, A.S. 1965. K metodike postanovki opytov po "metodu sklyanok" pri opredelenii potrebnosti prudov v udobreniyakh (On the methods of planning experiments on the "flask method" while determining the manure requirements of ponds). *Vopr. Ikhtiol.*, 5, 2, 35.

Chapter XIII

COMPLEX INTENSIFICATION

The practice of introducing an entire complex of intensification measures is appropriate only at the preparatory stage of the ponds in fish farms with technically correct management and a systematic follow-up. Ponds in which a complex of intensification measures is implemented are usually drainable and fishing in them relatively easy and complete. No trash fish or predators should be present. The pond should receive adequate water of the requisite quality; the supply should be consistent and uninterrupted and arrangements for the discharge of excess water present when such is necessary in the interest of the culture.

The complex of intensification measures includes: (a) preparation of the ponds, (b) feeding of fish, (c) rearing of live foods, (d) manuring of the ponds, (e) mixed stocking, (f) additional stocking, (g) necessary arrangement of hydroengineering installations, (h) well-organized breeding work ensuring rearing of high quality brood fish and stocking material, and (i) maintenance of the fish farm at the necessary level of sanitary-hygienic conditions.

Thus, the intensification complex in pond fisheries means the simultaneous and multifaceted treatment of ponds and the fish reared in them, using all the methods known for increasing fish productivity.

Intensification gives better results when the necessary measures are implemented *in toto*. However, the total complex may not always be practicable and in some cases a prior and thorough preparation of the ponds may not be essential. The measures which form a part of the intensification complex are developed and worked out in detail in accordance with zonal peculiarities, categories of ponds, their hydro chemical regimes, and so on. They are adjusted according to local conditions, such as soil, as well as domestic and economic requirements.

At present the complex of intensification measures has become a generally accepted method for increasing the yield of fish per unit of pond area. More often feeding of fish is combined with manuring of the fish ponds in the practices of pond fisheries, using additional stocking or a wider polyculture in this process. With a high level of stocking density it is necessary not only to increase the food for consumption, but also to increase the natural fish productivity by manuring the ponds and so forth.

To establish the efficiency of individual methods of intensification, in particular feeding, the index of food coefficient is used, while for manuring the index of fertilizer coefficient (FC) is recommended. The latter is calculated as follows. If, for example, manuring is done in fractioned portions over the water in the form of a solution in a quantity corresponding to the biological need of the pond, according to the findings of Lyakhnovich (1969), for 1 kg of increased fish weight a minimum of 2 kg of ammonium nitrate and 1.5 to 2.0 kg of superphosphate has to be applied. The total consumption of chemical fertilizers per kg of weight increase in the fish is known as the fertilizer coefficient.

As mentioned in an earlier chapter, the most commonly used combination of chemical fertilizers is nitrogen-phosphate in the ratio of 4 : 1. This corresponds to the equal quantity by way of ammonium nitrate and superphosphate with a fertilizer coefficient of 4.0. We may assume for the purpose of calculation that the fish are fed with any mixture in which the food coefficient is 3.0. It is also necessary to determine the natural fish productivity of the pond for this calculation.

To calculate the comparative efficiency of manuring and feeding, V.P. Lyakhnovich recommends that the following parameters be taken into account: (1) natural fish productivity, (2) consumption of manures and feeds over the vegetative growth period, (3) total production of fish, and (4) the food and fertilizer coefficients.

On the basis of the data obtained, the food and fertilizer coefficients are finally calculated.

For example, if the natural fish productivity of a pond is 250 kg/ha and over the vegetative growth period equal quantities of ammonium nitrate and superphosphate are applied at the rate of 450 kg/ha each (total 900 kg/ha) and 1,800 kg/ha of feeds, the resultant output of the produce in autumn will be 1,200 kg/ha.

To analyze the role of individual measures in the total output of fish from one hectare, the value of natural fish productivity should be deducted from the value of total output: 1,200 minus 250 = 950 kg/ha. This value can be attributed to the fertilizers and feeds consumed.

If the coefficient in the fish pond is equal to 3.0, the increase in weight of the fish as a result of the application of feed will be equal to the value obtained by dividing the quantity of feed consumed by the value of the food coefficient (1,800 : 3 = 600 kg/ha). If this value is deducted from the quantity used for total output under artificial feeding and manuring, one gets 350 kg/ha (950 minus 600 = 350). Finally, the increase in production due to chemical fertilizers with a fertilizer coefficient of 4.0 will be 225 kg/ha (450 + 450 : 4). If the value of increase in fish productivity because of fertilization is deleted from 350 kg, an extra growth of 125 kg/ha is obtained, which has not been accounted for in the foregoing calculations. In practice this may happen due to an imprecise determination of natural fish productivity, or not including other measures of intensification, or actually attaining a higher efficiency of feeding of fertilization than expected in the calculations. In this case the total weight (percentage) in the total increase in fish productivity is determined for each of the measures included and the extra fish productivity divided equally among the items.

In our example the natural fish productivity has been taken to be 250 kg/ha, which constitutes 20.8% of the total productivity toward autumn. The increase in productivity because of artificial feeding is 600 kg/ha or 60% and that due to the application of chemical fertilizers 350 kg/ha or 29.2%. If the extra growth is distributed proportionately, the values of 26.0, 62.5 and 36.5 kg/ha are assigned respectively. Adding these values to those calculated yields the following data: (a) initial natural fish productivity —276 kg/ha, (b) increase in productivity due to feeding—662.5 kg/ha, and (c) increase due to manuring —261.5 kg/ha.

The actually obtained food and fertilizer coefficients are calculated on the basis of these data. To calculate the former the value for food consumption is divided by the growth rate attributed to feeding—1,800:662.5 = 2.72. To calculate the latter the quantity of manures applied is divided by the growth rate attributable to them—900:261.5 = 3.44.

If the fish productivity is less than the value of the total fish productivity, then the growth rate for each of the intensification measures used should be reduced in accordance with its total weight in the total calculated increase in fish production.

These data together with the cost of the food items and fertilizers used make it possible to assess the economic efficiency of the intensification measures.

Raising of Marketable Under-Yearlings. Movchan (1958) has developed a method for the speedy raising of under-yearlings, which makes it possible to obtain in one season marketable fish weighing about 500 g or more. Young fish, selected on the basis of good characters, are first kept in hatching ponds under reduced stocking (1,200 to 1,500 fish/ha, southern part of Ukrainian SSR). Only the largest and most hardy are subsequently transplanted to foraging ponds, which have been carefully prepared beforehand. Throughout the year the young fish are intensively fed on daphnids and chironomids simultaneously reared in large quantities. Thus, in the very early stages of fry postembryonic development the best ecological conditions are created and the juvenile fish provided with abundant food. The possibility of raising under-yearlings to marketable weight in one summer period is explained by the fact that carp has a high growth potential, especially in the first year of its life. In the two-year rotation of pond fisheries this potential is checked by raising under-yearlings up to that weight accepted as standard for individual zones of the USSR. Yet it is known that in the southern regions under certain conditions under-yearlings over the first vegetative growth period can attain a weight of 400 to 600 g and more. However, these conditions must be observed: early spawning should be done at the end of April–beginning of May; food corresponding to the stage of development should be abundantly provided for the fry, and for this purpose live foods reared (see Chapter XI); larger juvenile fish should be selected for transplantation in the foraging ponds at the age of 18 to 20 days and a weight of 150 to 200 mg; the maximum area of such ponds should be 5 to 10 hectares or in well-maintained farms even more (it is important that catching of the fish should pose no difficulty in these ponds).

Stocking norms differ in different regions. Thus, for example, in western Georgia stocking of carp fry was done at the rate of 1,700, 3,000, and 4,000 fish per hectare. Ammonium nitrate and superphosphate were used as fertilizers and the fish fed a mixture containing soybean cake, formula feed, and meat bone meal. The under-yearlings achieved a weight of 672, 625, and 435 g after 154 days of rearing with a fish productivity of 12.94 to 23.43 q/ha and an output of 91.1%.

Experience has shown that at the optimum rate of stocking of 3,000 to 3,500 fry per hectare the weight of the fish toward autumn (second decade of October) may reach 500 to 600 g.

In Moldavian SSR for carp culture of one year, the recommended stocking norms are 2,500 to 3,000 fish per hectare with supplementary feeding and not more than 1,500 to 2,000 fish per hectare when they are maintained only on natural foods. The stocking rate in rice fields is 1,000 to 1,500 fish per hectare. The average weight of marketable under-yearlings stocked at the rate of 3,000 per hectare is 415 g, and stocked at the rate of 2,000 fish per hectare, 625 g.

In Uzbek SSR when marketable under-yearlings were raised with a stocking density of 1,300 fish per hectare, the fish productivity was 901 kg/ha with a unit weight of 731 g (losses 5.3%); at a stocking rate of 4,000 fish per hectare, 1,330 kg/ha with a unit weight of 546 g (losses 39.1%).

Foraging ponds are used for juvenile fish in the northern regions also but here the weight of marketable under-yearlings is comparatively less, even with a reduction in stocking rate. For example, in the Saratov district and in Bashkirian SSR with a stocking rate of juvenile fish of 1,000 to 1,200 per hectare the unit weight toward autumn is 250 to 350 g. The same results are obtained in the Voronezh and other districts of the RSFSR. Experiments done in Novosibirsk, which has a vegetative growth period of 110 to 130 days and a shorter period with optimum temperatures, indicate the possibility of raising marketable under-yearlings with an average weight of 114 to 250 g (maximum rate 435 g) under the stocking rate of 2,300 to 2,800 fish per hectare. Such a weight does not correspond to the standard weight of marketable carp. At the same time a higher weight of under-yearlings at the same stocking rates makes it possible to use them after hibernation as stocking material in foraging ponds during the spring of the next year, to obtain a higher unit rate of two-year fish.

Thus, the most effective results are obtained in the southern regions.

For the raising of marketable under-yearlings in monoculture the Ministry of Fisheries of the USSR has recommended the following norms for northern Caucasus, Moldavia, Trans-Caucasus, and Central Asia:

Age of young fry at time of stocking in ponds—12 days;

Stocking density of young fry in rearing ponds—2,500 to 3,500 per hectare;

Output of marketable under-yearlings—65 to 70%;

Average weight of under-yearlings—300 g.

Raising under-yearlings as marketable fish is also done in Bulgaria. Within a period of 155 days with a stocking rate of 2,500 to 3,750 and 5,000 young carp fry per hectare and manuring under an average temperature of 23.7°C, the unit weight was 433 to 621 g and the output 90%.

The advantages of raising marketable under-yearlings in a single vegetative growth period (i.e. changing from two-year rotation of pond fishery to one-year rotation) are as follows:

1. Reduction in the cost of raising fish.

2. Reduction in production costs since the need for expensive wintering ponds for under-yearlings has been precluded, in addition to maintaining rearing ponds for the same period. Hibernation is required only for brood fish and replenishing stock. In such a fish farm only the following ponds are mandatory: spawning, transplantation (nursery ponds), separate stock ponds for summer and winter maintenance, foraging, and quarantine-isolation ponds.

3. Reduction in quantity of carp yearlings as stocking material.

4. Reduction in food expenditures per kg of growth rate since the consumption of natural foods and artificial feed is cut back 20 to 25% compared to the maintenance of two-year fish.

From the foregoing it is obvious that raising marketable under yearlings, especially in the southern regions, is a promising and economically desirable venture.

Two Rearings in Foraging Ponds. The method of two rearings in foraging ponds in a single vegetative growth period of prolonged duration, possible in the southern regions of the USSR, has been developed by Movchan (1948). According to him two harvests of marketable fish are feasible within one season while simultaneously increasing the total produce and efficiency of the farm. For this purpose yearlings released in feeding ponds early in spring—after the necessary preparation, selection, feeding, and maintenance in hatching ponds, rearing ponds, wintering ponds, and foraging ponds—can achieve marketable weight toward midsummer. At the end of May the better fry are selected and released in the same pond.

The period of combined maintenance in the foraging pond of the two age groups is usually from May to end of July–beginning of August, when the two-year fish must be caught and sold as marketable produce. Only the under-yearlings remain in the foraging pond. In using this method the pond must necessarily be manured, food organisms for carp cultured, and the fish fed in accordance with the repetition of stocking. With this method V.A. Movchan achieved a fish productivity of 39 q/ha (26 q/ha of two-year fish weighing 500 g and 13 q/ha of marketable under-yearlings weighing 400 g).

In Uzbek SSR an experiment carried out by Puchaev (1967) has shown that five and six repeated stockings of yearlings and the same quantity of juvenile fish under artificial feeding and pond manuring (vegetative growth period 170 to 187 days) the weight of two-year fish toward the end of July–August was 500 to 810 g and the weight of marketable under-yearlings 400 to 520 g. The fry were released in foraging ponds immediately after catching from hatching ponds in one variant, and in another were released after prior rearing up to 5 g body weight. In the first variant fry with a unit weight of 0.03 g toward autumn yielded better results than those raised to 5 g weight. Their output was within the range of the commonly accepted norm for under-yearlings.

Two-year fish, before stocking with young fish commenced, were maintained alone in the foraging pond: in the first year of the experiment for 23 to 24 days and in the second year for 74 to 78 days.

Rearing two-year fish from yearlings of different initial weight (19 and 35 g) showed that better results are obtained with the use of larger stocking material. However, even from yearlings with a unit weight of 19 g the average weight of the two-year fish was 540 to 580 g and the weight of some specimens even higher. The total fish productivity with five repeated stockings was 1,030 to 1,530 kg/ha, and with ten repetitions 2,485 to 3,516 kg/ha. The production cost of marketable under-yearlings was lower than the cost of rearing marketable two-year fish by about 45 to 50%, which is associated with the reduction in the period and simplified scheme of fish culture.

Thus two rearings in a single vegetative period appears quite profitable in the southern regions. Healthy stocking material obtained from good brood fish should be used (yearlings and fry). Only ponds with good hydrochemical and hydrobiological regimes ought to be used and proper feeding and care of the fish must be ensured.

References

Ignat'ev, V.A. 1967. O vyrashchivanii tovarnogo segoleta karpa v Novosibirskoi oblasti (Raising marketable carp from under-yearlings in the

Novosibirsk district). Sbornik *Puti Uvelicheniya Produktsii Sel'skogo Khozyaistva. Zhivotnovodstvo.* Novosibirsk.

Ignat'ev, V.A. 1968. Tovarnye segoletki karpa v Sibiri (Marketable under-yearlings of carp in Siberia). *Rybovodstvo i Rybolovstvo,* no. 1.

Lyakhnovich, V.P. 1969. E'ffektivnost' udobreniya prudov (Efficiency of manuring ponds). *Rybovodstvo i Rybolovstvo,* no. 6.

Movchan, V.A. 1948. E'kologicheskie osnovy—intensifikatsii rosta karpa (Ecological Basis for Intensification of Carp Growth). Izd-vo Akad. Nauk UKrSSR, Kiev.

Movchan, Yu.V. 1960. Tovarny karp za odno leto (Marketable carp within one summer). *Rybovodstvo i Rybolovstvo,* no. 3.

Peskova, O.D. 1962. Vozmozhnost' vedeniya odnoletnei kul'tury v Gruzii (The possibility of one-year fish culture in Georgia). *Trudy Nauchn.-issled. Rybokhozyaistvennoi Stantsii Gruzii,* no. 7.

Puchaev, I.M. 1967. Sovmestnoe vyrashchivanie tovarnykh (stolovykh) segoletokov i dvukhletkov karpa v usloviyakh Uzbekskoi SSR [Combined raising of marketable (table) carp from under-yearlings and two-year fish under the conditions of Uzbek SSR]. Avtoref. Kand. Diss., Kiev.

Yaroshenko, M.F. 1965. O vozmozhnostyakh proizvodstvennogo vyrashchivaniya tovarnykh segoletkov karpa v prudakh Moldavii (The possibility of commercial culture of marketable under-yearlings in the ponds of Moldavia). *Izv. Mold. Fil. Akad. Nauk SSSR,* no. 4, p. 24.

Chapter XIV

SPECIAL TYPES OF WARM-WATER FISH CULTURE

Fish culture can also be organized in a complex with other branches of agriculture, and each combination can serve the purpose of a rational utilization of water-covered areas as ponds for fish rearing and other purposes.

Two forms of mixed farming are widespread—rice-fish and carp-duck culture. Peat bogs are used for fish rearing, which differ from usual ponds in thermal, hydrochemical, and hydrobiological regimes, as well as in natural fish productivity. Irrigation reservoirs, the water area of which is increasing, are another promising source for fish culture.

MIXED RICE-FISH FARMING

The area under rice cultivation is increasing every year and at present comprises hundreds of thousands of hectares. Culturing fish in rice fields is an ancient custom in a number of countries in Southeast Asia (China, Japan, Thailand, and Vietnam). Mixed farming is also done in Malaya, Indonesia, Ceylone, and some African countries. It has been successfully developed in Europe (Czechoslovakia, France, Hungary, Italy, etc.), the USA, and elsewhere. Paddy fields filled with water constitute water bodies which differ from the usual ponds in a number of hydrological and hydrobiological peculiarities, conditions of water supply, and hydroengineering installations, the design of which is determined by agronomical requirements.

In recent years the USSR has begun to utilize rice fields for fish culture (Krasnodar and Stavropol territories, Astrakhan, Kherson, Nikolaev, Crimean districts, Uzbek and Kirgizian SSR). However, rice-fish farming has yet to find a widespread application in this country.

At present two methods of fish culture in rice fields are followed in world practice: (1) together with rice in the same field, and (2) without rice, i.e., leaving the rice plots fallow under water.

Fish and Rice Culture. Besides culturing fish on fallow rice plots kept under water, fish farming may be combined with rice growing. Either method produces marketable fish. The stocking material is either brought in from fish nurseries or reared in the rice fields.

Rice fields are separate well-planned plots bordered by bunds of a height of about 40 cm. Their area is usually 0.5 to 5.0 hectares and sometimes more. Areas larger than this are less convenient for fish culture, especially at the time of fishing, because the fish hide between the rice stalks and do not enter the water channels.

The water enters the plots through a network of irrigation canals and therefrom collects into an outlet channel (Figure 92). Thus, a constant column of water is created in the rice plots to a depth of 10 to 15 cm and maximum 20 cm. Losses due to evaporation of the water layer during high temperature are kept constant with a complete replacement of the water over a period of 1.5 to 3.0 days.

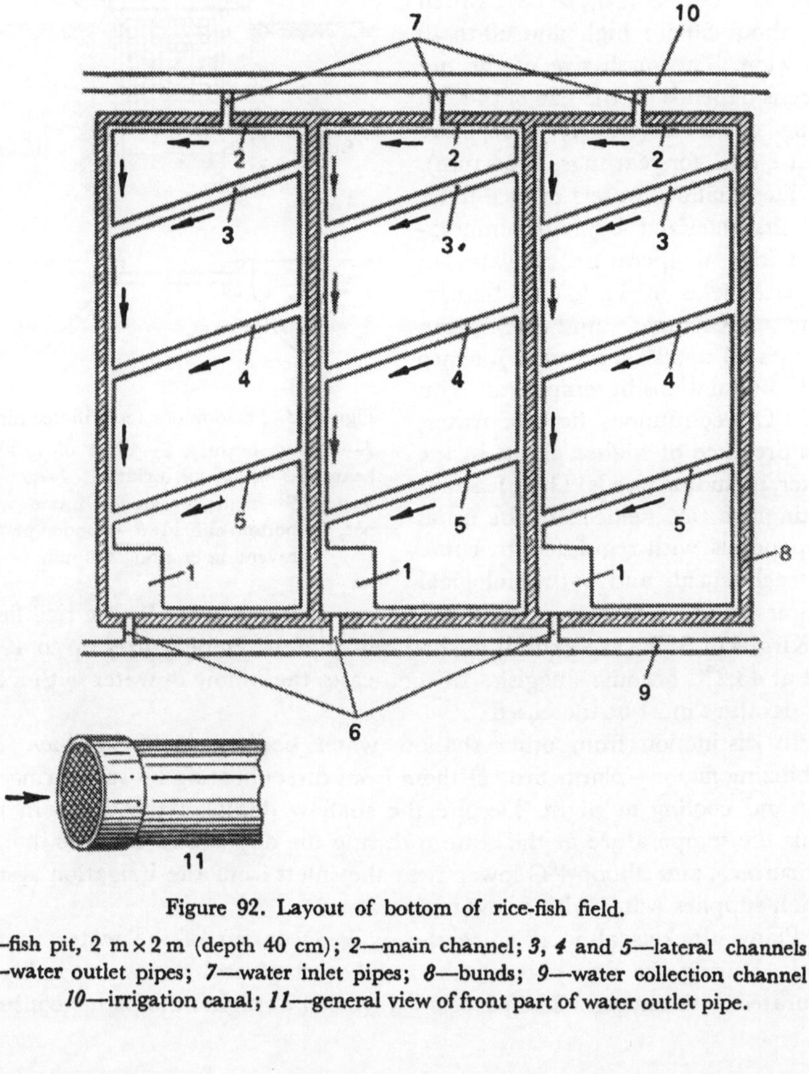

Figure 92. Layout of bottom of rice-fish field.

1—fish pit, 2 m × 2 m (depth 40 cm); *2*—main channel; *3, 4* and *5*—lateral channels; *6*—water outlet pipes; *7*—water inlet pipes; *8*—bunds; *9*—water collection channel; *10*—irrigation canal; *11*—general view of front part of water outlet pipe.

The water supply is interconnected. Water from the irrigating points enters the nearest plot, from whence it flows into a series of plots before passing into the outlet canal.

The adoption of rice fields for culturing fish is neither complex nor very costly. For this purpose small channels with a depth and width of 25 to 35 cm are cut alongside the bunds bordering the plots. The bunds are strengthened with the soil from the channels. These channels serve as a temporary shelter for carp at the time of water release from the plots. To prevent the egress of the fish under culture and the ingress of wild trash and predatory fishes in the fields, protective screens are fitted at the water inlets and outlets (Figure 93), which are about 50 cm high and 60 to 70 cm wide. The mesh size of the net screens depends on the size of the fish being cultured (for fry, not more than 1 mm; for yearlings, 3 to 5 mm).

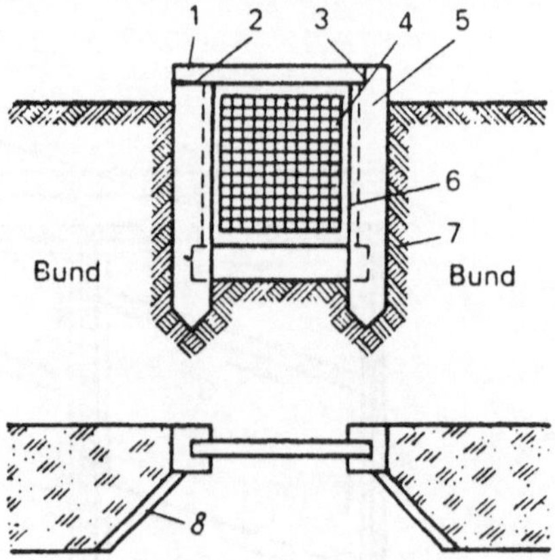

Figure 93. Fixation of screens in rice plots. *1*—locking beam; *2*—joints of locking beam; *3*—rings for locking; *4*—net on frame; *5*—stand; *6*—slot for frame with net; *7*—bottom of field; *8*—wooden planks preventing erosion of bund.

The shallow waters of rice fields are characterized by high temperatures (day temperature of water in the rice fields of Tashkent, Samarkand, Kokand, and Margelan reaches 36 to 38°C and more), sharp daily fluctuations in temperature (up to 19°C), continuous flow of water, and presence of higher plants in the water, including weeds. These features distinguish rice fields from the usual carp ponds with regulated thermal, hydrochemical, and hydrobiological regimes, and as such require unique farm management. In the rice fields in Kirgizian SSR carp exhibit normal behavior at temperatures up to 40°C, but at 43.2°C become sluggish. In such cases the inflow of water with a low temperature must be increased.

In distinction from other shallow water bodies, rice fields have one stabilizing factor—plants protect them from direct heating by sun during the day and cooling at night. Despite the shallow depth of the water in rice fields the temperature at the bottom during the day is 2 to 3°C less than at the surface, and about 4°C lower near the inlets from the irrigation system which supplies water with a low temperature.

Plants thickets also influence the oxygen content of the water in rice fields. During the day photosynthesis takes place and the water is highly saturated with oxygen. The process is reversed at night: the plants consume the oxygen and, moreover, the oxidative processes of decomposition of different organic substances continue at the bottom and in the water column. As a result, the oxygen regime deteriorates significantly during the night and early morning hours; however, it seldom reaches a lethal level. The remaining hydrochemical indexes, in spite of their specificity, are quite suitable for carp rearing.

The zooplankton and zoobenthos available in rice fields as food for fish have been assessed differently by various authors. Most believe that the food stocks in rice fields are not high and consequently the natural fish productivity of such fields is lower than in the usual ponds in most of the regions of the USSR (80 to 150 kg/ha; in Hungary 43 to 87 kg/ha). Contrarily, it has been asserted that the rice fields in Uzbek SSR are as good as the usual ponds and that natural fish productivity there can reach 150 to 200 kg/ha. In the rice fields of Ferghana Valley 183 species of planktonic and benthic organisms have been recorded, which constitute a significant quantity of biomass. The density of zooplankton reaches 180 to 600/m^3 of water, and that of zoobenthos 3,500 to 18,000 per m^2 of bottom area. The rice fields of northern Tadzhikistan are distinguished by a relatively high productivity of zooplankton and in high mountainous areas (Vorukh, Chorkukh) in the second half of summer and autumn in the rice fields of Vorukh have reached 157,400 to 437,500/m^3.

Other researchers on the other hand have pointed out that rice fields are relatively poor in live foods, and this condition compels carp fry in the process of growth to resort to different types of algae and macrophytes for food. In the rice fields of Uzbek and Kazakh SSR more than 200 species of algae have been recorded. As such, the fish could consume a large variety and quantity of phytoplankton. Fry are capable of eating ostracods, small cladocerans, copepods, desmids, and blue-green algae in large quantities.

Mass development of algae has beneficial and harmful implications. On the positive side algae serve as a source of energy for nitrogen-fixing bacteria; many species of blue-green algae are capable of fixing nitrogen from air and, after they perish, enrich the soil with organic matter and improve its structure. Finally, they serve as food for fish as well as for zooplankton and benthos. On the negative side most species of algae form a dense film on the bottom of the plots and prevent the normal development of rice. This film also absorbs nutrients from the water and reduces the effect of manuring. This undesirable phenomenon is obviated somewhat by culturing carp and other benthic-feeding fishes, which destroy the algal film in search of food.

Toward the end of the first vegetative growth period carp also consume weed seeds in rice fields. In their second year of life plant food (seeds of weed plants and phytoplankton) together with zooplankton and benthos constitute 80% of their food intake, especially in the last month of the feeding season.

Thus, the usual nature of carp feeding in rice fields changes: carp adapts to the given conditions of any rice field.

The differences prevailing in the composition of natural foods in rice fields are mostly the result of different agronomical practices and the conditions characteristic of different rice-growing areas.

Combined rice-fish farms are managed with one-year and two-year rotations. In some farms spawning is done in one set of plots, fry reared in another set, and a third set used for foraging. The construction of wintering ponds in the main rice-growing regions in the south is much cheaper than in the north; the ice cover is generally thin in the south and so the wintering ponds need not be deeper than 1 m. For hibernation purposes the irrigation canal may also be used, a large part of which does not operate during winters. Shallow earthen basins (not deeper than 1 m) with a constant supply of fresh water may also be used, in which case the irrigation canal is provided with screens to prevent the upward movement of fish.

The management of carp culture in rice fields almost does not differ from that at usual fish farms. Since it is closely associated with the agronomical base of the rice crop, however, certain deviations perforce are expected.

Cultivation of rice is undertaken at different times in different regions, depending on climatic conditions—from the beginning of March (Central Asia) up to 15 to 30 (northern Caucasus). The commencement and termination of fish culture coincides with these periods. One or two sets of spawners are kept for breeding in specially adapted rice fields, depending on their area. On the second day after spawning the spent fish are removed. The water outlet is covered with a gunnycloth or screen to a height 1.0 to 1.5 m and an area of about 1.0 to 1.5 m $\times$ 1.0 to 1.5 m, to prevent the spawners from jumping out. For the same reason the bunds bordering the spawning fields are built a little higher and more compact, and measures taken for supplying clear water to the field, especially when suspended particles abound. Turbid water can be passed first through other fields containing no fish, where the suspended particles will settle.

To prevent damage to the rice plants, fry from the spawning fields are transferred to rearing fields near the inlet of fresh water. At the point of entry of this water a small channel is dug and enclosed by a bund, leaving just a small passage; the water supply is stopped for two to three days and then fresh water released into the area enclosed by the bund. The fry congregate at the inlet of water through the passage made in the bund as well as in the channel. When they gather in sufficiently large quantities, the passage is closed and the fry caught with dip nets and transferred to the rearing fields. This operation is repeated until all the fry have been caught. They are counted and the quantity obtained from a rudd or pair of spawners determined.

Stocking norms for rearing in rice fields are calculated in the same way as those for standard rearing ponds in a carp farm. Autumn catching in the rearing and foraging fields is linked with the time of paddy harvest, the growth period of which varies from 85 to 120 days. In the southern regions harvesting is completed by September 1 to 15, which is too early for transferring the fingerlings to wintering ponds. Hence in these regions the rice fields are again filled with water after the paddy is harvested and the fingerlings kept in them until winter. The same practice is followed for two-year fish.

The output from the rearing fields varies from 45 to 78%, depending on the management level of the combined rice-carp farms. With a stocking rate of 5,000 to 10,000 fry per hectare the under-yearlings toward autumn attain a body weight of 40 to 50 g. Calculations for stocking norms in foraging fields are also done in accordance with the formula used for standard foraging ponds, taking the body weight of two-year fish toward autumn as 450 g. However, the output of 90 to 95% established for standard foraging ponds, in rice fields drops to 60 to 70%; this is attributed to the larger number of enemies of carp in such waters (larvae of dragonflies and beetles, frogs, water snakes, gulls, herons, jackals, fox, etc.). In practice the output is actually lower, not more than 50%, since complete catching of fish from fields of large areas is generally difficult. To obtain better results measures to control fish enemies should be implemented (fixing of traps, killing of birds and wild animals, surveying and destroying frog eggs, and fixing of small-mesh screens to prevent the entry of larvae of dragonflies and beetles into the fields from the irrigation system).

The approximate stocking norms of carp in rice fields for raising marketable fish are given in Table 47.

Table 47. Approximate rate of stocking carp in rice fields

Age	Weight of stocking material, g	Fish productivity from rice fields, kg/ha	Stocking norms, number of fish/ha	Weight of fish in autumn, g	Output of fish in autumn, %
15-day-old fry	0.3–0.5	125	1,000	250	50
40-day-old juvenile fish	5–10	125	600	300	70
Yearlings	30	110	250	500	85

In the course of fish culture in rice fields mainly two-year carp are grown from yearlings, preferably with a higher body weight (50 to 70 g). This makes it possible to obtain larger two-year fish toward autumn. Fish stocking material is better raised in a specialized fish hatchery or, as mentioned above, in rice fields adapted for this purpose. The advantages of specialized fish nurseries are as follows:

1. Overretention of stocking material up to the end of May–beginning of June is possible, when it is more appropriate to keep the fish for feeding on the basis of agronomical conditions. In the southern regions the fish hibernate in such fish nurseries (rearing ponds) and remain there until transferred to the rice fields for feeding.

2. If nursery ponds are available, after spawning and hatching the fry can be transferred to them and subsequently to the rearing ponds.

Thus, the advantages of rice-carp farming are not only obtaining two crops from one and the same area (rice and fish), but an increment in rice yield and an improvement in grain quality (greater absolute weight, lower percentage of husk, and fewer empty grains). For example, the yields of rice fields in the Ferghana Valley (Uzbek SSR) increased by 10 to 11% (9.2 q/ha), in Kirgizian SSR by 12 to 15%, in Ukrainian SSR on the average by 6.48 q/ha, and in Kazakh SSR by 6 to 8 q/ha.

The increase in rice yield from fields containing fish is explained by the fact that carp nosing through the soil in search of food destroy any biological film, consume the rice bug (a major pest of rice), and eat the weed seeds falling in the water. Moreover, the larval population of mosquitoes also declines considerably when grass carp are reared in rice fields.

In Uzbek SSR and other regions of rice cultivation rice farms have begun to expand the area of their rice fields (10 to 15 and even 20 hectares versus 1 to 2 hectares in the past); introduce high yielding rice varieties; mechanize field operations; incorporate crop rotations; and raise plant density (300 to 400 plants/m^2 versus 150 to 200 plants earlier). These measures have increased the rice yield by 25 to 30% but, sad to say, have unavoidably influenced the hydrochemical regimes in the rice fields, making them less suitable for rearing fish. As a result, fish productivity in such farms has declined markedly, falling to 30 to 60 kg/ha (Uzbek SSR).

At the same time, experimental data from Ukrainian SSR reveal that better results in carp rearing in rice fields could be obtained by planting yearlings of a unit weight of 15 to 20 g (and better up to 70 g). Within a period of 80 to 90 days, two-year fish had grown to 400 to 650 g and more. Even with these results, however, the output of fish decreased considerably and fish productivity dropped due to improper planning of the bottom and difficulties in complete fish catching.

In addition to carp, goldfish and probably silver carp, which feed on algae, could also be raised in rice fields. The fry of grass carp could likewise be released as they do not browse on the rice stems. Under the conditions of Kuban such fry have grown to 15 to 25 g at the time of harvest with an output of 30 to 50% of the stocked material.

In foreign countries tilapia (*Tilapia mossambica*) is raised in rice fields and, under conditions of additional manuring of the fields, from 170 to 315 kg/ha fish obtained together with an increase of 3.0 to 13.2% in rice yield.

In addition to extensive farming based on the use of natural foods, intensive farming using artificial feeds can also be done in rice fields to increase the yield of fish per unit area. The technique of feeding is analogous to that in the common culture of fish in ponds. Local resources can be used more appropriately as food: pupae of silkworm (waste) mixed with rye meal (one part) and the tadpoles, frogs, and fish (minced in a meat grinder) which are found in abundance in rice fields mixed with 10 to 15% flour or the sweepings from a grinding mill. Carp should be kept on natural foods up to the age of one month, for which it is essential to develop live foods. The development of natural fish productivity is decelerated by fertilizers applied for the main crop, rice. However, as revealed by studies conducted in Krasnodar, the quantity of these fertilizers is not sufficient for the desired development of phytoplankton, and the periods of fertilizer application do not correspond to those expected in fisheries. Applications must be more frequent than normally practiced in rice cultivation, which may disrupt the budget for the latter.

Combining rice and fish culture in rice fields ensures only a small increase in rice yields and provides not more than one quintal, maximum 1.5 to 2.0 q/fish, from the same rice fields. In a number of cases, especially when the rice fields are inadequately prepared, fish culture is hardly profitable. Nevertheless, in many Eastern countries different fish species are cultured together with rice (Siamese gourami, milkfish, burbot).

The practice of fish cultivation without rice or in fallow rice fields kept under water, is gradually being recognized as a more efficient method of mixed rice-fish culture.

Fish Culture in Fallow Rice Fields Kept under Water. Fallow rice fields are filled with water for the purpose of fish culture by following the common methods of intensification. This makes possible an increment in fish productivity, better utilization of mechanization, and greater clearance of the rice fields by foraging phytophagous fishes (in the first place, grass carp) together with carp, which consume weed plants (Figure 94). In Karakalpak ASSR (Rice State Farm "Maiyab") the rearing of two-year grass carp together with common carp after growing a rice crop for two consecutive years led to a rapid and more complete destruction of reed, algae, and pond weeds. Carp feeds very well on weed seeds and prefers the seeds of lesser millets over artificial feeds. The usual feeding behavior of carp changes in rice fields and it adjusts readily to the particular conditions of a given field.

Fish productivity increases insignificantly in mixed rice-fish forming. In Uzbek SSR it is expected that the annual separation of one-seventh of the total area under rice into fallow underwater fisheries will, under intensive management, raise the yields of rice 10 q/ha and produce 50,000 to 60,000 quintals of fish; in Krasnodar 300,000 quintals of fish ought to be obtained by the same method. The control measures taken against weeds include the

Figure 94. *A*—Fallow rice field under water overgrown with plants; *B*—Fallow field under water cleared by grass carp; plants remained in control plot where grass carp could not gain ingress.

application of chemical substances and carp rotation with crop alternation. In Uzbek SSR rice is alternated with such valuable crops as maize, lucerne, barley, etc.; in Krasnodar it is rotated with lucerne and legumes (wheat, vetch-oat mixture).

In the seven-year rotation program followed at the State Farm "Krasnoarmeisky" in Krasnodar crops are rotated as follows: rice, rice, rice, lucerne, lucerne, rice, vetch-oat mixture. In using one-seventh of the area for fish culture the vetch-oat mixture was replaced by fish, while the other crops and their sequence remained unchanged. The replacement of vetch-oat mixture by fish was found to be quite profitable.

Proper planning is necessary for fallow rice fields kept under water for fish rearing and for successive rice crops. The water level in each plot is planned in such a way that the maximum difference in depth is not more than 10 cm and the deviation from the average horizontal area of the feeding pond not more than ± 5 cm. Additional elevation of the bunds is also necessary so that the water level does not fall below 50 cm. The usual water collection channels and fish pits must be made in each field to facilitate fish catching. To prevent fish egress from the field, and to forestall ingress of unwanted fishes protective screens with a metal mesh 1.0 m × 1.5 cm are installed at the inlet and outlet points. Similar screens are fixed in each field to prevent the transfer of fish from one field to another. The obvious advantage of fish rearing in fallow fields kept under water is that the period of fish culture is not tied to the period of rice cultivation. The fish can be released much earlier, as soon as water appears in the irrigation system. This makes it possible not only to increase the vegetative growth period of the fish being cultured, but to avoid overretention of the stocking material in wintering ponds, which frequently leads to infectious diseases and waste.

Stocking norms for carp are calculated in the same manner as for standard fish ponds. Grass carp and silver carp are also raised to improve the conditions of the rice fields (Table 48).

Table 48. Stocking norms of fish in fallow rice fields kept under water

Fish species	Average unit weight, g	Stocking density, number of fish/ha
Common carp	50	1,000–1,200
Grass carp	120–140	50–60
Silver carp	60	600–700

In the conditions of Uzbek SSR it is possible to obtain 10 to 12 q fish/ha and in Krasnodar territory 6 to 8 and even 10 q/ha. The weight of two-year fish by autumn in Uzbek SSR reaches the following values: carp—525 g, grass carp—800 g, silver carp—500 g; in Krasnodar: carp—up to 550 g, silver carp—430 to 530 g, and grass carp—700 to 900 g. Calculations on the economic efficiency of fish culture in fallow rice fields kept under water for an area of 315 ha in Uzbek SSR, have shown that the cost of the fish was 53.65 rubles, gain per hectare 209.04 rubles, and net gain 65,846.32 rubles. The farm realized a net profit of 33 kopecks per ruble. In Krasnodar the per hectare profit was even higher, i.e., 400 rubles. These data prove the economic benefits of fish culture in fallow rice fields kept under water.

References

Chizhik, A.K., S.A. Dunets et al. 1967. Rezul'taty opytnykh rabot po vyrashchivaniyu ryby v risovykh chekakh (Results of experimental work done on raising fish in rice fields). Sbornik *Rybnoe Khoz-vo.* Urozhai, Kiev, no. 3.

Chizhov, N.I. and A.I. Anoshin. 1969. Vyrashchivanie ryby na risovykh polyakh, vyvedennykh pod vodny par (Fish raising in fallow rice fields kept under water). *Trudy VNIIPRKh.* Moscow, vol. 16.

Meien, V.A. 1940. Razvedenie ryby na risovykh polyakh (Fish Rearing in Rice Fields). Pishchepromizdat, Moscow.

Mukhamediev, A.M. 1961. Razvedenie ryb na risovykh polyakh (Fish Rearing in Rice Fields). Gosizdat UzbSSR, Tashkent.

Obukhova, V.M. 1961. Al'goflora risovykh polei nekotorykh raionov Kazakhstana (Algal flora of rice fields in some regions of Kazakhstan). *Trudy Inst. Botaniki Akad. Nauk KazSSR,* vol. 10.

Obukhova, V.M. 1961. Znachenie vodoroslei v rezhime risovykh polei (Significance of algae in conditions of rice fields). *Izv. Akad. Nauk KazSSR, Ser. Botaniki i Pochvovedeniya,* no. 1, p. 10.

Piskarev, K.V., E. Omarov and I. Ibrahimov. 1965. Opyt vyrashchivaniya karpa na risovykh polyakh Yuzhnoi Kirgizii (Experience in carp raising in the rice fields of southern Kirgiziya). *Uch. Zap. Oshskogo Gos. Ped. Instituta,* no. 6.

Rekomendatsii po vyrashchivaniyu ryby na risovykh chekakh (Recommendations for Fish Raising in Rice Fields). Krasnodar, 1970.

Tezisy dokladov konferentsii po voprosam rybovodstva na risovykh chekakh i irrigatsionnykh sistemakh (Abst. Conf. on the Problems of Fish Rearing in Rice fields and Irrigation Systems). Krasnodar, 1968.

Yudin, V.L. 1966. Rybovodstvo na risovykh polyakh Uzbekistana (Fish rearing in the rice fields of Uzbekistan). Sbornik *Biol. Osnovy Rybnogo Khoz-va na Vodoemakh Sr. Azii i Kazakhstana.* Alma-Ata.

Yudin, V.L. 1967. O razvedenii ryby na risovykh polyakh Uzbekistana (Fish rearing in the rice fields of Uzbekistan). *Uzbekskii Biol. Zhurn.,* no. 3.

Yudin, V.L. 1968. Belyi amur v riso-rybnom sevooborote (Grass carp in

rice-fish rotation). Sbornik *Novye Issledovaniya po Ekologii i Razvedeniyu Rastitel'noyadnykh Ryb*. Akad. Sci. USSR, Ikthiol. Komissii, Moscow.
Yudin, V.L., A.A. Yudlere and Ya. Borisov. 1968. Rekomendatsii po biotekhnike rybovodstva na risovykh polyakh v usloviyakh vodnogo para (Recommendations on the Biotechnique of Fish Rearing in Fallow Rice Fields Kept under Water). UZINII, Tashkent.

MIXED CARP-DUCK FARMING

The grazing of ducks in carp feeding ponds is a profitable proposition. Ducks moving over the silt loosen it and improve soil aeration. Moreover their droppings are a valuable organic manure; 100 kg of duck excreta contains 0.8 kg nitrogen, 1.5 kg phosphorus, and 0.4 kg potassium. The increase in fish weight as a result of manuring ponds with duck excreta is 60 to 70% higher than with poultry manure. Ducks are excellent conditioners of fish ponds as they feed on aquatic plants. The food value of several such plants is given in Table 49.

It can be seen from the data presented in the Table that compared to the best fodder crop, lucerne (*Medicago sativa*) aquatic plants are comparatively rich in raw proteins and mineral substances. Duckweed is nutritionally a rich fodder (see Chapter XI). Various kinds of aquatic animals and their

Table 49. Food value and chemical composition of coastal and aquatic plants

Plant	Percentage					
	Coefficient of feeding	Raw protein	Fat	Nonnitrogen soluble matter	Cellulose	Ash
1	2	3	4	5	6	7
			Coastal plants			
Lucerne	1.00	18.40	3.30	49.70	18.80	9.80
Clover	1.77	19.00	4.10	41.90	24.60	10.40
			Semiaquatic plants			
Water smartweed	0.34	24.26	3.85	27.76	34.80	9.33
Arrowhead	1.35	21.64	3.68	18.81	36.72	19.15
			Floating plants			
Common duckweed	3.41	25.75	4.65	27.24	24.57	17.79
Greater duckweed	3.76	20.94	2.66	35.18	26.60	17.79
Ivy-leaved duckweed	2.76	30.40	2.70	24.00	20.80	22.10
Oedogoniales (algae)	2.76	19.25	3.19	21.60	32.26	27.40

1	2	3	4	5	6	7
		Submerged plants				
Water thyme	1.60	18.30	2.50	42.50	16.00	20.10
Hornwort	0.82	17.38	2.64	38.86	27.99	13.33
Sago pondweed	1.15	20.94	2.64	36.55	26.07	13.00
Sharp-leaved pond-weed	1.60	21.78	2.56	28.16	26.75	26.25
Clasping-leaf pond-weed	1.35	18.26	2.54	27.26	46.34	7.25
Curly pondweed	0.36	18.89	2.31	46.80	16.94	14.66

Notes. 1. The term "coefficient of feeding" means the quantity of green mass of aquatic plants eaten by ducks compared to the quantity of lucerne taken as a unit.

2. Vegetative growth phase of all the plants up to flowering.

larvae live in the plants, which ducks prefer even to lucerne. Adult ducks of the Peking breed consume up to 1 kg of aquatic plants in 24 hours.

The usefulness and profitability of rearing ducks in foraging carp ponds is determined by the following considerations:

1. Ducks are not competitors for food with carp. They consume soft submerged plants and plants floating on the water surface; they nibble on the rhizomes of coarse plants and help in their destruction. Less concentrated feed is needed therefore and the expenditure involved in the removal of excessive soft, coarse, and floating plants is obviated.

2. Ducks consume tadpoles, frogs, and their eggs, as well as some aquatic insects—all pests of fish.

3. Ducks grow better while grazing on water and their productivity is higher.

4. Duck droppings, although small in quantity, are highly concentrated and constitute the cheapest organic manure. Such manure significantly improves the natural food base of the pond and the fish yield. Experience has shown that natural fish productivity increases from 28 to 91%.

5. Two products are obtained from one and the same pond area: fish and duck meat.

Data presented in Table 50 show the advantages of duck rearing under water grazing.

The fish productivity from two-year carp in foraging ponds also increases as a result of duck grazing (Table 51).

In Lithuanian SSR 40% of the total poultry meat produced comes from fish farms and that, too, from mixed carp-duck farming (rearing of 250 to 300 ducks/ha). Fish productivity has also increased from 28.5 to 113.1%. In 1969 the yield of duck meat in terms of live weight increased to 39,788 quintals versus 1,430.6 quintals in 1961. The cost of production was 87.87

Table 50. Comparative results from maintenance of ducks with and without grazing in ponds

Production parameters	Without grazing in ponds		With pond grazing	
	Khakhi breed	Peking breed	Khakhi breed	Peking breed
Egg-laying per duck (number of eggs laid in 36 days)	19	13	31	21
Laying percentage	53.8	28.8	33.1	37.2
Percentage of output of ducklings from eggs laid	36.0	34.9	46.2	54.3
Percentage of output of ducklings from fertilized eggs	55.5	44.4	68.7	74.4
Food consumption per kg of growth of young ducks (g)	670	840	457	570

Table 51. Fish productivity of carp foraging ponds as a result of duck grazing

Ponds	Average unit weight of 2-year carp during autumn catches (g)	Fish productivity, kg/ha
Control (without grazing)	353	200
Experimental ponds with duck grazing	411	383
Same	372	256
Same	425	361

rubles per quintal, percentage of survival of the ducklings 88, and labor consumption per quintal of weight increase 1.6 man days.[1]

Due to the increase in the natural food base in ponds under the influence of duck manure, it becomes possible to increase the stocking rate of carp. In the process of calculating the stocking norms according to the formula for foraging ponds (see Chapter IV), the index of natural fish productivity increases on the average 30 to 40%.

All the favorable aspects of mixed carp-duck farming are realized when these conditions are observed: (a) raising of 25- to 30-day-old ducklings together with carp yearlings in the foraging ponds; (b) over-growth of higher aquatic plants in the ponds covers 30% or more of the area (ponds without

[1]In 1970 the survival of ducklings increased to 91.1%, total production of duck meat to 48,359 quintals, and labor consumption was 1.3 man days per quintal of increase in duck weight.

these plants are less suitable); (c) oxidizability of the natural water not over 20 mg O_2/l since at a higher oxidizability the ponds become so polluted that conditions favorable for the development of epizooty are created (branchiomycosis, etc.); and (d) observing correct stocking norms for ducks. As a tentative norm, 200 to 250 ducks/ha of pond area with a depth up to 1 m is recommended. The magnitude of this rate depends on the extent of overgrowth of higher aquatic plants in the ponds. If less than 30% of the pond is covered with plants, the stocking norm for ducks must be reduced accordingly. An increase in the norm can have an adverse effect. If, for example, 2,000 to 3,000 ducks/ha of pond area were stocked, the bird droppings would accumulate to such an extent that they would not be fully mineralized. Anaerobic processes would then develop leading to the release of poisonous substances and mass bird and fish poisoning. The release of 1,000 to 1,200 ducks/ha is permitted only in those cases when the water is meant for irrigating fields.

Recent studies (Chizhik, 1962) indicate that the stocking norms for ducks may soon be significantly raised.

In organizing fish culture in lowland peat bogs the best results were obtained under a stocking norm of 100 ducks/ha (maximum 125 to 150 ducks). The current rates of releasing ducks in a combined carp-duck farm are applicable to fish ponds of small areas. With reference to larger ponds, 200 to 500 hectares, farm management is faced with certain adaptations. In Ukrainian SSR, according to the data of Shpet and Kharitonova (1965), the stocking rate of ducks for such large ponds should not be calculated for the total area. Ponds of equal area may have different rates of water exchange. Hence stocking rates should be determined only for the maintenance of ducks near duck colonies and on the basis of water volume, and not on the basis of pond area. Nearly 240 to 300 m^3 water per year should preferably be allowed for each duck raised.

In the management of mixed carp-duck farming one must ensure that the ducks go around all the corners of the pond, thereby spreading their manural effect uniformly. Excessive accumulation of duck manure in patches can lead to pollution and an outbreak of branchiomycosis. Hence feed pans are distributed uniformly along the coastal areas of the pond, in the pond itself or in the form of floating rafts. Ducks are not released in the head pond or the one which supplies the entire farm with water as a safeguard against epizooty.

The water inlets in the ponds, areas where carp usually congregate, are protected from birds. In the initial stages of stocking the yearlings tend to remain in the coastal areas and swim at the surface. Such fish may be preyed on by the ducks. Hence the ducks are not released in the ponds during the first week of yearlings stocking. If, however, the ponds are stocked with healthy, strong, standard stocking material, 25- to 30-day-old ducklings pose no threat to the fish. However, even in this case the ducklings are fed abundantly to reduce their search for food. Marketable fish are caught in autumn after the ducks have been removed from the ponds; otherwise they might injure the carp when the water is reduced in the ponds. Ducks are also removed at the time of control catches.

Under the conditions tenable in most of the regions in the USSR, ducks of fast-growing breeds (for example, the Peking) kept for grazing in water become ready for marketing at the age of 65 to 70 days, when their weight reaches 2 kg or more. After marketing the first batch of ducks, the second batch is released into the ponds and allowed to remain up to the end of the vegetative growth period. In the southern regions raising a third batch of ducks is also possible.

Mixed carp-duck farms have significantly developed in a number of Eastern European countries (Czechoslovakia, German Democratic Republic, Hungary, etc.). In Hungary three breeds of ducks are raised together with carp—Peking, Hungarian, and British—characterized by white feathers and high productivity. In the usual foraging ponds mostly 200 ducks are released per hectare (the stocking norm can be increased to 500/ha) using three-week-old birds, which at the age of 50 to 59 days have attained a weight of 2.3 to 2.5 kg and are ready for the table. Under the climatic conditions of Hungary generally three batches of ducks (sometimes even four) can be raised in one season, giving a total yield of 2 to 4 tons of duck meat per hectare. In this process one man feeds and serves 10,000 to 12,000 ducks. Feeding is done from automatic feed pans (Figures 95 and 96) fitted around

Figure 95. Feeding platform for ducks at the experimental demonstration farm "Sarvash" in Hungary.

Figure 96. Feed pans (in background) installed in ponds around a filled island where ducks take rest (fish farm "Sarvash" in Hungary).

an island created by filled earth in the center of the pond as a place for the ducks to rest. The feed pans for the fish are installed at some distance from the feed pans for the ducks. The latter are filled with a limited quantity of feed at any time of the day; on the average 3.3 to 3.5 quintals of feed are required for one quintal of duck meat. Geese are also raised in some fish farms of Hungary.

In Hungary a special nine-year rotation is also followed where rice is rotated with carp and duck culture. After three years of fish and duck rearing the pond is dried, plowed, etc. for the cultivation of rice for three years. Thereafter it is again plowed, levelled, and converted into a dry field in which legumes and leucerne are planted. Subsequently the field is once more readied as a pond for fish and duck rearing.

As a result of maintaining ducks at fish farms the total increase in fish productivity from one acre increases by 20 to 25%, and because of improved soil fertility the rice yield by 30 to 35%. Under the influence of cultivating rice, legumes, and other crops for several years, as well as drying, the area is disinfected, which protects fish from diseases in the course of subsequent fish rearing.

In the German Democratic Republic mixed carp farming has been practiced for some years. Toward the end of 1970 the production of carp was 105,000 quintals and of duck meat from the same area, 100,000 quintals. The stocking norm per hectare of pond area is 200 to 300 ducks (sometimes more), which increases the fish productivity on the average by 1.0 to 1.5 q/ha. One laborer serves a pond of 100 hectares and feeds 10,000 ducks.

In Peitza—one of the best mixed farms in the German Democratic Republic—2,162 tons of marketable fish and 1,200 tons of duck were obtained in 1970. The ducks are released in the ponds at the age of 19 to 20 days and fed for 51 days. By that time they have grown to 2.4 to 2.5 kg in weight. A longer duration of grazing is not considered appropriate since, besides the consumption of feed, the fat content increases in the meat. The ducks are protected from predators (jackdaws, crows, cats, foxes, etc.).

In Czechoslovakia ducks are fed for 54 to 58 days.

Geese are also grazed in ponds in the USSR, and a stocking norm of 20 to 25 geese per hectare of coastal pasture is recommended. The pond serves mainly as a swimming pool for geese in which they dip, drink, and breed. Geese are better for manuring the pond; however, their consumption of aquatic plants is negligible and peripheral meadows must be provided for them (100 m from the water line along the shore).

Geese droppings as a manure for the ponds increase natural fish productivity. For example, in the steppe zone of Ukrainian SSR natural fish productivity increased 40.6% as a result of geese grazing in ponds.

References

Baranov, S.A. 1965. Ryaska kak kormovoe rastenie i vozmozhnosti ee massovogo kul'tivirovaniya (Duckweed as a fodder plant and the possibilities of its large-scale cultivation). *Sbornik Nauchno-tekhn. Inform. VNIRO*. Moscow, no. 10.

Chizhik, A.K. 1962. Rybovodnoe ispol'zovanie kolkhoznykh prudov lesostepnoi zony USSR raznogo khozyaistvennogo naznacheniya (na primere kolkhozov Volochisskogo r-na Khmel'nitskoi oblasti) [Exploitation of the ponds in collective farms in the forest-steppe zone of Ukrainian SSR for different production purposes (based on the example of the collective farms of the Volochissk region of Khmel'nitskii district)]. Avtoref. Kand. Diss., Kiev.

Chizhov, N.I. 1966. Rybovodstvo v Vengerskoi Narodnoi Respublike (Fisheries in the Hungarian People's Republic). *Rybovodstvo i Rybolovstvo,* no. 4.

Letichevskii, M. 1961. Ryba i utki s odnogo vodoema (Fishes and ducks from the same reservoir). *Sel'skoe Khoz-vo Povolzh'ya.* no. 12.

Martyshev, F.G. 1956. Kombinirovannoe karpo-utinoe khozyaistvo (mixed carp-duck farming). *Ptitsevodstvo,* no. 5.

Menzel, H.V. 1967. Über die entenzüchtung in ungarischen teichwirtschaften, *Dtsch. Fisch. Ztg.*, no. 6.

Schuberth, L. 1967. Der stand der entenzüchtung die weitere rationalisierung der zuchtarbeit, *Dtsch. Fisch. Ztg.*, no. 6.

Shpet, G.I. and N.N. Kharitonova. 1965. Vliyanie vygula tutok na povyshenie ryboproduktivnosti krupnykh karpovykh prudov (Effect of duck grazing in increasing the productivity of large carp ponds). Sbornik *Rybn. Khoz-vo,* Urozhai, Kiev, no. 2.

Frček, M. 1957. Vliv chovu kachen na priozenu vyrobnost rybniku, *Českosl. Rybařsvi,* no. 6.

Valenta, B.A. 1969. Proizvodstvo myasa utok i nekotorue voprosy intensifikatsii utkovodstva v rybkhozakh Litovskoi SSR (Production of duck meat and some problems of intensification of duck farming in fish farms of Lithuanian SSR). Avtoref. Kand. Diss., Kaunas.

Vavilkin, A.S. 1954. Sovmestnoe vyrashchivanie ryby i utok v usloviyakh trofyanikh kar'erov (Mixed farming of fish and ducks under the conditions of peat quarries). *Referaty Dokladov TSKhA,* no. 19.

FISH CULTURE IN PEAT QUARRIES

The number of peat quarries is increasing due to the development of the peat industry. At present, in the RSFSR alone there are about 400,000 hectares under peat quarries. It is expected that by 1980 the total area under quarries in the USSR will be one million hectares. There are three types of peat bogs: highland, transitional, and lowland. Categorization is based on the composition of the vegetation, the location of the bogs in relation to the relief of the locality, and the availability of mineral matter and nitrogen for plants. The important feature of peat is its genesis. The quantity of major biogenic elements in the ash of highland and lowland bogs is given in Table 52.

Peat in lowlands is almost three times richer in mineral elements[2] than that in highland; it contains more phosphorus, potassium, calcium, and nitrogen.

Table 52. Quantity of major biogenic elements and ash content as percentage of dry matter (from K.N. Shishkov)

Quantity	Highland bog					Lowland bog				
	Ash	CaO	K_2O	P_2O_5	N	Ash	CaO	K_2O	P_2O_5	N
Maximum	4.7	1.40	0.11	0.98	2.56	70.0	36.90	0.60	1.42	3.94
Minimum	1.8	0.24	0.01	0.01	0.48	6.0	0.34	0.01	1.11	1.14
Average	3.5	0.65	0.05	0.23	1.29	10.3	3.48	0.14	1.43	2.69

[2]Bogs with a mixture of sand and silt as well as lime are characterized by a higher ash content.

However, the phosphorus and potassium in both types of peat, taking into consideration fishery requirements, is sufficient. This deficiency, especially that of phosphorus, is one reason for low natural fish productivity.

The highland or moss type of peat marsh is characterized by higher acidity and poorer decomposition. Fish survival in water of higher acidity is precarious. To make highland bogs suitable for fish culture reclamation works are essential; among these lime application is the most important.

The transitional type of peat quarry, as its name implies, occupies an intermediate position between highland and lowland quarries. Its properties vary from those closer to the highland to those more characteristic of lowland quarries.

The depth of peat quarries depends on the depth of peat deposition. For the purpose of fish culture those quarries are selected which have a depth corresponding to that of the requisite category of fish ponds in standard farms. In using fallow peat quarries for fish culture it should be remembered that the depth is not uniform due to the pattern of peat excavation. The bottom of such quarries is generally covered with a slushy layer of peat to which, with the passage of time, a silt layer is added. In shallow peat quarries, shallow waters are created (depth from 20 to 80 cm) with an unstable water regime in which every atmospheric change is reflected. Such an unstable hydrochemical regime likens peat quarries to other shallow water bodies, in particular rice fields. The water temperature in shallow water quarries is, in most cases, higher than the atmospheric temperature or close to it. This is a characteristic property of shallow water bodies.

Peat quarries filled with water differ from ponds located on nonpeaty soils by the formation of a unique hydrochemical and hydrobiological regime. The existing norms for evaluating factors in standard ponds are purely relative in characterizing the medium of peat ponds; the same norms may be harmful or beneficial depending on several other elements. The number of excavations is reflected in the chemical properties of the water of peat ponds. Ponds three to four years old or more are quite suitable in their hydrochemical parameters for carp farming. Changes which occur in peat ponds with the passage of time ultimately lead to their "maturation", i.e., the chemical nature of the water improves for culture.

It is important to know the pattern of the water regime in a peat pond in the process of its biological utilization. The maturation of the ponds may proceed differently due to the presence of various types of peat or, with just one type may depend on the nature and conditions of the peat deposits. Hence the physico-chemical nature of different types of peat ponds should be taken into account during the course of pond pisciculture.

Natural fish productivity, the primary index, depends on a number of factors: nature of the formation of peat deposits influencing the hydrochemical and hydrobiological regimes; preparation of the peat ponds for stocking fish;

number of small trash fish in them, competitors for the food of the main object of culture, carp; and maintenance of the ponds and fish throughout the vegetative growth period. Furthermore, aquatic plants, especially toward the middle of summer, quite often form a continuous green carpet which shades the pond, thereby reducing the free water surface and obstructing fish catches. Such a carpet usually forms with the luxurious development of duckweed and filamentous algae.

Nevertheless, the potential of peat ponds for fish culture has been recognized and efforts directed toward its realization.

Peat ponds created from lowland bogs are more promising for pisciculture, since peat deposits comprise about 70% of the total area, creating more favorable conditions for the growth and development of fish. In the nonchernozem zone of the USSR the soil of bogs of this type is characterized by higher ash content, low moisture retention, and a higher level of peat decomposition. Depending on plant composition, nutrients in the water, chemical properties of the water of the feeding rivers, and reaction of the soil solution, the nitrogen content of the soil in lowland bogs varies from 1.6 to 4.0% and calcium from 1.2 to 6.0%. These soils are much richer in phosphorus and differ little from other soil types with regard to potassium content. Microorganisms are also more abundant in soils of lowland bogs.

Lowland bogs are quite suitable for the culture of carp, tench, grass carp, bighead, silver carp, goldfish, Crucian carp, pike, sander, and under especially favorable conditions, even trout and whitefish.

The techniques of fish culture in peat ponds are similar to those applied in ponds located on nonpeaty soils but include certain variations. In releasing spawners in the spawning ponds it must be kept in mind that the growth of aquatic plants in peat ponds is slightly delayed because of the later thawing of this soil compared to mineral soils. Peaty soils freeze to a lesser depth than mineral soils but remain frozen for a longer time. The absence of meadow plants and the delay in development of aquatic plants as a substratum for egg-laying of carp in the spawning ponds slightly shifts the spawning periods, although it has been noted that spawners recruited from peat ponds breed one to two weeks earlier. In the case of poor development of aquatic plants (the substratum for eggs) artificial media are provided.

In selecting spawners for breeding in peat ponds preference is given to those raised at a farm in which the conditions do not differ sharply from those in peat quarries. For a rational management of fish culture in peat ponds it is appropriate to develop a new breed of carp, one adaptable to marshy water bodies.

The fertilization of eggs is nearly 83 to 92%, i.e., similar to that in standard hatching ponds.

Standard fingerlings of carp can be obtained from rearing peat ponds with fairly minimal losses. The natural fish productivity in these ponds is

75 to 220 kg/ha. In recent years, fingerlings of grass and silver carp and bighead have been successfully reared in combination with common carp in peat ponds (Ivano district). Grass and silver carp, even under a fairly high density of stocking (common carp 53,000/ha, grass carp 19,000/ha, silver carp and bighead 7,200/ha) exhibit a better growth rate than common carp alone. However, the ratio of individual fish species in the case of stocking in different categories of peat ponds, considering their unique nature, should differ from standard ponds and hence requires study. The feeding role of

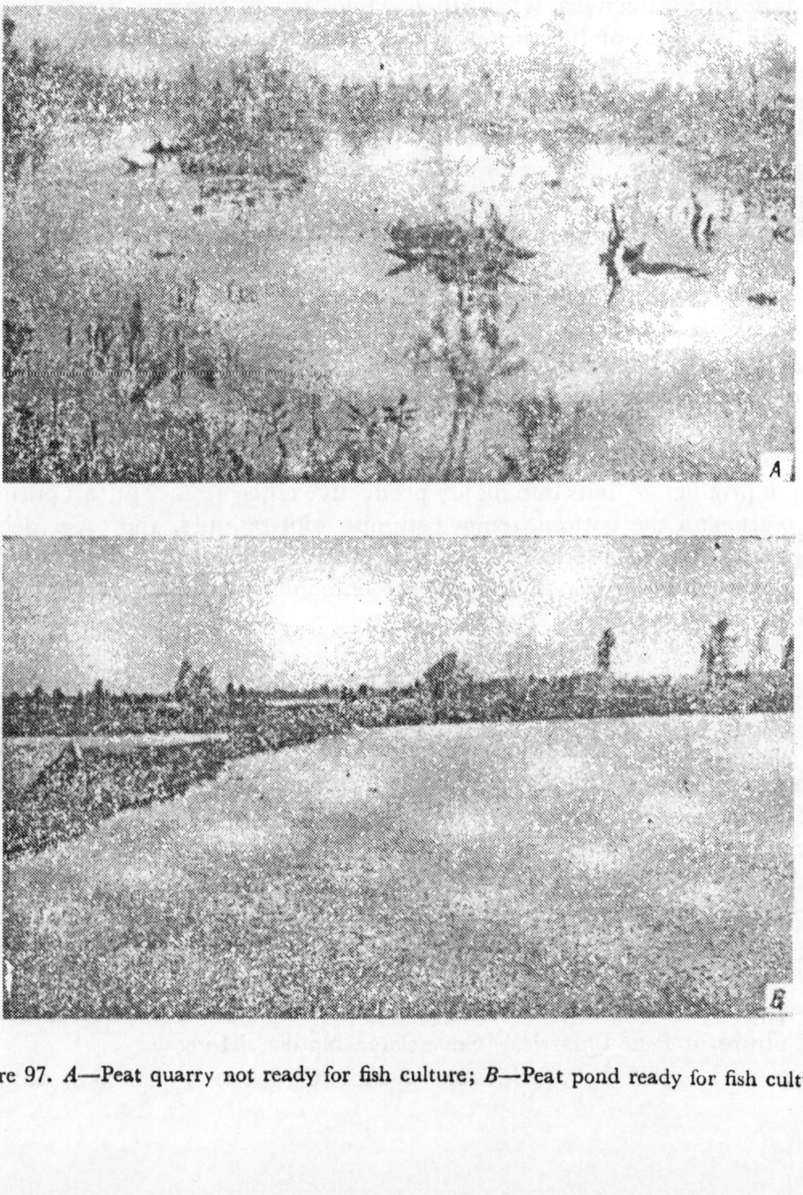

Figure 97. *A*—Peat quarry not ready for fish culture; *B*—Peat pond ready for fish culture.

grass carp, when reared in foraging, replenishing, and stock ponds with well-developed higher aquatic plants, becomes significant.

Two-year carp grow well in peat foraging ponds and their output is generally within the norms for standard fish ponds. The natural fish productivity of peat foraging ponds does not exceed 100 to 120 kg/ha. With intensive farm management, including artificial feeding of the fish and manuring, the yield of fish can be increased to 600 to 750 kg/ha. At the fish farm "Biserovo" in the Moscow district constructed on a peat deposit (area 365 hectares) fish productivity is even higher. In the Ivanovo fish farm the average fish productivity is 7.7 q/ha, while that of foraging ponds at the Shatursk Experimental Farm of the All-Union Research Institute of Pond Fisheries is 14 to 16 q/ha and of rearing ponds 9 to 10 q/ha. The output from wintering ponds is nearly 90%. Carp are raised in peat ponds in the Gorkii district, Lithuanian SSR, and in other Republics. In Lithuanian SSR a recently excavated peat bog of 1,000 hectares has been earmarked for the construction of a fish farm.

In addition to single-year foraging farms, it is recommended that full-system fish farms be organized with hatching, rearing, stock, and wintering ponds. The latter should have a higher rate of water exchange since the sources of water supply during winter are characterized by an extremely low oxygen content, especially in January–February, when it drops to 0.2 to 0.1 mg/l. Hence special and continuous aeration of the water entering the wintering ponds is required.

Under proper management peat quarries in most cases can be converted from unproductive units into highly productive enterprises. For this purpose one must plan the bottom, remove stumps, root residues, and other debris which would hinder production processes, arrange the water outlets properly, remove intermittent deposits, make the embankments more compact, and so forth (Figure 97).

References

Fatchikhina, O.E. 1948. Dinamika soderzhaniya fosfora i zheleza v chernom ozere (Dynamics of phosphorus and iron content in the Black Sea). *Gidrokhimicheskie Materialy*. Izd. Gidrokhim In-ta Akad. Nauk SSSR, vol. 15.

Khairulina, E.N. 1962. Puti povysheniya ryboproduktivnosti vyrostnykh prudov, postroennykh na torfyanykh kar'erakh (Ways to increase fish productivity of rearing ponds in peat quarries). *Voprosy Prudovogo Rybovodstva. Trudy VNIIPRKh*. Moscow, vol. 11.

Martyshev, F.G. 1957. Razvedenie ryby v torfyankh kar'erakh (Fish Culture in Peat Quarries). Sovetskaya Nauka, Moscow.

Mints, A.G. 1962. Vyrashchivanie dvukhletok karpa v prudakh, postroennykh na kar'erakh Gidrotorfa (Raising of two-year carp in ponds in the peat quarries of Gidrotorph). *Voprosy Prudovogo Rybovodstva. Trudy VNIIPRKh.* Moscow, vol. 11.

FISH CULTURE IN IRRIGATION NETWORKS AND MULTIPURPOSE RESERVOIRS

In the "directives" of the XXIV Congress of the Communist Party of the Soviet Union for the 1971–1975 Five Year Plan for the development of the national economy, and in the subsequent "guidelines" developed by the Central Committee of the Communist Party of the Soviet Union and the Council of Ministers of the USSR, several measures were proposed for more extensive irrigation networks and expansion of projects relating to the national economy and land improvement.

The expanded construction of irrigation networks has provided a golden opportunity for extending fish culture. Phytophagous fish free irrigation canals of choking weeds, improve their efficiency, and expedite water flow. The exploitation of the Karakum canal for grass carp has almost totally eliminated aquatic weeds while, obviously, yielding a food product. The usual mechanical method of eradicating obstructive plants is less effective and comparatively more costly.

In the case of an extremely thick plant growth in canals (development of emergent aquatic plants up to 2.6 to 3.0 kg/m^2 and submerged plants 2 kg/m^2), the stocking density of grass carp should be 300 to 400 fish/ha with an average weight of 800 to 1,000 g, which gives an average growth of 150 to 300 fishes/ha. Grass carp two to three years old check the growth of aquatic plants better.

Because of the long duration of the vegetative growth period in such regions as northern Caucasus, southern Ukraine, Moldavia, Trans-Volga, etc., these areas are especially interesting for fish culture, more so because in the arid south pond fisheries are few due to extremely limited water resources.

According to the findings of Chizhik (1969), the fish productivity of an irrigation canal may reach 50 to 100 kg/ha. Besides grass carp, common carp, silver carp, and bighead can be cultured. However, in flood areas where aquatic plants develop in great abundance, preference should be given to grass carp. In some cases the enclosure method of fish culture can also be undertaken in canals.

The sources of water supply for irrigation are rivers, small rivulets, wells, springs, lakes, and artesian waters. According to the nature of the water supply the hydrological and hydrochemical conditions for rearing fish change. Each source has its own peculiarities. Water from artesian wells, for example, is quite suitable for trout culture; after heating it in water storage tanks to 18 to 20°C it may be used for irrigation.

Irrigation networks can be used for a full-system fish farm, but here we are assuming one-year production or foraging farms for which fish hatcheries would be constructed separately. The management of fish culture in irrigation networks differs from that in standard fish farms in such matters as the selection of fish for rearing, periods of water supply and discharge, peculiarities in fishing, and methods of intensification.

The water supply during summer ought to come from manured fields and in winter include drainage and highly mineral water. One peculiarity of the hydrological regime of drainage waters is that under the conditions in southern USSR the release of fish is not possible in winter, but during autumn there is no need for wintering ponds. It becomes possible to use marketable fish in this case much earlier, in July and August. Annual reclamation procedures can be implemented thereafter.

Lack of adequate experience in fish culture in irrigation networks does not permit generalized recommendations at the present time. Each water body needs to be approached individually. Methods pertaining to the construction of fish nurseries, rearing of fish, and intensification measures adopted in pond fisheries are applicable in irrigation water resources wherever the requisite conditions prevail. It is important to use the waters discharged after irrigation for the development of pond fisheries. However, if due to agronomical requirements toxic chemicals are used in the fields before irrigation, such water could lead to the mortality not only of the fish but also the organisms which serve as their food. Hence, before using irrigation water reservoirs for fish rearing one must ascertain the agronomical practices followed in the cultivation of crops—the composition of the fertilizers and toxic chemicals applied to the fields, their dosages, and time of application.

Tanks and reservoirs of water maintained for the irrigation of agricultural crops and/or as drinking water for cattle can also be adopted for fish culture. In reservoirs of multipurpose utilization, however, one-year fish culturing would be more meaningful, with the stocking material supplied from fish nurseries. The methods of management of fish culture in such unique reservoirs would differ distinctly in each zone. Of paramount importance is managing the fluctuations in the quantity of water over the vegetative growth period. In calculating the quantity of water in reservoirs of multipurpose exploitation, one considers the effective volume left after assessing the quantity of water to be consumed by the local population for raising farm animals, for fish culture, for irrigation and other production purposes, and loss of water through seepage and evaporation (especially in dry years). The latter can be quite high in irrigation canals, necessitating the implementation of compensatory measures.

Agricultural crops respond differently to soil moisture. The pattern of irrigation depends both on soil conditions and meteorological factors such as quantity of rainfall and air temperature. For example, sandy and sandy-loamy soils require more irrigation compared to heavy soils. All the factors influencing the water balance of a pond (soil, wind speed, shade, average amount of rainfall, norms of irrigation in different periods of vegetation growth and for different crops, etc.) should be taken into consideration. It would be foolhardy to organize fish culture in reservoirs of multipurpose utilization without considering these factors.

Surface water runoff and water absorption greatly affect the water balance. Surface runoff should be minimalized. Data on the relationship between soil moisture and width of forest belt in the Kamennya steppes reveal that the moisture content in a 7 m forest belt is 35.2%, in a 32 m belt 48.4%, and in a 64 m belt 51.2%. Surface runoff and soil erosion are almost eliminated in the case of wider forest belts (32 and 64 m).

Waters with characteristic surface currents in passing through forest and grassy belts are cleared, to a great extent, of suspended silt particles before they reach the ponds. A study of the water-absorbing role of forest belts in the Kamennya steppes has shown that in the period of snow melting forests absorb more than 12 times the quantity of water absorbed by fields. Water streams passing through a forest belt 10 to 12 m wide are totally cleared of suspended particles. Thus forest belts and meadow areas are not only water-absorbing regions but also regulators of surface flow through subsurface intake. Experience has shown with an increase in forest area of 1%, the surface flow of water reduces 0.2 to 1.5%.

Forest belts greatly influence the evaporation of water from the surface by reducing the surface runoff during spring and increasing the underground current in low-water periods. Data collected by the hydrogeological station of the Kamennya steppes over several years reveal that evaporation is 25 to 30% less in ponds constructed in forest belts compared to ponds not surrounded by forests. Thus, afforestation around the ponds increases the effective volume of water by reducing losses through evaporation, which increases the period of their exploitation. The average magnitude of evaporation over several years from an open surface of small water bodies in arid and semidesert regions, according to Veselovskii (1956), varies from 1,000 to 1,100 mm in a year.

In pond fisheries the surface flow of water is important not merely as a source of water supply; it also contributes a large quantity of nutrients and improves natural fish productivity. A total stoppage of surface flow leads to a reduction in the natural productivity of the fish ponds. Thus, manuring fish ponds with substances which are deficient is of greater significance in reservoirs of multipurpose utilization, in order to replenish any loss caused by the slackening or termination of surface flow.

Change in the nature of surface flow is not the only factor which influences the production processes of ponds of multipurpose utilization. The management of fish culture in these water resources has its own peculiarities and the most important aspect is the quantity of water. Fish may be placed under unfavorable conditions of external environment during the period of their most intensive growth (June, July, and August) due to diversion of water for irrigation purposes. In dry years, with an increase in loss due to evaporation, this may give rise to premature catching and disposal of the fish. The depth of reservoirs maintained for irrigation is considerably reduced in summer and hence the area which can be irrigated necessarily shrinks. In such reservoirs so-called "dead storage" or minimum water of a depth of 1.5 to 2.0 m (according to the findings of A.N. Kostyakov, 2 to 3 m) must be kept for sanitary purposes; in certain cases this depth is increased. The minimum water depth suffices for the needs of pond fisheries, but in practice is not always maintained.

In the course of construction of ponds of multipurpose such a reserve of water should be planned which toward the end of the irrigation period will not be less than 30% of the total pond area at the time of dead storage.

To maintain a sufficient quantity of water, especially in dry areas, it is necessary to fill ponds of multipurpose (fish rearing-irrigation) with an "extra" water supply. Hence their depth compared to that of standard foraging ponds is slightly more. After the first and second irrigations they are kept at a level generally followed in pond fisheries, which corresponds to the period of the most intensive fish growth (mid-June).

Ponds of multipurpose differ in construction. Some are constructed on rivers and rivulets by erecting dams and others on ravines and gullies. They may be drainable or nondrainable and filled with water from a perennial source or by natural precipitation. In such ponds specific biological and hydrological conditions develop which are reflected in the production regime of the fish farms.

The four most typical schemes for organizing a complex farm involving agriculture and fish culture are described below.

First Scheme. Ponds constructed on ravines may be arranged in the form of a cascade (chain) with a dependent water supply (Figure 98), and are mainly supplied with water accumulated during the spring floods.

Because of the large areas under ravines, which in the European part of the USSR alone comprises two million hectares, this system has great practical significance. Cascaded ponds for pond fisheries can be used as feeding areas. The water from the first two ponds is used for lift irrigation and the water from the third pond in gravitational irrigation. The first and third ponds of the cascade are provided with bottom water outlets, and the second equipped with siphons. At the place of the water outlet in pond No. 3 and the water collection units of ponds Nos. 1 and 2, screens are provided.

At the time of operating the pumping station a current is created near the water collection units which attracts fish. These screens prevent damage to the fish by the blades of the pump and the falling of fish into the scoop and pump, and from there into the water collection system, and finally their release into irrigated fields through the main irrigation canal.

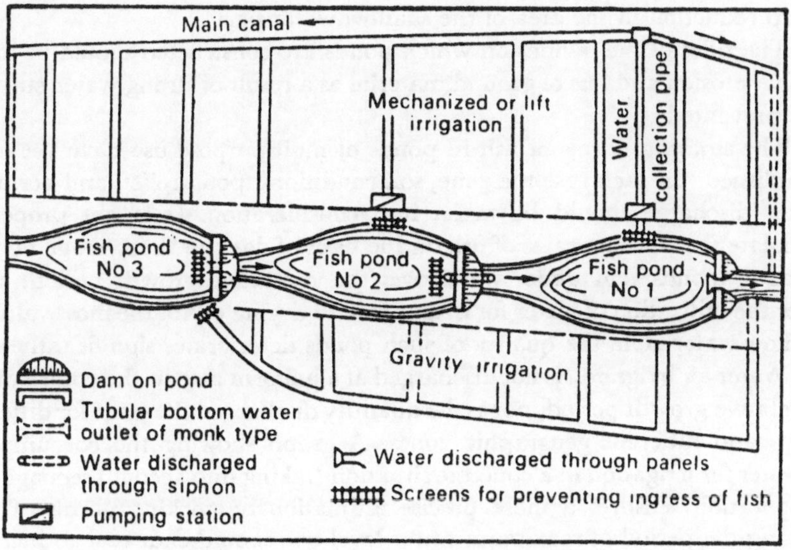

Figure 98. Cascade of ponds for fish culture filled with water by atmospheric precipitation for gravitational and lift irrigation.

For example, in the Generalov irrigation system, connected to the Tsimlyansk hydrocomplex, in the absence of the necessary protective screens in just one month the quantity of fry which had gone into the fields and died exceeded the installed capacity of the fish farms with regard to production and recruitment of fry for many years. In the irrigation canals in the delta of the Tereka River more than 500,000 fish were let out per day. Similar cases have occurred in other regions where the pumping unit threw out tens of thousands of valuable young fish which ultimately died in the fields.

Experiments conducted by the Laboratory of Ichthyology at the Institute of Animal Morphology of the Academy of Sciences of the USSR in collaboration with the Main Fish Directorate (Glavrybvod) have shown that larger quantities of fry (70 to 90%) enter the pumping units at night. Changing the irrigation time to other hours of the day considerably reduces fish mortality. However, it is important that more effective measures be developed to preclude this situation.

As a result of water discharge for irrigation the tank area is gradually depleted in the vegetative growth period and phytoplankton and zooplankton carried away. The shallow water left in the ponds is heated by the sun, i.e., a water surface develops which is most valuable from the viewpoint of fish culture. Depending on the structure of the basin in the ponds the reduction in water level proceeds in different ways. In ravines with steep slopes and depth (5 to 7 m) the water level decreases slightly, while in water bodies created in low-lying areas, even a small discharge of water causes a sharp reduction in the area of the shallow water part.

The sides of the ravines on which ponds are constructed should be made so that erosion and loss of ground material as a result of strong water currents are prevented.

The stocking norms of fish in ponds of multipurpose use have yet to be established. In each case the zone, soil conditions, pond relief, and period of water discharge should be taken into consideration. It is not proper to calculate the stocking rates of fish on the basis of data on the average flooded areas of ponds kept under water over the vegetative growth period, since after the utilization of water for irrigation and drying out of the most valuable shallow water zone the quality of such ponds deteriorates significantly.

Water for irrigation is not discharged at a uniform rate in the course of the vegetative growth period, nor is the quantity discharged the same for different crops and different geographic zones. A graph showing the consumption of water for irrigation in a concrete situation (taking into account seepage and evaporation) ensures a more precise estimation of stocking norms of fish. During the period of maximum water level one should endeavor to increase the growth of fish in order to obtain the planned unit weight toward autumn. If the hydrochemical regime of the pond, especially the gas regime, does not undergo sharp changes as a result of reduction in water level, estimates for stocking of carp can more appropriately be done on the basis of the initial area. In the process of water discharge and, consequently, increase in density of stocking, the fish must be fed more intensively.

In autumn the water is discharged from all the ponds, the fish caught, and the commonly recommended melioration measures implemented. Water discharge commences with pond No. 1. When its upper part adjoining pond No. 2 is dry, the siphon in pond No. 2 is opened and the water from this pond flows into pond No. 1. After drying the upper part of pond No. 2 the panels at the discharge point of pond. No. 3 are raised and the water from this pond drained into pond No. 2. Eventually all the water discharging devices in the ponds are operated simultaneously.

Total water discharge is expedited by planning the bottom of the ponds properly. In constructing ponds according to this scheme it is essential to create lesser depth in the upper part of all three ponds to speed up water discharge from ponds Nos. 2 and 3.

Second Scheme. Ponds meant for fish culture and gravitational irrigation (Figure 99) are filled with water from atmospheric precipitation. The bottom of such ponds is planned in the usual way except that bottom discharge outlets are installed. At the place where the main canal joins the pond, screens are placed to prevent the egress of the fish stocked. The water for gravitational irrigation is supplied through the main canal in which protective screens are also fitted. The water is released through the bottom discharge system and farther along the canal by the natural slope of the gully.

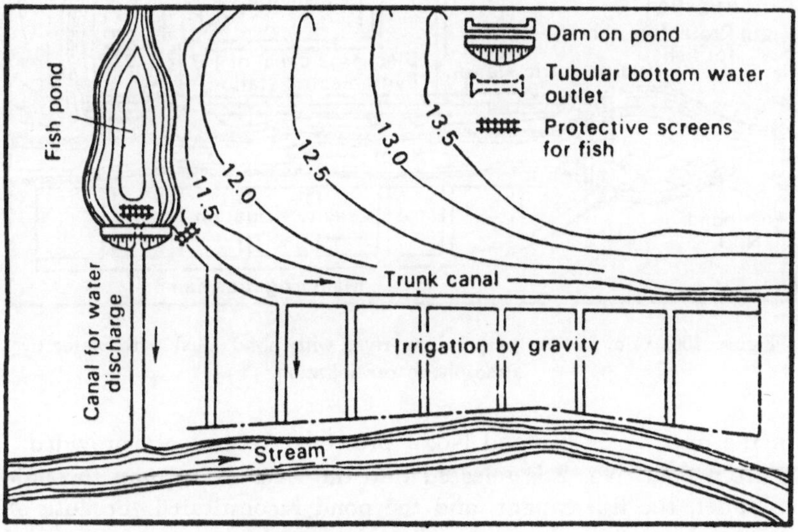

Figure 99. Pond for fish culture filled with water by atmospheric precipitation and used for gravitational irrigation.

Fish stocking and culture management are the same as in the first scheme. The only difference is that in the second scheme only one pond is involved. In autumn, after water discharge and fishing, the necessary reclamation is done. The pond is again filled with water in spring, mainly using floodwaters, and readied for fish culture and irrigation.

Third Scheme. This constitutes a combined system in which pond No. 1 is constructed on a rivulet by erecting a dam and the water used by a hydroelectric station for generating power; pond No. 2 is constructed in a dry valley and fed by atmospheric precipitation. The water from pond No. 2 is used for lift irrigation (Figure 100). The water level in both the ponds can be maintained at the required level during the vegetative growth period—in pond No. 1 through the continuous inflow of water from the river, and in pond No. 2 by water received from the hydroelectric power station. Here, it should be kept in mind that in the so-called ponds for hydroelectric power stations daily fluctuations in the water level occur, which are associated with the working conditions of the electric power station.

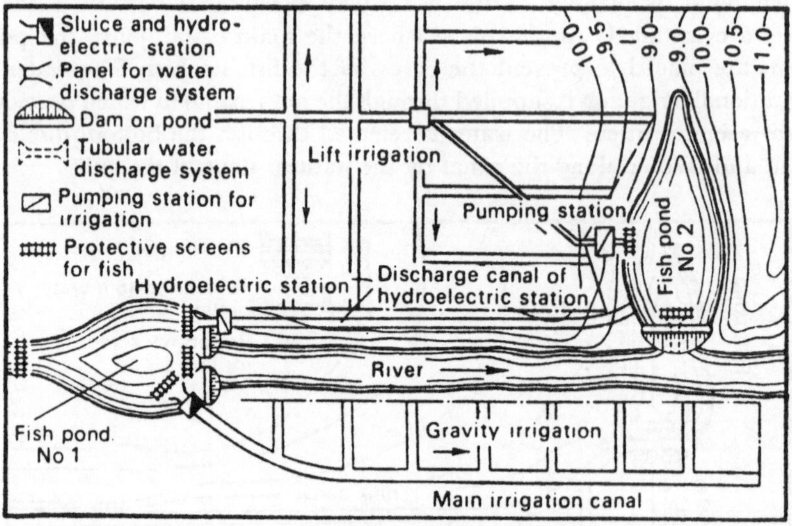

Figure 100. Combination of pond on river with pond filled with water by atmospheric precipitation.

In the upper part of pond No. 1 protective screens are provided. The water from pond No. 2 is released into the river in autumn through the water outlet, the fish caught, and the pond reconstituted. Because of the working regime of the hydroelectric power station, water discharge from pond No. 1 is not possible and hence fishing is done in water. In constructing this pond, therefore, the river bed should be carefully cleared of any obstruction to thorough fishing. The winter regime of the pond also totally depends on the operation of the hydroelectric power station. If it is possible to discharge the water for some time up to the usual river level (through the water discharge system), the dry places of the pond should be reclaimed.

FOURTH SCHEME. The pond is constructed on a small rivulet, fed with ground and rainfall water, and used for fish culture and irrigation. A dam is built and protective screens installed at the top to prevent the ingress of fish from the rivulet into the pond and the egress of the fish cultured. A water discharge system and a tubular water outlet for releasing water for the purpose of gravitational irrigation are incorporated in the dam. The water reaches the trunk canal via the water discharge system and reaches the fields through a network of distributors (Figure 101).

At the time of autumn catches the water is drained through the water discharge system, remaining only in the narrow stream of the rivulet, which poses no problem for fishing. The dry part of the pond must be reclaimed.

The level of water is constantly maintained in the pond by water from the rivulet and undergoes little or no change throughout the vegetative

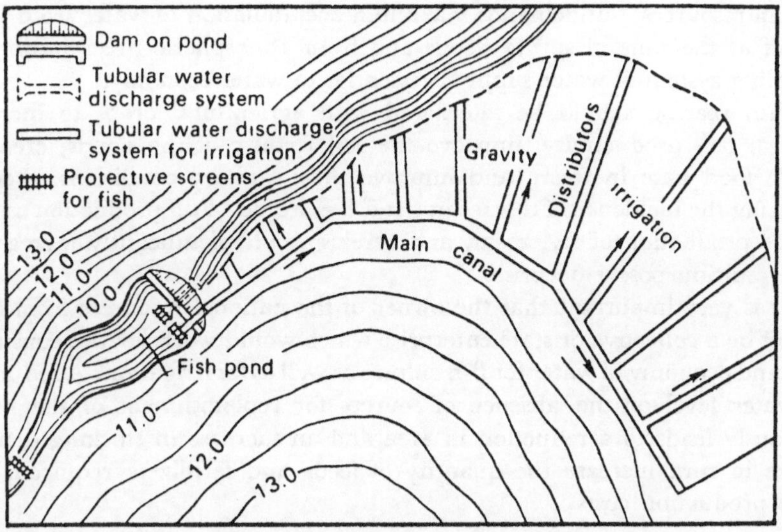

Figure 101. Pond for fish culture constructed on a small rivulet with ground and rainfall water supply and used for gravitational irrigation.

growth period, depending on the quantity released for irrigation (the difference between water discharge and inflow at a given moment).

Water discharge in autumn for catching the fish cultured and undertaking other necessary measures are not always tenable in ponds utilized for both irrigation and fish culture. For example, in the steppe zone, where the magnitude of the spring water flow varies highly from year to year and excessive water from one year is rarely conserved for utilization in the next, a total discharge of water from the ponds would sharply deteriorate the irrigation conditions in areas which otherwise have a guaranteed yield level. In such cases fish culture must be done in the same way as in nondrainable ponds. With a cascade type of pond arrangement, in the steppe zone the foregoing method of fish culture may be followed, complete catching of fish effected, and the subsequent reclamation measures implemented. However, even in this scheme water must be stored in one of the ponds in that quantity necessary to meet the irrigation requirements of the next year. Hence the upper pond of the cascade should be large enough to meet this need.

To accelerate the production of farm animal products in the country the construction of large state, collective, and intercollective complexes on an industrial basis has been proposed by the XXIV Congress of the Communist Party of the Soviet Union.

Pond fisheries can very appropriately be organized in combination with other branches of agriculture (fisheries-animal husbandry, fish-crop cultivation). This makes possible a more rational utilization of water, land, and human resources, but demands maximum accumulation of water from local runoff at the time of spring floods and from atmospheric precipitation. A recycling system of water supply is essential to water economy.

Fish rearing should be alternated with agricultural crops to increase natural fish productivity, improve the soil fertility of the ponds, create a stable food base in them, and improve their sanitary conditions, thereby reducing the incidence of disease in the fish cultured. Fish production as well as the production of fry, meat, and cereals would significantly increase in such multipurpose enterprises.

It is very important that the owner of the national resources from local runoff be a collective or state enterprise which would work for the consumption and economy of water for fish culture as well as for irrigation. A reduction in water level or the absence of sources for replenishment of the ponds obviously leads to a reduction in area and an increase in stocking density. These in turn increase the quantity of feeds and fertilizers required, i.e., raise production costs.

References

Chizhik, A.K. 1969. Osobennosti organizatsii prudovogo rybovodstva v zone orashaemogo zemledeliya (Peculiarities of organizing pond fisheries in the zone of irrigated agriculture). *Rybnoe Khoz-vo,* no. 4.

Isaev, A.I. 1969. Oroshaemoe zemledelie i rybovodstvo (Irrigated agriculture and fish culture). *Rybovodstvo i Rybolovstvo,* no. 2.

Lipatov, D.A. and A.V. Uvarov. 1964. Kak organizovat' vyrashchivanie ryby na mestnykh i irrigatsionnykh vodoemakh (How to Organize Fish Farming in Local and Irrigation Reservoirs). Pishchevaya Promyshlennost', Moscow.

Nikol'skii, G.V. 1966. Razvitie orosheniya i melioratsiya zemel i zadachi biologicheskikh rybokhozyaistvennykh issledovanii (Development of irrigation and land reclamation and the problems of biological studies in fisheries). *Vopr. Ikhtiol.,* 6, 3, 40.

Veselovskii, N.V. 1956. Prudy v zasushlivykh raionakh i ikh gidrokhimiya (Ponds in the Arid Regions and Their Hydrochemistry). Izd-vo Akad. Nauk SSSR, Moscow.

Chapter XV

COLD-WATER (TROUT) FARMING

In cold-water ponds cold-loving fish are cultured, primarily brown trout (*Salmo trutta morpha fario*) and rainbow trout (*S. irideus*). Lake Sevan trout (*S. ischan*) is also reared, which is characterized by rapid growth; other culturable fishes include "gegarkuni"*, the cold-loving Caspian inconnu, its northern subspecies nelma, and peled, broad whitefish, vendace, ripus, etc. Rainbow trout (Figure 102) is the most widely cultured species in cold-water fish farming. It is a hybrid form obtained in North America by crossing the migratory trout (*S. irideus*) sometimes treated as a subspecies of *S. gairdneri*, and the river trout (*S. shasta*). Rainbow trout was introduced into Russia from Germany in 1890.

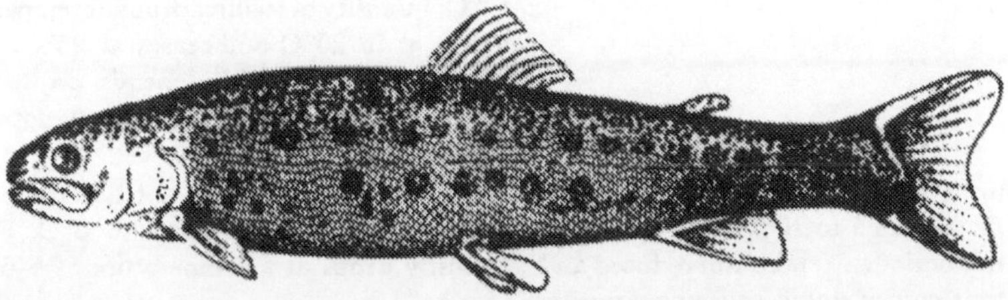

Figure 102. Rainbow trout.

In recent years more attention has been given to the culture of rainbow trout. Acclimatization and raising of rainbow trout has increased its area of distribution. It is now present in all the continents of the world except Antarctica.

The wide distribution of rainbow trout in pond culture is explained by its lower sensitivity and demands in terms of external environment, good growth, and high food qualities. The edible part of this trout constitutes 78 to 80% of the fish; 100 g of muscle protein contain 18 amino acids, of which 10 are essential for man: threonine—4.15, valine—4.86, leucine—9.17, isoleucine—4.75, methionine—3.28, histidine—1.78, arginine—6.72, phenylalanine—3.97, lysine—11.79, and tryptophan—0.75 g.

*A fast-growing form of Sevan trout—General Editor.

Major differences in the biology of carp and trout—the two leading fish for culture in two different kinds of fish ponds—are given in Table 53.

The biological peculiarities of trout, which is more exacting in conditions of external environment, temperature regime, and quality of water, have led

Table 53. Biological peculiarities of common carp and rainbow trout

Carp	Trout
1. Fish of stagnant warm-waters. Thrives in small ponds characterized by soft water, well-heated by sun, protected from winds, with a soft clayey-silty bottom, rich in organic matter. Quiet and lethargic fish.	1. Fish of cold and fast-flowing hill streams with plentiful shade, hard water, and coarse sandy and stony bottom. Thrives in flowing ponds, the deeper preferable to shallow ones. Very active and agile fish.
2. Feeds on small aquatic animals—crustaceans, chironomid larvae, worms, etc., as well as seeds. Pacific; does not attack its own kind.	2. Feeds on larger aquatic animals—amphipods, mollusks, larvae of dragonflies, small water beetles, tadpoles, etc. Fond of water bugs and aquatic insects important as food. Young trout pacific under normal conditions; larger fish sometimes predatory.
3. Withstands water temperatures up to 35°C. Optimum temperature for feeding 20–27°C.	3. Lives in temperature range of 0–30°C. Optimum water temperature for feeding 10–15°C; intensity of feeding drops at temperatures higher than 20°C and ceases at 25°C. Upper temperature threshold depends on temperature conditions to which a given population has adapted.
4. During winter does not feed; quiescent at 4–6°C. Looses 5 to 10% of body weight during this period. Metabolism reduced and proceeds at cost of stored substances during winter period (fat, etc.).	4. During winter does not hibernate unless temperature of water close to 0°C. Feeding intensity drops at a temperature of 4–6°C.
5. Oxygen content in water for intensive feeding 6–7 mg/l, for reduced feeding about 3 mg/l. Oxygen content critical for this fish, 0.7 mg/l.	5. Oxygen content in water for intensive feeding 10–11 mg/l. Lowest oxygen content tolerated 1.0–2.6 mg/l and for embryos around 7 mg/l. Respiration of adult trout suppressed at about 4–5 mg/l.
6. Spawning takes place in spring-summer, commences in temperature range of 17–20°C. Fecundity about 180,000 eggs per kg of live weight of female fish. Eggs small, 1.3 mm in diameter, yellow, adhesive, and stick to plants.	6. Spawning in southern regions (Abkhazia) takes place in January–February, in northern regions (Leningrad district) at the end of April and even in May and beginning of June. Fecundity 1,600–2,500 eggs per kg of body weight. Eggs larger, 4–5 mm in diameter, and nonadhesive.
7. Emergence of larvae from eggs depends on water temperature; usually occurs 3–6 days after spawning.	7. Emergence of larvae from eggs depends on temperature; usually 330 degree days* at a temperature of 9°C required.

*Degree day = product of number of days and average water temperature.

to the adoption of certain unique methods for its rearing and breeding. In the first place trout will not reproduce in ponds, which makes it necessary to maintain brood fish in ponds up to gonadal maturation, followed by artificial fertilization of the eggs and their incubation in special apparatus at the fish farms.

The type of pond required for fish culture is determined by the biological peculiarities of carp and trout. Ponds in a cold-water trout farm are of the flowing type, deeper, have a pebbled-sandy bottom, and must be filled with hard spring water saturated with oxygen.

Construction of a Trout Farm. Great attention is paid to the quality and quantity of the source of water supply while constructing a trout farm. The water must be clean, transparent, rich in oxygen, and free from dirt, suspended particles, and toxic chemicals. An oxygen deficiency in the original source of water supply can be compensated by means of aeration. For this, water supply troughs or channels are laid at a height of at least 15 to 20 cm above the water level in the pond. Usually waters from springs, small rivulets, and rivers are used in trout farms, which have a temperature range from 3°C during winter to 20°C in summer. Experience has made it possible to rear rainbow trout in ponds utilizing water from artesian wells also.

A primary condition for rearing trout in ponds is flowing or running water. The requirements of flow vary under different kinds of farm management (extensive, semi-intensive, and intensive). A trout farm is more often managed in an intensive manner, involving dense stocking. The degree of intensification depends, first of all, on the rapidity of water replacement and the quality of water. Researchers in foreign countries have determined that the inflow of water in trout ponds, depending on the extent of intensification, should range from 300 to 1,400 l/sec per hectare. Soviet experience has established that 50 to 100 kg of marketable trout can be obtained with a water consumption at the rate of one l/sec.

An increased water exchange in trout ponds, especially under high stocking density, is essential for meeting the requirements of trout with reference to hydrochemical regimes. A high rate of water flow generally maintains the oxygen content at the required level, removes accumulated metabolic products and organic matter, and creates conditions closer to those found in natural waters.

The bottom of trout ponds plays a less important role than in carp ponds, since at high intensity it is not the natural food but the extraneous feed that is more important. Extensive trout culture based on the level of natural fish productivity is not in vogue. Trout ponds are not constructed on soils rich in organic matter, porous, or marshy.[1]

Hydroengineering installations are similar in trout and carp farms. The

[1]Marshy lands can be used only after thorough reclamation.

water supply should be independent; however, in the case of a water deficit a dependent source of water supply can be used. In the latter case, in addition to a water discharge system equipped with two rows of panels, recycling units are installed in the body of the dam. To render the ponds completely dry a slight slope in the direction of the water outlet is created (gradient usually 1:100 to 1:200), and protective screens or boxes with a fine mesh net at the bottom fitted in the water channels to prevent trout egress from the ponds.

Trout farms may be of two types: full system or partial system. In the full-system farm trout are raised from eggs to marketable size, which is attained in two summers, i.e., after 14 to 16 months from hatching. A full-system trout farm (Figure 103) includes an incubation section, tanks for fry, live fish storage tanks, cooling units for storage of feeds, feed preparation unit, base for rearing live foods, ponds of all the necessary categories (rearing, foraging, nursery, and replenishment), space for processing, residential quarters, etc. A partial-system fish farm is either a hatchery or a foraging (production) farm. A fish hatchery includes the same facilities listed above except for foraging ponds and live fish storage tanks. A production farm needs only foraging ponds, a cooling unit for storage of feeds, feed preparation unit, and residential and service accommodations. The stocking material (yearlings) in production farms is obtained from fish hatcheries.

Full-system trout farms and trout hatcheries must maintain brood fish in sufficient quantity to ensure the production of the required quantity of eggs and stocking material.

Thus the systems in trout farms and the problems faced by each are similar to those of carp farms. However, the organization of trout farms and the characteristics of individual links differ, because governed by the biological peculiarities of trout. For example, spawning ponds would be redundant since trout do not reproduce at the farm. Trout feed in winter so wintering ponds are also superfluous; the fish are left during winter in the summer ponds.

A trout farm is usually managed on a two-year rotation scheme. Rainbow trout in this period attains a weight of 150 to 200 g and more. For a higher unit weight, three-year rotation is followed, in which case they grow to 600 g and more. Trout farms of two-year rotation prevail in the USSR and Western Europe.

Trout ponds differ from carp ponds of the same category. The area of each rearing pond (raceway) in trout farms is not very large, ranging from 100 to 500 m^2 with an average depth of 1.0 to 1.2 m, and water consumption of 50 to 60 l/hr per kg of live weight of under-yearlings. The area of the foraging pond is 300 to 1,000 m^2 with an average depth of 1.5 m, and water consumption 35 to 40 l/hr per kg of live weight of two-year fish. It is easier to control fish rearing in such small rearing and foraging ponds in an intensive trout farm. These ponds can be quickly drained and filled with water, which

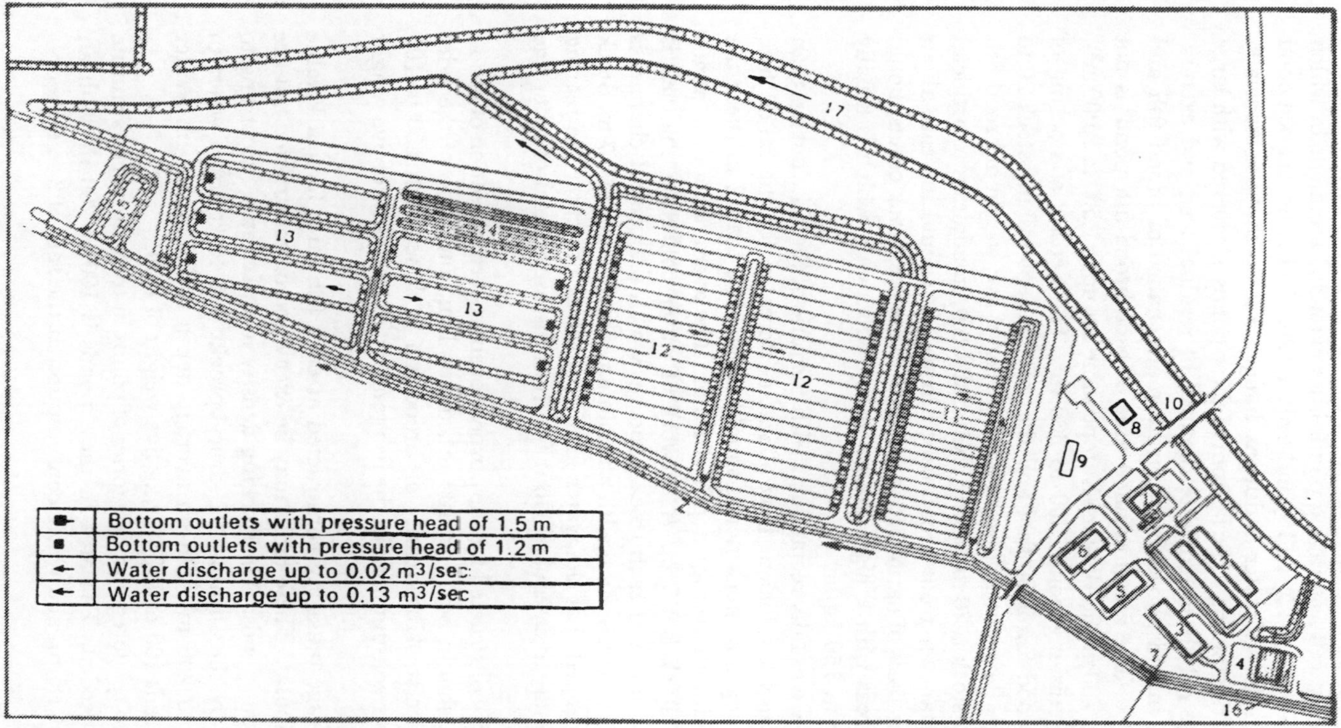

Figure 103. Plan of trout farm at the State farm "Urozhiny" (Altai territory).

1—office-laboratory; *2*—tanks for fry; *3*—incubator; *4*—tanks for brood fish; *5*—store for dry feed and equipment; *6*—feed section and cooling unit; *7*—water supply from wells; *8*—boiler section; *9*—cesspool; *10*—ferro-concrete bridge; *11*—rearing ponds; *12*—foraging ponds; *13*—stock ponds; *14*—replenishing ponds; *15*—quarantine ponds; *16*—water supply canal; *17*—new course of river; *18*—road.

is very important in terms of production processes. The most convenient shape for rearing as well as foraging ponds is an extended rectangle in which the ratio of the sides is: 1 : 5, 1 : 10, and even 1 : 20. A channel-shaped pond facilitates a rapid flow of water and its exchange. Such a shape is especially important for rearing ponds. The bottom of these ponds is covered with large pebbles or stone rubble to expedite removal of residual feed and excreta.

Stock ponds should allow the separate maintenance of brood fish and replenishing stock. The recommended area of each brood fish pond[2] is not more than 500 to 700 m^2 (in Western Europe some are up to 1,500 to 2,000 m^2), and for the replenishing ponds 200 to 300 m^2. The density of stocking of rainbow trout brood fish is 30 fish, and that of the replenishing stock 20 to 50 fish per 10 m^2. The water consumption per kg of live weight of brood fish and replenishing stock is 70 to 90 l/hr. Natural foods, in addition to artificial feeds, are important for trout brood fish; fish weight should increase about 20 to 25% on the basis of natural foods alone. Therefore, stock ponds should be located on areas with a high natural fish productivity, which is usually taken to be 100 to 150 kg/ha.

Stock ponds are usually rectangular with the ratio of the sides not less than 1:5; in such ponds a sufficient flow of water is created for the brood fish. The maximum depth of stock ponds should not be less than 2 m (near the bottom outlet) with an average depth of 1.5 m. The average depth in ponds for replenishing stock is 1.5 m. A network of water collection channels is installed in the bottom of each stock pond, in which the central channel is 1.0 to 1.5 m wide at the bottom with a minimum depth of 0.5 to 0.7 m. Stock ponds are so arranged that the water from them does not enter the rearing and foraging ponds, to safeguard trout fry from disease typical of older age groups.

Fish tanks are constructed for the purpose of short-term retention of brood fish. Their area should be 75 to 100 m^2, depth 1 m, and ratio of the sides 1:5 to 1 : 10. The stocking density of brood fish in these tanks is 50 to 100 fish/10 m^2. Water consumption in the fish tanks is 70 to 90 l/hr of live weight of brood fish.

Live fish storage tanks are constructed at each fish farm for marketable fish awaiting disposal. These tanks may be excavated or constructed but are similar to the trout ponds. The stocking density in these tanks (according to the norms fixed by the Institute "Gidrorybproekt") is 45 kg/m^3 of water in an area of 150 to 200 m^2 with an average depth of 1.5 to 1.7 m. Water consumption here is 100 to 120 l/hr/kg live weight of trout.

In addition to the foregoing categories of ponds in trout farms, quarantine ponds are also needed. The area in these ponds is 100 to 200 m^2 with an average depth of 1.5 m, and water consumption at the rate of 3 to 6 l/sec.

[2]According to the norms fixed by the Institute "Gidrorybproekt," 300 to 700 m^2.

The duration of filling of each pond with water, depending on the area, may be taken as follows (hr): rearing pond—3 to 12, foraging pond—3 to 15, replenishing pond—3 to 8, stock pond—4 to 48, quarantine pond—2, and fish tanks—2.

The following percentage of areas for different categories of ponds generally prevail in Western Europe for trout farms working on a two-year rotation: rearing ponds—18, foraging ponds—65, and stock ponds—7 of the total farm area; 10% of the area is reserved for other purposes. Depending on the objective of the farm, this ratio may change. If there are ponds for replenishment and quarantine, these provide favorable conditions for breeding work and improve sanitary and hygienic conditions. Gracheva (1953) has recommended the following percentage of area for ponds of different categories: rearing ponds—60, finishing ponds—20, stock ponds—5, replenishing ponds—10, quarantine and other ponds—5%. These percentages, which differ notably from those followed in Western Europe, incorporate a provision for a sufficient increase in the area of rearing ponds and a considerable reduction in the area under foraging ponds. These changes are based on different methods of trout rearing:

(a) Release of fry from incubation apparatus directly into the rearing ponds (see below);

(b) Rearing of fry in rearing ponds up to the end of July—beginning of August only on natural foods, after which artificial feed is provided.

In the scheme of trout culture discussed earlier the rearing ponds may be similar in morphological characteristics to the corresponding ponds in carp farms. In the latter, however, such ponds have a depth of 1.5 to 2.0 m and occupy 25 to 30% of the total area, and also differ from typical trout farms in water exchange and configuration, and are closer to foraging carp ponds. With regard to foraging ponds, these in trout farms have a higher water exchange and are not very large in area, and their corresponding depths ensure fish hibernation.

The requirement for ponds of different categories is determined on the basis of the following parameters: projected yield of produce, water consumption, average unit weight, and norms of output from each type of pond. For example, the total area under ponds and tanks would have to be 10,257 m^2 to realize the following goals: 100 quintals of marketable two-year fish (67,000) with an average weight of 150 g and 90% survival (of the yearlings planted), with 90% survival of the yearlings (of the fingerlings stocked), 60% survival of the fingerlings (of the fry stocked), 90% survival of the fry (of the larvae or early fry released over the period of initial rearing), and 80% survival of the larvae (from eggs over incubation period). In this case the percentage of the total area covered by rearing ponds would be 27.3, foraging ponds—48.7, stock ponds—14.6, replenishing ponds—8.8, live fish tanks—0.3, and nursery tanks—0.3.

MAINTENANCE OF TROUT BROOD FISH. The qualities of brood fish and replenishing stock are assessed on the basis of these characters: general health and condition, absence of deformities, good external appearance, rapid growth (associated with a better storage capacity and conditions of food), good coloration, activity, and swiftness of reaction to external stimuli. After attaining sexual maturity other relevant characters are added to these indexes: age, spawning time and quantity, and size and color of the eggs.

Female rainbow trout spawn best at the age of four to seven years and males at the age of three to six years.[3]

The fecundity of rainbow trout changes with age.[4] According to Steffens (1966), a four-year-old female produces 2,300 to 2,500 eggs with a diameter of 5.1 mm, and a seven-year-old produces 4,200 to 4,400 eggs with a diameter of 5.5 to 5.7 mm. The relative fecundity (for 1 kg of weight) of four-year-olds is about 2,300, of six-year-olds 1,770, and of seven-year-olds 1,680 eggs. It is better to use female fish weighing 750 to 2,500 g since they produce 2,000 to 4,000 eggs with a diameter of about 5 mm (range of 4.7 to 5.6 mm).

Feeding conditions notably influence the fecundity of rainbow trout. Studies done by Scott (1962) reveal that food shortages reduce fish fecundity. A reduction in ration to one-third leads to a reduction in fecundity by one-half, and a concomitant decrease in egg size.

The eggs of older fish contain more water and fat, which reduces their viability. Males mature one year earlier than females. The spermatozoa of two-year trout lose their motility in water after 33 to 35 seconds, of three- and five-year fish after 51 to 53 seconds, of five- and six-year fish after 62 to 65 seconds, and of seven- and eight-year fish after 45 seconds. Spermatozoa separated from the milt by centrifugation maintain motility and fertilization for 7 days at a temperature of 5 to 10°C and for 9 days at 4°C.

Brood fish are maintained in stock ponds or in a fast-flowing rivulet enclosed with screens and nets. The latter simulates the conditions of trout found in nature (rapid water flow). The quality and fertilization of eggs of rainbow trout maintained in fast-flowing ponds is higher. A desirable level of water current is achieved with a consistent consumption of water at the rate of 5 to 10 l/sec for every 100 brood fish (according to the data of American researchers, at the rate of 1,500 l/sec per hectare.

The food regime of brood fish is important. Some authors hold that to obtain sex products of better quality trout should be reared on natural foods. However, experience with female rainbow trout fed artificial feed during winter and the prespawning period has shown favorable results: In these

[3]The maximum longevity of rainbow trout is 11 years.

[4]It is not advisable to use eggs collected from females spawning for the first time, since the roe is comparatively small and yields a higher percentage of waste during the period of egg development.

conditions the quantity of eggs was greater and the extent of losses during incubation cut almost in half. One may therefore consider supplementary feeding of trout brood fish with readily assimilable and balanced feed a rational measure. A sharp changeover from natural foods to artificial feeds and back to natural foods can have an adverse effect (sterility or poor quality of progeny).

Brood fish are fed throughout most of the year, depending on temperature, at the rate of 1 to 2% of body weight. In the last two months before spawning feeding is done only two to three times a week and stopped immediately before spawning. The average weight of two-year replenishing stock should be 300 to 400 g and the weight of three-year fish 500 to 600 g. In the replenishing stock under-yearlings comprise 55%, two-year fish 30%, three-year fish 10%, and four-year fish 5%. The number of trout spawners required is calculated on the basis of the average working fecundity of a single female fish, taken as 1,500 eggs. In this case the reserve stock is kept at 25 to 50%. The ratio of females to males is taken as 3:1 (75% females and 25% males). The annual loss in brood fish is assumed as 5%, and the quantity of replenishing stock to total number of brood fish as 200%.

Stripping Trout for Eggs and Milt. Artificial Fertilization. The incubation section is prepared long before work on artificial fertilization commences. The area is disinfected, the incubation apparatus and other equipment washed, the ration of the spawners changed, and the fish maintained in flowing water. The quality of the water in this period must be the same as in the case of natural spawning (clear, clean, rich in oxygen, and flowing cool water). The requirements of trout in terms of hydrological and hydrochemical regimes also increase in the period of embryonic and postembryonic development.

If the water happens to be polluted with suspended particles, it is filtered; in the case of an oxygen deficiency or the presence of large quantities of carbon dioxide and salts of ferric oxide, the water is aerated.

When the fish manifest a tendency to move together in pairs, the better looking, larger, and well-fed ones are selected and separated on the basis of maturation stage for release (males and females separately) in earthen or cement tanks or cisterns (with a water exchange of three to four times per hour), or in ponds with holding norms of 10 to 15 fish/m^2 (when the area of the fish tank is 50 to 75 m^2). Fully mature females are taken for artificial fertilization since their eggs are readily stripped with a gentle flexing of the body or a light (never firm) stroking of the abdomen. Mature eggs are spherical in shape; the diameter of eggs of small fish is 3.0 to 3.5 mm and in older and larger fish reaches 5.3 mm. Immature females are kept in fish tanks and are examined periodically every three to five days.

The external appearance of males and females changes during the spawning period. The abdomen of the female has a fuller profile and usually remains white with swelling and redness around the vent. The body coloration of the male becomes more brilliant and the abdomen, although it remains compressed, appears darker, almost black sometimes; the lower jaw elongates and sometimes curves upward like a hook. Large eggs, highly pigmented, are better for fish propagation. The ripe eggs of brown trout are transparent and pale orange in color, while those of rainbow trout are pale yellow to orange due to the presence of carotenoids.

Yarzhambek (1966) and Iriskina (1968) have established that eggs containing a large quantity of carotenoids are more hardy during development and losses during incubation fewer. Two types of carotenes accumulate in the muscles, liver, ovaries, skin, and fins of trout maintained on natural food: red carotene (astaxanthin) and yellow (lutein). Rainbow trout is also capable of accumulating canthaxanthin, which has a favorable influence on growth. Carrots and the oily preparation of β-carotene should be used as carotenoid supplementation, since they are not assimilated by the fish. The larvae of chironomids, gastropod mollusks, and crustaceans are quite suitable among live food organisms (Yarzhambek) as a source of carotenoids for juvenile fish. The diet of brood fish must include cyclopoid copepods, daphnids, diaptomids, amphipods, meals prepared from shrimps or crabs, and gammarids, which are rich in the pigment astaxanthin. Chironomids and mollusks do not contain this pigment, and oligochaetes are extremely poor in carotenoids. It has been established (Galkina, 1959) that poorly pigmented eggs have a lower resistance to high temperature; even with a small increase in water temperature (up to 10°C) losses three times higher than in brightly pigmented eggs have been recorded. Larvae obtained from very small eggs (about 35 mg) have a smaller yolk sac and lower weight and length. Such eggs, obtained from females spawning for the first time are 3.9 mm in diameter and unsuitable for the purpose of fish propagation.[5] Young fish obtained from brightly colored eggs are characterized by higher weight and fat content than those from poorly pigmented eggs (Brizinova, 1969).

The milt quality can also be determined on the basis of external appearance: high quality milt has quite a dense consistency and is cream colored, while milt of inferior quality, aside from containing fewer spermatozoa of lower motility, is more watery with a bluish tinge. Poor quality milt should not be used for the fertilization of eggs.

To determine the time of ripening of the sex products brood fish are systematically observed in the fish tanks. Overmaturation of the eggs should never be allowed; it occurs with protracted retention in the body cavity usually due to incorrect water temperature. Egg retention in the body cavity for two to three days is permissible but up to six to eight days risky. Under

[5]According to N.P. Novozhinin, only eggs weighing more than 40 mg should be kept for incubation

the conditions of prolonged retention the eggs lose their capacity for fertilization. Overripe eggs yield a higher percentage of male fish, malformed fish, and nonviable fish. Phased maturation in the males makes it possible to milk them two to three times with an interval of 8 to 14 days for fertilizing the eggs of a smaller number of females.

When the eggs and spermatozoa have ripened, the spawners are carefully caught in the fish tanks and transferred to the incubation section. The eggs are fertilized by the dry method introduced in the nineteenth century by the Soviet scientist, V.A. Vrasskii, which is widely known as the "Russian method of fertilization".

Before stripping the female is gently wiped with a clean towel to prevent dirt falling on the eggs; the eggs are collected by applying a mild pressure on the abdomen. Generally, the eggs from three to five females are collected in the same clean, dry, enamel-coated, porcelain, or glass container (Figure 104). Stripping must be done very carefully, taking precautions to see that

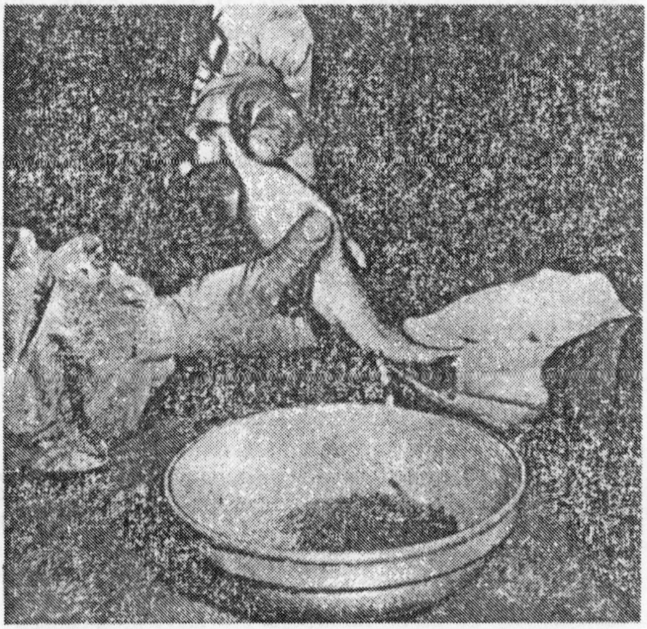

Figure 104. Stripping female trout for the purpose of artificial fertilization.

the eggs do not drop into the container but rather slide down its side. Pressure is applied on the abdomen of the fish commencing from the lower part of the ovary (near the vent) and gradually moving upward. Normally the eggs emerge freely in the form of individual grains. Eggs exuded in the form of clusters with blood clots are still not ripe and hence not suitable for fertilization. No water should fall into the container during egg extraction.

The stripped eggs are mixed in the same container with milt collected from two or three males.[6] The contents are moistened with water and stirred gently with a duck or hen feather.[7] After five minutes, water is poured carefully down the side of the container until all the eggs are covered with a 2 cm layer. The eggs are allowed to stand for three to five minutes during which time fertilization takes place. Thereafter the eggs are washed with water poured down the side of the container, which is continuously but gently shaken, the milt poured off, and the process continued until the water becomes transparent. The addition of water activates the spermatozoa. The process of egg swelling continues for about two hours.[8]

The eggs should neither be fertilized nor washed in bright sunlight. Neither should there be a notable difference between the temperature of the water in which the spawners are kept and the air temperature at which fertilization of the eggs is done.

Stripping should be done quickly since the spawners cannot remain alive outside water for very long. Some authors suggest that the eggs from a female fish can be collected in installments and good results obtained; usually eggs are extracted by stripping only once. The spawners from which the sex products have been collected (spent fish) are released into fish tanks under flowing water for a few hours, then transferred to the stock ponds where they remain until the next spawning season.

The method of egg and milt extraction given above has some shortcomings, arising first of all from stripping very large trout specimens. Such specimens flop about or struggle during stripping, which often damages the sex products. The use of anesthesia, although it rectifies this situation, does not totally exclude the possibility of injury to the eggs, nor their admixture with urine and feces. In the mid 1950's Worton (Australia) improved this process by using compressed air and increased the percentage of fertilization from 50 to 90. This method is used in the USA at the fish farm Skamaniya (Washington) and at the fishery station of Benner-Spring (Pennsylvania); the process has now been mechanized (Figure 105). In Worton's method a laboratory type pump delivers and absorbs compressed air; it is equipped with a 1.6 hp electric

[6]The best results are obtained from fertilizing eggs with milt collected from a few males, or the milt mixture from males cultured in different farms, or the milt mixture of local and imported males.

[7]Some pisciculturists believe that less damage occurs when the eggs are mixed carefully with the hand.

[8]V.P. Vrasskii originally kept the eggs and milt in separate dry containers after extraction. The milt was subsequently diluted with water and this mixture added to the eggs. The entire operation required no more than five minutes. This method was eventually modified as described in the text.

motor. The fish are anesthetized with methyl pentothal, washed until the anesthesia is totally removed, and rolled in a soft cloth saturated with water. A hypodermic needle of 16 caliber is then inserted into the cavity near the pelvic girdle and the inner barrel of the syringe replaced by a 15 cm section of rubber pipe from the air compressor. The pump is then turned on which passes compressed air into the abdominal cavity at a pressure of 0.4 kg/cm². The fish is then inverted and gently shaken to release the eggs. When the eggs stop flowing from the vent, the pump sucks out the air. The entire process of egg extraction is finished in about 15 seconds. The eggs pass through a net sieve and the waste material collected thrown out. Eggs which are clean, intact, and without fecal or other foreign matter are transferred to a dry container, fertilized, and placed in the incubation apparatus.

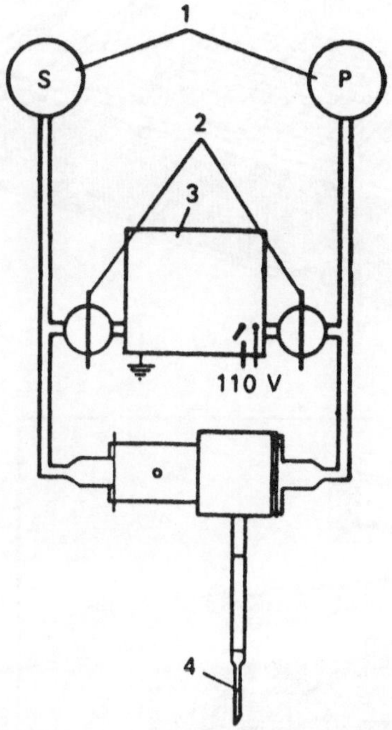

Figure 105. Diagram of device used for egg extraction with the help of compressed air. *1*—manometers; *2*—regulators; *3*—pump for egg extraction under pressure; *4*—hypodermic needle.

Eggs fertilized by one or the other method are counted after swelling and then carefully and uniformly arranged in the frames or trays of the incubation apparatus located in the incubation-larval section. It is better to arrange the eggs in one layer with such a density that four occupy about 1 cm². The Californian apparatus of Shuster's system is commonly preferred for incubation (Figure 106). Its dimensions are 36 cm × 20 cm × 13 cm, egg capacity 5,000 to 6,000, and water consumption 1.0 to 1.5 l/min. Water consumption may be more—when the dimensions of the outer box are 100 cm × 60 cm × 18 cm and those of the inner 90 cm × 59 cm × 13 cm. In such an apparatus 15,000 to 17,000 eggs can be arranged in one layer. Apparatus with smaller dimensions are more convenient and facilitate egg care during incubation. According to the data of "Gidrorybproekt", the rate of loading of the incubation apparatus should be 40,000 to 60,000 eggs/m². The extent of losses in eggs increases in an incubation apparatus in which the bottom is made of galvanized iron: the harmful effect of the poisonous properties of zinc is responsible for this. If such apparatus are made of steel, they should be coated with asphalt paint.

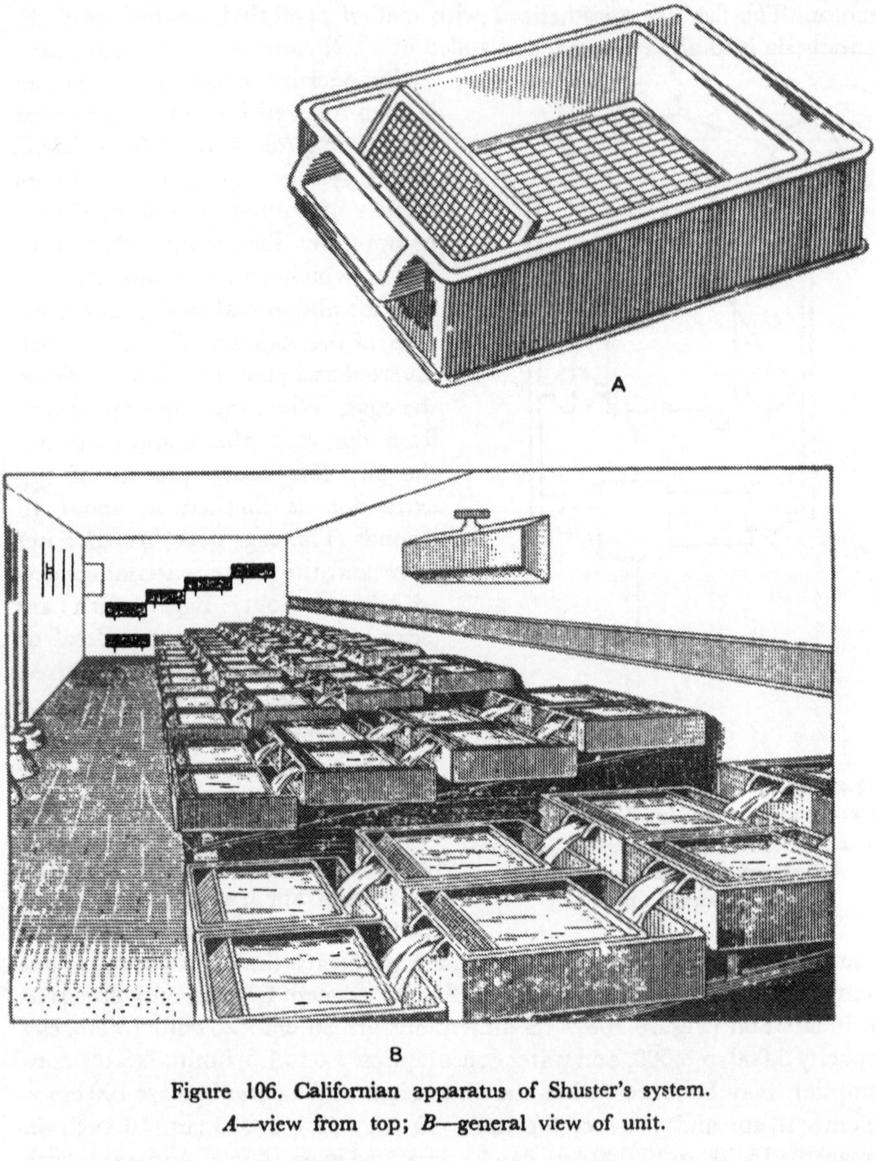

A

B

Figure 106. Californian apparatus of Shuster's system.
A—view from top; *B*—general view of unit.

Egg counting is commonly done by weight or volume. In the weight method, for example, 1,000 eggs are visually counted and their weight recorded. After this the total quantity of eggs is weighed. By dividing the total weight of the eggs by the weight of 1,000 the total number of eggs is obtained.

In the volume method for counting eggs a measuring cylinder is used. First the number of eggs in 1 cm^3 is recorded and then the total quantity of eggs placed in the cylinder. The total number of eggs is calculated by multiplying their number in 1 cm^3 by the total volume of the measuring cylinder and the result expressed in cubic cm.

A special count is done of the eggs laid on the frames and these counts recorded in the register.

Care of Trout Eggs During Incubation Period. For the normal development of fertilized eggs in the incubation apparatus certain conditions are essential:

1. Continuous and uniform flow of fresh cold-water must be supplied to the apparatus from the same source which supplies water to the entire fish farm. The water temperature is maintained at 5 to 9°C for rainbow trout[*] and not higher than 2 to 3°C for brown trout. At a water temperature below 1.5°C and above 14°C mortality of eggs of rainbow trout is observed. It is better to supply spring water to the apparatus. If the water temperature increases, it is essential to take measures for its reduction. Low temperatures but not subzero temperatures are a characteristic feature of the incubation section. Temperature affects the embryo differently at different stages of its development, which can be seen from the data presented in Table 54.

The flow of the water should be carefully regulated. A strong water flow can lift the eggs up, causing them to emerge from the period of dormancy

Table 54. Effect of temperature on embryogenesis (in days) (from Timoshina, 1968)

Developmental stages	Temperature, °C				
	13	10	7	5	2
Beginning of cleavage (32 blastomeres)	0.92	1	1.5	2	3
Midcleavage period (128 blastomeres)	1	1.5	2.5	3	5
Blastula	2	2.5	4	4	7
Epithelial blastula	3	4	8	9	17
Half growth of yolk	8	9	16	21	29
Closing of blastopore	10	11	24	28	40
Beginning of eye pigmentation	15	20	34	40	59
Emergence of larva	25	30	50	65	102
Half absorption of yolk sac	37–92	45–48	67–72	86–92	128–135
Complete absorption of yolk sac	50–52	60–65	86–92	102–110	147–155

[*]Some researchers assert that the optimum temperature for the development of rainbow trout is 5 to 7°C. The development of eggs of rainbow trout takes place at a temperature of 2 to 13°C, but the outer limits are not favorable.

essential to embryonic development. Fish eggs are very sensitive to any kind of physical jolt in the initial period of cleavage. The formation of air bubbles between the bottom of the apparatus and its inner wall should be forestalled since such bubbles impair the respiration of the development embryos. To avoid this phenomenon wooden floats are placed at the point of water entry into the apparatus. A reduction in the flow of water is not advisable.

In the case of tap or spring water supplied under pressure, air bubbles are created in the apparatus due to the reduction in pressure when the water emerges from the pipe. Such bubbles cover the eggs and pose serious threats for the larvae. The smaller bubbles fuse into larger bubbles, which exert a pressure on the gill arches and pericardium of the larvae. This pressure impairs blood circulation and leads to larval mortality. S. Kupka (1957) has recommended that water be supplied to the apparatus through a sedimentation-distribution tank. The rate of water consumption in the incubation apparatus is taken as 0.2 to 0.3 l/min for 1,000 eggs. At a higher oxygen content the water consumption is reduced, which reduces the adhesiveness of the eggs.

2. The optimum oxygen content in the water of the apparatus should be 6 to 8 mg/l, for which the necessary analysis must be done not less than once a day. A watch is maintained on the hydrochemical regime of the water and metabolic products removed. The oxygen requirement varies during the embryonic period, increasing in the period of the formation of the blood circulatory system. A deficiency of oxygen results in various types of deformities. The requirement of developing eggs for oxygen is less than in fully formed young fry. Thus, in the period of development and formation of the vitally important organs and systems the embryos are more sensitive to the conditions of external environment than in the process of growth in the subsequent period.

3. The water is cleansed of fine silt particles which could settle on the eggs and interfere in the normal gas exchange process and create the risk of an infection. The permissible quantity of suspended material in the water is 1.5 mg/l. To prevent the settlement of silt particles special, usually sand-pebbled, filters are used. Since the filters become polluted rather rapidly they should be systematically examined and cleaned.

The eggs are washed to remove silt, taking extreme care that no mechanical injury occurs. In the initial period of cleavage, at the time of gastrulation and closing of the blastopore, as well as in the beginning of pigmentation of the eyes and immediately before hatching of the larvae, washing of the eggs should not be done, since during this period they are particularly sensitive to external mechanical injury. The frequency of washing depends on the quality of water as well as on the working efficiency of the filter. If clean water containing only a small quantity of suspended particles is supplied to the apparatus, the eggs are washed not more than once or twice during the entire incubation period. Different methods of washing have been devised but the one more often used is lifting and submerging the trays or frames of eggs while simultaneously spraying them with water from a sprinkler. Jolting the trays must be avoided and the spray of water fine. The outer sides of the frames as well as the apparatus are carefully scrubbed with a brush. The clean frames of eggs are replaced in the apparatus since the eggs again enter into a period of high sensitivity. A special device for washing the eggs has also been designed (Figure 107). In this case the frame of eggs is removed from the incubation apparatus and placed under a gentle shower of water issuing from a transporter (conveyor) for one minute.

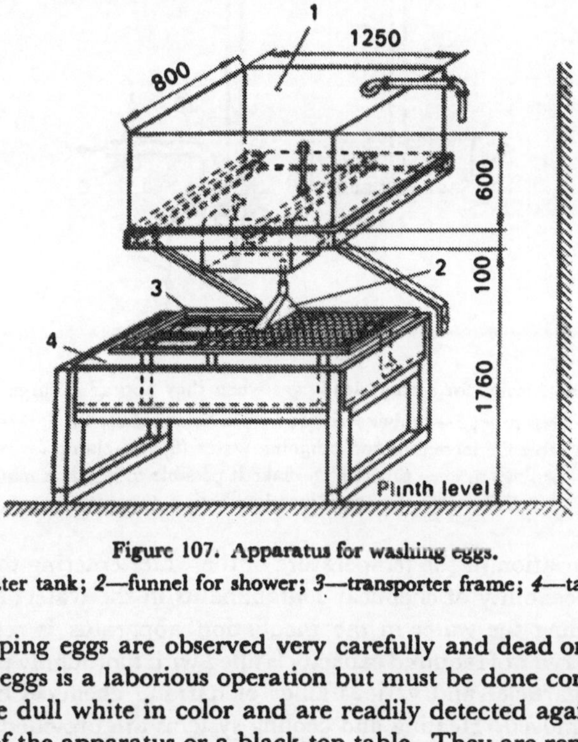

Figure 107. Apparatus for washing eggs.
1—water tank; *2*—funnel for shower; *3*—transporter frame; *4*—table.

4. Developing eggs are observed very carefully and dead ones removed. Sorting dead eggs is a laborious operation but must be done conscientiously. Dead eggs are dull white in color and are readily detected against the dark background of the apparatus or a black-top table. They are removed with a forceps or a pipette with a rubber syringe. The frames or trays of eggs are removed from the apparatus to the black-top table and the dead eggs counted in every survey eventually totalled to obtain the loss sustained throughout the incubation period. This loss should not exceed 15 to 20% in rainbow trout. All operations relating to washing the eggs and removal of dead ones are done in the periods of least sensitivity of the eggs to rubbing and mechanical impacts. The first such period ranges from the first hour after fertilization up to no more than a day, and the second period is at the "eye spot stage". In the case of a high level of egg degradation or the development of infection,

washing and removal of dead eggs should be done in accordance with need (Figure 108).

To avoid the shortcomings mentioned above (supply of water from rivers, streams, lakes, etc. which has to be cleaned and stored for sedimentation;

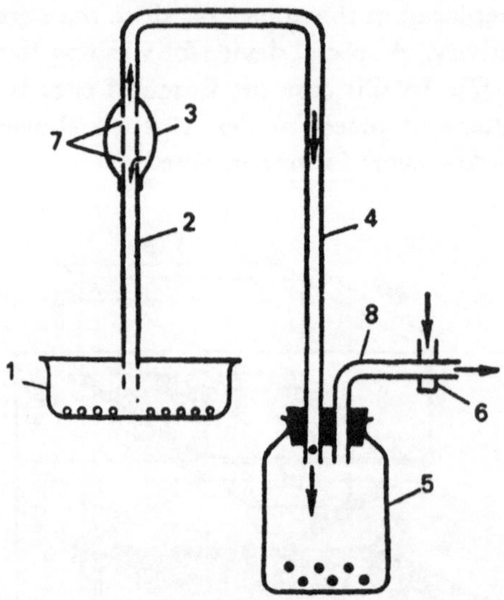

Figure 108. Device for sorting dead eggs when they appear in large numbers. *1*—egg tray; *2*—glass tube; *3*—rubber syringe; *4*—plastic or rubber tube; *5*—bottle with cork stopper and two tubes for incoming and outgoing water (8); *6*—clamp for regulating water supply; 7—two openings in glass tube which make it possible to regulate movement of water and release live eggs into incubation apparatus.

significant variation in the temperature of the water entering the incubation apparatus; possibility of chemical contaminants in the water; low pH; and so on), recycling the water in the incubation apparatus is recommended. For this a reservoir of required capacity is filled with thoroughly purified water from which particles and various kinds of harmful chemical contaminants have been removed. Heating and cooling systems are provided to maintain the water temperature at the required level. The water is aerated and pumped into the incubation apparatus from which it returns to the reservoirs by gravity flow, thereby completing the cycle. The loss of water due to evaporation is supplemented from the sedimentation/storage tank or in the case of total water exchange (replacement) from the reservoir. Less water is required in this system and the water is of better quality.

5. Fish eggs must be protected from the harmful effects of sunlight. Daylight and strong electric light are also harmful to developing embryos: development is retarded which is subsequently reflected in growth. To avoid

such consequences the apparatus are covered with lids and the windows of the incubation section protected from light.

The incubation period of the eggs of brown trout is 460 degree-days at 4°C and for rainbow trout 320 to 350 degree-days at 6 to 10°C. As soon as the prolarvae begin to hatch the water current through the apparatus is increased 1.5 to 2.0 times. Hatching is usually completed in 5 to 7 days but may extend up to 10 days. Dead prolarvae and egg-shells are removed since this material collects in the protective screens and thereby disturbs the water exchange, and undergoes decomposition leading to a deterioration of the gas regime.

Care of Larvae and Fry. The prolarvae after hatching are kept in the same incubation apparatus after removing the trays in which the eggs were arranged. On hatching the prolarvae of rainbow trout are 1.5 to 1.8 cm long and weigh 45 to 50 mg. Between 50 to 65% of the weight is due to the large yolk sac, the contents of which sustain the prolarva in its initial stage of development. During the time that the prolarvae remain quiescent at the bottom of the incubation apparatus, undisturbed surroundings are essential because the hatchlings are very sensitive to external conditions. They are protected from bright light and the apparatus covered with a lid. It is essential to thin out the number of larvae by carefully transferring them to larger apparatus.

The yolk sac of the prolarvae of rainbow trout is absorbed within 15 to 18 days. When two-thirds of the yolk sac have been consumed the larvae move toward the water surface, become motile, and change over to active feeding. The larvae should be regularly fed in the period of mixed feeding (see below).

The larvae are transferred to trough or tanks of different shapes (rectangular or round) where they are kept up to the age of 1.5 months. A sketch of the trough of the Cherfas-Kozlov system is shown in Figure 109.[10] The size of the trough is 200 cm × 100 cm × 40 cm with a water supply from below. Before the larvae change over to active feeding they are released into the tanks at the rate of 15,000 to 20,000 prolarvae/m² under a consumption of water at the rate of 0.3 to 0.5 l/min/1,000 larvae. The output is 14,000 to 19,000. The average weight of larvae up to the moment of their changeover to active feeding is 0.1 g. Light now becomes an important factor as the fry grow better under good illumination. Until their transference to the rearing ponds the larvae are reared in troughs or tanks at a density of 5,000 larvae/m², a water consumption of 0.5 to 0.6 l/min and an optimum oxygen content of 8 to 10 mg/l. The average weight of the fry is 1.5 to 2.02 g and more. The loss during the rearing period is generally 10 to 15%.

[10]An oval-shaped trough gave good results at the Chernorechenskii trout farm. It provided better water exchange and a higher index of fish rearing, making it possible to reduce the area of costly tanks.

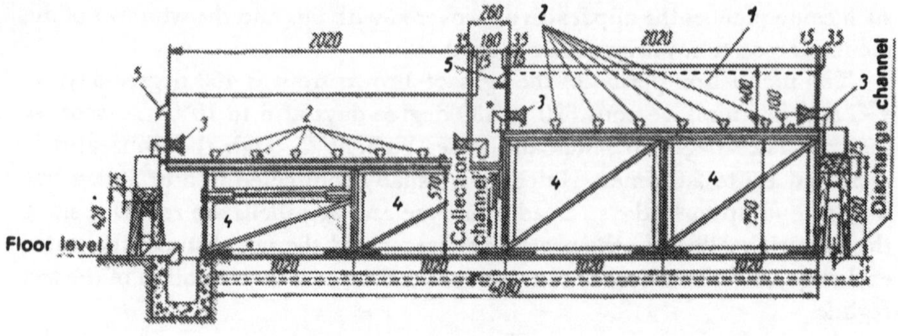

Figure 109. Trough of Cherfas-Kozlov system for rearing trout fry.
1—galvanized iron tank; *2*—net caps; *3*—water outlets; *4*—stand for fixing galvanized iron tank; *5*—overflow.

The fry in troughs or tanks are fed intensively for 30 to 45 days. They grow normally, with the larger and stronger fish predominating over the smaller and weaker ones with regard to food intake. With fry growth overcrowding occurs in the troughs and tanks and hence the fish are separated into two or three groups, depending on the number of ponds available at the time of transplantation, and each group transferred to a separate rearing pond.

The oxygen content of the water exerts a considerable influence on the growth of juvenile rainbow trout. At a lower oxygen concentration growth slows down to half of what it should be and the absolute and relative indexes of food consumption and recovery factor drop. This is partly explained by the deterioration in the digestibility of proteins.

The stocking density of fry in rearing ponds varies from 25 (the desired norm) up to 100 fish/m^2 (permissible maximum level) with a water exchange of 5 to 10 times in 24 hours. According to the norms of "Gidrorybproekt" the quantity of stocking material in rearing ponds may be 150 to 200 fish/m^2 with a water consumption of 50 to 60 l/hr for one kg of live weight of fingerlings.

The care of juvenile fish in rearing ponds includes correct and systematic feeding, maintenance of proper sanitary conditions and water exchange, control on growth, removal of diseased and dead fish, and so on.

The fingerlings remain in the rearing ponds up to autumn and, according to the standards of "Gidrorybproekt", should attain an average weight of 15 to 20 g with an output of 70 to 80% of the fry stocked. After every 20 days the fingerlings are graded into size groups—small, medium, and large. The devices for grading fingerlings before they are transplanted for hibernation have already been described (see Chapter IV).

At the time of catching fingerlings the water is discharged from the ponds and fish traps installed behind the bottom outlets to facilitate catchment and grading. After grading the fish of each group are placed in separate ponds.

The last grading in autumn is usually done on the basis of length (cm); from 5 to 8, 8 to 11, 11 to 14, and 14 and above.[11]

Stunted fish (length 5 to 8 cm) should be rejected as underdeveloped. The catching of fish in the ponds is done at fairly low temperatures (not freezing) and in the warmer parts of the year during the coolest hours of the day. The fish traps used for catching are depicted in Figure 29.

If the rearing ponds are quite suitable for the winter maintenace of trout under-yearlings, the fish are left there for hibernation and the yearlings transferred to the foraging ponds the following spring. The density of stocking of under-yearlings in rearing ponds is 100 to 150 fish/m^2 with an average weight of yearlings of 40 to 70 g and an output of 90%. The consumption of water for one kg of live weight of yearlings is 50 to 60 l/hr. The stocking density of yearlings in foraging ponds is 50 to 70 fish/m^2. The average weight of two-year fish is taken as 130 to 150 g and more with an output of 90 to 95% of the total number of yearlings stocked, and water consumption at the rate of 35 to 40 l/hr for one kg of live weight of two-year fish.

Brood fish have to be caught with special care in the case of autumn catches of trout of older age groups. To avoid loss of eggs catchings should not be delayed. If other fish species are raised in the same pond with trout, the trout should be caught first.

Feeding of Rainbow Trout. The main method of intensification of trout farming is feeding under higher density of stocking than done at carp farms. If the growth of fish in carp farms depends on natural foods to some degree, in trout farms natural foods are not so important and can even be completely replaced by supplementary feed. This being so, the type of feed given determines the success of the trout farm. The biological peculiarities of trout, their predatory mode of life, and their requirements for live foods distinguish the nature and methods of feeding rainbow trout from those described earlier for carp. In the first place trout need products of animal origin: meat, fish, and still better live foods. In trout farms usually the meat of farm animals is used (horse, donkey, goat, etc.) as well as the different organs of these animals such as lungs, spleen (fresh as well as dried), liver, brain, heart, and thoroughly washed guts. The following are also used as feed: slaughterhouse waste, meat powder, prawn meal, silkworm pupae, rye flour, blood-yeast in the form of a dry preparation, coagulated milk, serum, dry fatless milk powder, eggs of birds, snails, frogs, tadpoles, mollusks, May beetle, etc. The most economical flesh feed is made from cow liver, the dry matter of which

[11]Other group sizes are followed in Czechoslovakia, the Federal Republic of Germany, and other countries.

contains 79.71% protein, 12.22% fats, 3.2 carbohydrates, 4.75% ash, 42.69 mg% iron, and 1.29 mg% phosphorus. Cow liver is rich in vitamins of the B group. Slaughterhouse waste is widely used in the Federal Republic of Germany and the pupae of silkworms in Japan.

Marine and fresh-water fish, as well as the waste material obtained as a result of their grading, are also commonly used for feeding trout, both in fresh form and in fish meal. In Denmark the main component of trout feed among marine fish is herring and other less valuable fishes. Feeding trout fish increases the growth rate and the survival rate of fry. Fresh-water and marine fish with a high fat content (greater than 5 to 7%) are not suitable for feeding trout. To avoid lipoid degeneration in the trout liver young fish which contain less fat should be used. Fresh-water trash fish (bleak, ruff, gudgeon, loach, verkhovka or moderlieschen, minnow, etc.) are considered a better natural food, especially in the case of trout brood fish which avoid feeding of marine fish. It is better to use moderlieschen (*Leucaspius delineatus*) for feeding trout. This species is characterized by phased spawning, minimal vital demands, and early maturation, which makes it possible to obtain large quantities of progeny within one vegetative growth period. These fish can be stored in live form in special small ponds or fish tanks.

Insects are no less important in feeding trout. Trout prefers aquatic bugs but aerial insects also occupy a significant place in its ration; their occurrence in the gut contents reaches up to 33% in May, 60% in June, and 40% in July. Aerial insects constitute 53% of the ration of rainbow trout in the Trans-Carpathian rivers.

Trout adjusts quite readily to a change in food, which makes raising easy. However, its food value in such cases also changes. According to Gorovaya (1968) the muscles contain less fat but more protein if rainbow trout is fed its natural foods. The color of the muscles also changes. In fish fed mainly on insects and plankton a characteristic pink coloration is seen, while in fish fed on artificial feeds the muscles are white.

Trout fed on fish differs in taste from trout raised on dry feeds. The type of feed also influences the color of the meat. Cataxanthine—a product of oxidation of yellow carotenoids—added to the feed gives a pink color to the trout meat.

Rainbow trout is a predatory fish and as such cannot grow exclusively on plant food (Privol'nev, 1968); it consumes and assimilates plant proteins but its intensity of growth on such foods is much lower. This is partially explained by the fact that in the proteins from plant sources a deficiency of the less important amino acids such as lysine and methionine usually occurs. Trout develop shorter intestines when fed animal foods compared to plant foods. The rate of digestion of proteins in fish of the same age, but different in size, is not the same; it is higher in smaller individuals. In the case of excessive proteins the food mixture is digested at a slower rate. The intensity of protein digestion and its assimilation are increased by adding 10% phosphatides.

The nutritional composition of some food items of trout is given in Table 55.

Ballastic materials (bran, wood sawdust, etc.) are also important in the ration of trout; these have a catabolic effect and facilitate the effectiveness of the digestive enzymes. Such substances (cellulose, etc.) are not digested by trout and only a small quantity is added to the ration.

Table 55. Nutritional composition of some food items of trout (according to data from the Fodder Research Institute in Brno and other sources)

Type of food	Nitrogen substances,%	Digestible proteins, %	Concentrated nonnitrogen substances,%	Fat, %	Ash, %
Spleen	17.70	—	—	3.00	0.90
Liver	20.00	—	—	4.00	2.00
Calf brain	17.50	—	—	10.00	0.80
Heart	17.50	—	1.00	10.60	2.50
Fresh blood	15.19	14.43	0.48	0.89	1.36
Boiled blood	19.19	18.55	0.01	0.07	0.80
Blood serum	38.23	36.28	—	0.28	0.96
Meat meal	43.88–77.42	29.04	0.61–4.75	3.37–24.27	7.59–17.29
Meat-bone meal	49.20–63.50	23.20–46.70	2.80–3.70	1.60–15.60	21.40–24.90
Bone meal	9.10–35.50	4.90–18.20	3.70–4.80	1.60–4.60	48.50–79.30
Blood meal	70.70–85.20	60.40–77.90	0.80–4.90	0.70–4.40	3.50–8.60
Fish meal	42.00–26.30 [*sic*]	32.00–56.00	0.50–7.80	1.60–13.50	13.10–35.80
Wheat bran	12.80–16.10	9.00–11.00	54.10–54.50	4.10–4.50	5.50–12.40
Rye bran	12.50–16.20	8.10–10.10	55.20–59.90	3.50–3.90	5.40–10.70
Sprouted barley	26.30	12.97	42.30	2.20	6.50
Cereals	34.50	29.30	37.50	7.90	7.49
Feed meal:					
rice	12.20	6.60	44.50	12.50	10.70
wheat	10.90–17.00	9.90–14.50	60.60–73.70	1.80–3.90	1.00–3.50
malt barley	10.50	6.40	76.20	1.90	1.40
rye	13.80	8.90	65.80	3.20	3.00
Meat of pond mollusks:					
boiled	13.00	—	6.40	1.20	6.70
raw	6.90	—	3.10	0.50	2.80
Mollusks	6.10	—	2.80	0.50	41.90
Beef	20.00	—	1.50	3.50	2.70
Horse meat	20.50	—	2.00	3.20	2.50
Pupae of mulberry silkworm (on dry weight basis):					
raw protein	59.4–66.4	—	9.0–12.7	18.0–24.0	2.8–5.8

The food coefficients of some of the common food items of trout according to the data of a number of Soviet and foreign authors are presented in Table 56.

Table 56. Food coefficients of some items of trout food

Type of food	Food coefficient
Fresh daphnids	6–7
Larvae of chironomids	4
Fresh spleen	5–8
Fresh liver	7
Meat meal	1.5–3.0
Meat-bone meal	3.5
Blood meal	1.5–3.0
Fish meal	1.05–3.0
Feed meal	6
Meal prepared from pupae of mulberry silkworm	2
Bran	6–7
Slaughterhouse waste	5–6
Meat of warm-blooded animals	5–6
Horse meat	5–8
Fresh fresh-water fishes	5–10
Fresh marine fishes (herring)	5–7
Dried fish	5–6
Fish admixtures	3–5

The pupae of mulberry silkworm contain a large quantity of fat and have a harmful effect on fry; large-scale mortality commences eight weeks after feeding due to fat accumulation in the liver and kidneys. The mortality of fingerlings of carp and trout is significantly decreased by defattening the pupae with ethyl either for 10 hours or until the fat content drops to 0.1%. Experiments conducted by Tsirkova (1966) have shown that defattening the pupae to 0.5% and adding them to the food mixture KRT-VI yields even better results.

In composing the formula for trout rations one must incorporate all the vitally essential components, among which proteins take first place since the supply of different amino acids must be ensured. Lysine, cystine, and tryptophan are very important for rainbow trout. The following amino acids have been found to occur in the muscular tissues of rainbow trout: methionine, cystine, tryptophan, arginine, histidine, leucine, isoleucine, threonine, valine, tryosine, lysine, and serine.

With regard to protein quality, marine and fresh-water fish in fresh form, natural foods, and the meat and viscera of homoiothermal animals occupy first place. Under natural conditions trout consumes about 48% of the dry weight of proteins from which it assimilates 92%. When trout is maintained in ponds on artificial feeding, its requirement for protein must perforce be met. Studies have shown that raw protein should constitute 28 to 40% of the food mixture. According to Steffens (1968) the requirements of raw protein to obtain one kg of food protein in trout, pigs, and broilers is determined by the same value.

In the Federal Republic of Germany formula feeds for trout contain 33, 37, 40, and 46% proteins. Increasing the protein content from 26 to 30% up to the foregoing values made it possible to reduce food consumption from 2.5 to 1.3 kg per kg of increase in body weight.

Carbohydrates do not play an important role in the nutrition of rainbow trout. Trout does not digest many of the complex carbohydrates since it is not naturally endowed to secrete the necessary enzymes. Most ichthyologists hold that carbohydrates should not exceed 8 to 9% of the trout ration (although their quantity in natural foods is two times greater[12] than that in dry matter) and that no more than 4.5 g of assimilated carbohydrates per kg of body weight of trout should be provided per day. In terms of the physiological activity of trout fat has the maximum value. Nevertheless, its quantity should not exceed 5% of the ration even though in relation to the dry matter in the natural food of trout the fat content is up to 15%. One must remember that the composition of the food also influences the composition of the fatty acids in the body and liver. Any excess fat is deposited in the kidneys, creates edema, and leads to increased waste. Spawners in such cases lay poor quality eggs. Reduction in the fat content (below 2 to 3%), containing mainly unsaturated fatty acids, leads on the other hand to a deterioration in growth and an increase in fish loss.

Minerals are especially important during the period of growth. As indicated in Chapter XI the sources of minerals for fish are food and water entering the gills and skin. In the case of trout the mineral content of the pond water must be controlled. Excessive minerals in the food or water may cause diseases and even mortality.

Bone or fish meal or calcium phosphate is added to the ration of fry. The salt content (common salt) for adult trout should not exceed 3% of the ration. Any increase in salt leads to inflammation of the digestive tract and causes a high rate of mortality. Trout fry given food containing just 1% of common salt usually die.

Elements such as calcium and phosphorus should be present in the food in a ratio of 2:1. Trace elements are also important: iron, cobalt, magnesium, manganese, iodine, boron, silicon, and others act on the metabolism and participate in the enzymatic processes. However the nature of the influence of trace elements on trout has been studied rather poorly. Hence they should

[12]According to N.O. Christensen (1966) a carbohydrate content in the food of trout greater than 9 to12% causes disorders of the liver leading to mortality.

be restricted in commercial fish culture until their influence on fish is better known and well-founded information available on norms of usage for different age groups. Vitamins are also included in the ration of trout. Those of group A (retinal, axerophthol) are essential in the oxidation processes in the tissues, prevent delay in growth, and protect against skin diseases. Vitamin A is produced from the carotenoids provided in food and is deposited in the liver. According to the findings of American authors the ration of trout should contain vitamin A in the form of fish oil at the rate of 2% of the total weight of the feed.

The biologically active substances of the vitamin B complex are very important as regulators of protein, fat, carbohydrate, and water metabolism. According to the findings of a number of researchers the trout ration should contain the following quantities of vitamins[13] (mg/100 g of ration):

Thiamine (B_1)	0.1–1.0	Inositol	25–50
Riboflavin	0.5–1.5	Choline	5–10
Pyridoxine (B_6)	0.1–1.0	Biotin (vitamin H)	0.005–0.025
Cyanocobalamin (B_{12})	0.009	Folic acid	0.1
Nicotinic acid (PP)	0.1–1.5	Pantothenic acid (B_3)	1.00–2.0

A deficiency in these vitamins causes a loss of appetite, poor growth, impaired digestion and metabolism, degeneration of the liver, neurosis accompanied by anemia, general physical debility, and eventual fish mortality. To avoid such consequences substances are added to the ration of trout which are rich in the vitamin B complex—yeast, cereal, bran, and liver. To enhance vitamin B_{12} in the spleen 10% powdered cow dung or poultry excreta is added to the ration in Czechoslovakia. The growth rate of fry is increased by 46.48% with the addition of cow dung, and by 36.43% with the addition of poultry dung, with no loss of fish. Dry cow dung is recommended as a source of vitamin B_{12} as an additional component in the ration in Yugoslavia and Bulgaria.

Ascorbic acid (vitamin C) activates metabolism in the cells and is especially important for older fish. Presumably fish synthesize vitamin C and in most cases its addition to the ration is not necessary.

Vitamins of the D group regulate phosphorus-calcium metabolism and stimulate trout growth. Their deficiency leads to underdevelopment of the operculum and abnormal deposition of calcium and phosphorus in the bones, sometimes even causing their softening. This phenomenon can be prevented by the addition of mineral substances to the feed—powder of shells of mollusks, phosphorus-calcium powder, and fish oil.

[13]The daily requirement of some vitamins by trout (and carp) according to Mann (1960) is as follows (per kg of fish weight): A—100 to 500 iu; B_1—0.15 mg; B_2—0.11 to 0.33 mg; C—100 mg; pantothenic acid—0.1 to 1.4 mg; nicotinamide—0.55 mg.

Vitamin E (tocopherol) is no less important. A deficiency of this vitamin results in poor egg fertilization, sterility of the brood fish, fat degeneration in the liver, and diseased kidneys. Vitamin E also prevents muscular dystrophy. This vitamin is added at the rate of 3 mg/100 g of feed.

Vitamin K prevents diseases of the mucous membrane of the intestines, anemia, and thrombosis. It is added to the food at the rate of 5 mg per 370 g of fish live weight.

Diverse opinions have been expressed regarding the addition of antibiotics to trout feed; this problem needs further investigation.

Data are summarized below on the effect of phosphatides on the metabolism and physiological condition of rainbow trout with the addition of concentrate phosphatides at the rate of 5 to 10% by weight of food mixture, using lecithin of raw sunflower (a by-product in the preparation of sunflower oil):

(a) Prevention of cirrhosis of the liver in trout, which causes anemia, and the cure of sick fish (5% phosphatides suffice for four-year fish);

(b) Regulation of fat metabolism and release of excess fat from the liver and its deposition in the muscles and body cavity;

(c) More productive utilization of food which improves growth;

(d) Increase in blood formation and increase in hemoglobin count.

The foregoing obviously improve the physiological condition of the fish, reduce losses, and increase the productivity of trout ponds.

According to the results obtained at the All-Union Research Institute of Lake and River Fisheries (1969) the addition of 5 and 10% edible yeast to the ration increases the growth of two-year trout by 22 to 31% and improves their physiological condition.

Yeast proteins contain all the amino acids essential for the animal organism. In a ration poor in vitamins of the B group at least 10% edible yeast is added; when these vitamins are sufficient yeast is added at the rate of 5%. An addition of 5% yeast increases the total fat content of rainbow trout and the fat content of the liver. Dry yeast, according to the results obtained by British pisciculturists, can be replaced by the proteins of soybean meal and by a mixture of the vitamins present in yeast, which makes the mixed feeding of trout cheaper.

The effect of carbamide on fish has been poorly studied. Its concentration at 0.25 to 0.50% (Knyazeva, 1967) has a suppressive effect on the larvae and yearlings of rainbow trout. Disruption in the development of larvae, deformation, and mortality occur after 18 days at a concentration of 0.50%, and after 25 days at a concentration of 0.25%. Yearlings of trout given the same concentrations die in three days' time. The larvae of carp are more resistant and live for a longer time in a carbamide solution of 0.06 to 0.35%, and fingerlings can withstand a solution of 0.5 to 1.5% for 45 days, i.e., they are less sensitive to such concentrations than the yearlings of trout. Urea injected into the body cavity of fish causes poisoning but its intake in moderate doses through the digestive tract does not.

The food for different age groups of trout is not similar and is divided into three groups: (a) food for feeding fry at different stages of their development (which should be highly nutritious); (b) food for grown-up under-yearlings and two-year fish; and (c) food for brood fish.

Fry feeding commences when the yolk sac is reduced to half. Fresh blood or blood-containing tissues obtained from the spleen are added to the apparatus containing larvae in a passive state and they gradually adapt to the consumption of this food. Regular feeding commences from the moment the yolk sac has reduced two-thirds and the larvae assume the normal position of a fish—backside upward and head directed against the current.

Initially the fry are fed fresh cow spleen, which is rich in vitamins and easily assimilated; cow liver, a source of biotin, is even better in terms of growth rate. Other foods which are good for larvae include zooplankton, fresh blood, egg, cottage cheese mixed with blood, emulsion prepared from egg and milk, etc. After two months the fry will feed on slaughterhouse waste, fresh low quality fish (intact), liver, frogs, tadpoles, and mollusks. For the enrichment of these basic foods with easily assimilated proteins, mineral matter, and vitamins, the ration is reinforced with blood, fish and bone meal, meal prepared from the pupae of mulberry silkworm, proteinic paste prepared from the material washed off after the hydrolytic processing of whale liver, spleen, and liver, as well as edible yeast, germinated seeds of cereals, etc. To increase the consistency of the foregoing, rye or wheat flour or a mixture of both is added. A high quality food for trout fry is cow heart. Feeding fry a mixture of cow liver and heart in equal quantities increases growth and reduces mortality. This food is costly, however. Hence one month after feeding commences the spleen, which is cheaper, is minced with a small quantity of fresh fish and slaughterhouse waste, or high quality fish meal and prawn meal. The spleen contains (percentage on dry weight basis): proteins—83.21, fat—9.52, ash—7.16, carbohydrates—0.11, calcium—1.17, phosphorus—1.66, and iron—129.43 mg. According to the findings of American authors fresh fish should not constitute more than 20 to 25% of the ration.

Natural food is very significant in the rearing of fry. In the course of retention of fry up to six months of age on natural foods or even by the addition of 20 to 40% live foods to spleen, the fish become more viable and their growth improves. Daphnids, gammarids, chironomids, larvae of the mosquitoes *Culex* and *Corethra,* tubificids, enchytracids, larvae of the blowfly, etc. serve as good live foods for trout.

To increase the quantity of natural foods fertilizers are sometimes applied to trout ponds. This has a special significance in the rearing of fry. Although this method of intensification is less effective because of the higher rate of water flow in trout ponds than in carp farms, nevertheless under certain definite conditions it makes it possible to improve their natural food base significantly. Chemical fertilizers (phosphorus, superphosphate) are generally applied at the rate of 60 kg P_2O_5/ha after 14 days from the application of calcium. Potassium salt is also added. Since trout is very sensitive to the quality of water, in particular high acidity, organic manures are used sparingly and are generally applied at places in the pond where the water flow is less. The water current is stopped in manured ponds until the fry which have changed over to active feeding are released in them.

Experiments carried out in Lake Sevan (Armenia) with ash manure (0.4 to 0.5 kg of grass per m^2 of area) have shown that the addition of plant material to ponds in small quantities and at short intervals of time (six to seven days) facilitates the maintenance of optimum concentrations of micro-organisms and favorable physico-chemical conditions for the development of food fauna in the zones of manure application where trout fry tend to congregate.

Similar to carp, trout is also fed food mixtures composed in such a way that all the elements make a physiologically complete ration. Such mixtures are prepared either at the farm or at a central place. When the food mixtures are prepared at the farm, mainly local resources are used or only a small part of the components (in particular vitamins) obtained elsewhere.

Recently, in the USSR and abroad standard feeds and dried food concentrates have been introduced. In the USA standard feeds are prepared mainly from waste material from food industries. The so-called "Oregon feeds (Oregon pellets)" have been very successful. These feeds, prepared from 60% dry and 40% moist components, are in the form of frozen pellets which must be thawed before use. Their composition is as follows (%):

Cotton seed meal (steamed and extracted beforehand, containing 44% protein)	23
Herring meal	21
Crab extract or prawn meal (equal fractions) and condensed fish broth, dried together	6
Flour from germinated wheat	3.60
Waste from maize distillation in dry form	2.40
Multivitamins	1.50
Albacore viscera	20
Paste made from internal organs of salmon	20
Corn oil	1.80
Choline chloride (70% liquid)	0.65
Antioxidant	0.05

On the average 2.33 kg of this feed is consumed per kg weight increase and the annual feed cost reduced up to 34%. In addition to this ration, in the

USA dry granulated feeds are mainly used for feeding rainbow trout, which are prepared by special firms for fry and fish of older age groups. Dry feeds for trout are now widely used in a number of countries (Denmark, German Democratic Republic, USSR, etc.) and their centralized production being organized. These feeds are well assimilated by fish, do not dissipate in water, and are about 30% more economical than mixtures prepared at farms. They can be easily transported and stored and their delivery to the fish mechanized. According to the findings of Steffens (1968) dry feeds can be fed to rainbow trout with no harmful consequences for about three months. A more prolonged use of dry feeds causes anemia in trout; hence the food ration must be changed. The periods may vary: after three months of dry feeds fresh foods are given for a few weeks (for example, cod mince), or fresh foods are given one to two days per week.

Based on studies by a number of authors and the experience of trout culturists, the following feed mixtures are recommended at present for different age groups.

1. For larvae, live zooplankton, hard-boiled egg yolk mixed in water, emulsion made from egg and milk, and blood-containing tissues of the spleen pressed through a sieve or muslin cloth.

2. For fry, paste-like mixtures and pellets containing the following components (%):

A. Blood-containing tissues of the spleen	85.0
Meat-bone meal or fish meal	14.0
Fish oil	1.0
B. Blood-containing tissues of the spleen	55.0
Fresh and thoroughly ground fishes	30.0
Fish meal	14.0
Fish oil	1.0

3. For under-yearlings (%) (from L.I. Pyder, 1963):

A. Blood	27.0
Spleen	28.0
Fresh fish	27.0
Offal from fish-processing industry	0.5
Blood meal	0.5
Fish meal	14.0
Bran	3.0
Poultry eggs	0.02
Fish oil	0.01
B. According to I. Smisek and I. Kupka (1961):	
Meat-bone meal	19.6
Dry animal proteins	4.9
Blood meal	19.6

Pupae of silkworm	29.4
Protein mixture for pigs	9.8
Brewery yeast	4.9
Peas	9.8
Precipitate	2.0

C. According to I. Smišek and I. Kupka (1961):

Meat-bone meal	30
Blood meal	30
Protein mixture for chicken	30
Feed meal	10

Malikova (1969) developed the feed mixture KRT (blood, fish, pupae) for young salmon which, over the years, has been improved and at present is recommended in the altered form, KRT-VI (Table 57).

Table 57. Proportions of components of KRT feed

Components	Quantity per kg of feed
Blood meal	200 g
Fish meal	120 g
Silkworm pupae meal	120 g
Algae meal	50 g
Mustard powder	20 g
Chalk powder	2–3 g
Edible yeast	30 g
Fish oil	5 g
Concentrate of vitamins:	
A	15,000 iu
D	7,500 iu
Penicillium mycelium	200 mg (which can be replaced by 50 mg of penicillin)
Biomycin	50 mg
Furazolidon	50 mg
Ammonium molybdate	5.5 mg (which corresponds to 3 mg of molybdenum)
Sodium pyrosulfate	15 g
Water	Used as necessary for making total weight of 1 kg

This mixture is physiologically a complete ration and is recommended for larvae in which the yolk sac is still not fully absorbed. Until the fry change over to external feeding, the ration should consist of 75 to 80% spleen and 20 to 25% KRT. Thereafter the dose of KRT is gradually increased up to 70 to 80% when the fry have attained a weight of 0.5 g, and the quantity of spleen reduced to 20 to 30%. The mixture is prepared as a paste and spread on pots or conical feed pans.

According to the recommendations of the State Research Institute of Lake Fisheries (1969), for trout yearlings, two-year fish, and three-year fish the following feed mixture should be used (%):

For farms located near industrial centers where cattle slaughtering is done:

Spleen of farm animals	55
Fish meal	20
Meat-bone meal	10
Bran	5
Edible yeast	5
Phosphatides	5

For farms located near places of processing fish:

Fish waste or small trash fish	80
Edible yeast	10
Bran	5
Phosphatides	5

The highest fat content in trout yearlings is recorded when phosphatides are increased to 10% in the ration. The addition of 10% carbamides (synthetic urea) accelerates digestion and in combination with phosphatides and yeast reduces the quantity of fat deposited in the body cavity.

The amount of food mixture supplied depends on the water temperature. The State Research Institute of Lake Fisheries (1969) has recommended the following periodicity of feeding rainbow trout (which still requires further verification at fish farms) (Table 58).

Table 58. Periodicity of feeding trout, depending on temperature and age

Water temperature, °C	Age of fish	Periodicity of feeding
15	1+	1 to 2 times a day
16	2+	Once a day
9	1+	Once in 2 days
9	2+	Once in 2 days
3–4	1+	Once in 3 days

The passage of food along the alimentary canal of rainbow trout also depends on water temperature and age of the fish. For two-year fish (Privol'nev, 1969) the changes are as follows:

Temperature, °C	Hours
8–9	38–42
10–12	32–35
18–20	24

In the case of three-year fish the passage of food at the same temperatures takes place in 71 to 85, 57 to 60, and 35 hours respectively.

The digestibility of proteins in various foods is given in Table 59 (according to data compiled by Nouz, Kitamikado, and Inabo).

Table 59. Digestibility of food proteins by rainbow trout

Food component	Quantity, %	Digestibility of protein, %	Author
1	2	3	4
Defatted pupae of silkworm	50	6.6	Nouz, 1960
Starch	50		
Defatted pupae of silkworm	70	8.3	
Starch	30		
Meat of flatfish	60	83–85	
Wheat flour	30		
Dextrine	10		
Dried mysids	90	82	Inabo, 1960
Wheat flour	10		
Pupae of silkworm	90	82, 89	
Wheat flour	10		
Whitefish meat	90	89, 92	
Wheat flour	10		
Cachalot meat	90	75	
Wheat flour	10		
Meat of fin whale	90	65	
Wheat flour	10		
Soybean pods	90	81, 87	
Wheat flour	10		
Cotton seeds	90	75	
Wheat flour	10		
Whitefish meat	90	81	Kitamikado, 1964
Potato starch	10		
Whitefish meat	80	82	
Potato starch	20		
Whitefish meat	60	78	
Potato starch	40		
Whitefish meat	40	74	
Potato starch	60		
Legume waste	80	44	
Wheat flour	20		
Casein	75	97	
Potato starch	15		
Olive oil	10		

(Contd.)

1	2	3	4
Casein	76.5		
Potato starch	13.5	97	Kitamikado, 1964
Olive oil	10		
Casein	68		
Potato starch	12	96	
Olive oil	20		
Pacific saury meat	70		
Wheat flour	20	75	Nouz. 1965
Dextrine	10		
Sardine meat	70		
Wheat flour	20	78	
Dextrine	10		
Anchovy meat	70		
Wheat flour	20	75	
Dextrine	10		
Herring meat	70		
Wheat flour	20	77	
Dextrine	10		

Proteins are poorly digested when large quantities of carbohydrates are present; fat exerts little influence on the digestibility of proteins. The digestibility of saturated fatty acids is much higher than that of unsaturated fatty acids.

According to the findings of Japanese pisciculturists the digestibility of casein from milk by trout is 99.5%, of egg proteins 83.9%, whitefish meal 91.9%, and defatted soybean meal 92.4%.

Rainbow trout weighing 2.6 kg consume food for 35 minutes at a water temperature of 19 to 22°C and digestion continues for 24 hours. At a water temperature of 14°C digestion continues for two days. In the process of growth this trout consumes less food and the average level of saturation is about 3.0 of the body weight.

To determine the daily norms of food for trout weighing from 100 to 200 g the following parameters have been recommended by Steffens (1965):

Water temperature, °C	Feeding norms (% of body weight)
Above 12	1.5
10–12	1.2
8–10	1.0
6–8	0.8
Below 6	0.6

Trout brood fish are best maintained on natural foods which, unfortunately, are not always available. They should be provided food twice a day at the rate of 2% of their body weight, which norm before spawning is reduced to 0.5 to 1.0%. A good diet for brood fish consists of fresh small fish and meat of homoiothermal animals and mollusks. Dry concentrated feeds, blood, fish and meat meal, pupae of silkworm, proteinic paste mixed with flour, sweepings from a grinding mill, or powdered cake can be used as supplementary items. High fat content in the ration has a harmful effect on the production of sex products. Dried prawn powder or prawn meal and phosphatides have a favorable effect on egg fertilization. Shells and snails also constitute a good food for trout brood fish; they impart a red color to the eggs. As revealed by experience in the Federal Republic of Germany crustaceans, especially crabs, impart a brighter color to the eggs; a more stable progeny is obtained from such eggs.

Preparation of Feeds and Methods of Feeding Trout. The feed given to fish should be fresh and of high quality. Low-quality feed causes diseases and mortality. With long storage the fats in the feed are oxidized, leading to the development of liver diseases. Feeds should be steam-processed only when there is some doubt about their quality as such processing reduces their digestibility, dramatically deteriorates the vitamins, and has an adverse effect on some of the essential amino acids.

Feeds for fry must be prepared very carefully in the initial period of their feeding. The spleen, which is full of blood and very sticky, is cut into small pieces, pressed through a small-meshed sieve to remove the membranes, and fed to the fry in the form of a paste. Feeding is done three to four times a day, although according to American pisciculturists it should be done more often (every alternate hour). Grown-up fry (approximately one month after feeding commences) are fed in addition to spleen, liver, fresh fish, and slaughterhouse waste passed through a meat mincer with disks 1 to 2 mm in size. The coarse and sticky parts such as membranes, tendons, and bones are removed. The blood pressed out while mincing the liver and so on is preserved by adding bran to the mixture which absorbs it. Fish meal, meat meal, and other kinds of meals are sifted through a small-mesh sieve to remove large particles, added to the foregoing mixture, and the mixture passed again through the meat mincer.

The food for fry is served in troughs or basins usually placed in small flower pots, or in small-mesh nets suspended in the water column. Small quantities are given at a number of places to avoid large congregations of fry at one point. For fry maintained in ponds it is appropriate to use small boxes in the form of a pyramid with a netted bottom. A long rod is fixed to the boxes to facilitate uniform food delivery throughout the pond at different points from the shore. The boxes are carefully submerged up to two-thirds of their height in the water, lifted out of the water, submerged again, lifted out, and this procedure repeated several times. The food at the bottom of the box passes through the small-mesh net (1 mm mesh), falling into the water in the form of a fine suspension.

Two months after feeding commences the prepared feed is changed. Powdered feed is added to the feed ground in the meat mincer and this mass thoroughly mixed by passing it through a meat mincer with disks 2 to 3 mm in size. After air-drying the mixture is supplied to the fry directly in the water, uniformly broadcasting it over the area. Feed given in this manner in small portions is eaten by the fish before it can settle to the bottom. Feed should be provided two to three times a day only in that quantity which can be consumed by the fry.

The main feed is passed through a meat mincer and flour, bran, and blood meal added as binding material. It is also appropriate to add up to 5% edible yeast as a source of vitamins of the B group. The mass thus obtained is thoroughly mixed and again passed through a meat mincer with disks 4 to 5 mm in size. Trout when transferred to foraging ponds have thus become accustomed to taking feed suspended in the water and lying on the bottom; this obviates the need for feed pans. The feed is distributed in small quantities directly in the water in certain areas of the foraging pond twice a day.

In the USA the process of trout feeding with granulated feeds has been mechanized. The mechanical feeder consists of a 23-liter bunker with a lid which protects the pellets from water. An electric motor rotates the rotor for sprinkling the feed and is equipped with an automatic relay. One relay serves ten feeders (feed pan). The relay is regulated in such a way that feeding takes place at hourly intervals for 20 seconds. The feed pans are suspended on poles and can be pulled to the shore for filling. The pellets are tossed up and down while passing through the openings of the lid to the rotor. Feed is provided every hour in summer from dawn to dusk and after several-hour intervals in winter.

In the course of trout feeding it is essential to regularly check on the sanitary-hygienic conditions of the tanks, troughs, and ponds, the quality of water, edibility of feed, condition and growth of the fish, and water temperature. Fish which have died or are suspected to be diseased are removed. The quantity of daily ration is changed in accordance with the water temperature.

As in the case of carp the feed for all adult groups of trout should not be changed abruptly; the new feed should be gradually introduced to allow the fish sufficient time to adapt to it.

The weight of the daily ration is calculated on the basis of the average weight of the fish as a result of periodic control catches. The daily quantity of ration is increased with an increase in fish body weight.

The annual quantity of feed is spread over different months[14] in accordance with the average monthly temperatures (%):

March	1	August	18
April	4	September	17
May	7	October	14
June	13	November	10
July	16		

Culture of Rainbow Trout in Fish Cages. Like carp, rainbow trout can be cultured not only in ponds but also in fish cages. This method is practiced in the USSR as well as abroad but not on a large scale.

In cage culture the density of stocking is considerably increased and is analogous to the stall maintenance of animals in animal husbandry. Such a method of fish culture is a higher form of intensification of pond pisciculture since the fish do not expend energy in moving about in search of food; there is also no wastage of the feed. In this case the quality and temperature of the water must correspond to the biological peculiarities of the fish (the water must be clean and free of pollutants from industrial effluents and domestic sewage), contain the requisite oxygen content, and have a total hardness of 12 to 14 German degrees.

In rearing rainbow trout in cages the oxygen content should be higher than 5 mg/l because with a lower oxygen concentration trout suffer distress, which ultimately leads to a deterioration in their feeding and a reduction in growth rate. The most favorable temperature is 15 to 18°C. At a temperature higher than 20°C feeding intensity drops and at 25°C ceases.

The State Research Institute of Lake and River Fisheries (Privol'nev, 1968, 1970) has successfully reared rainbow trout in cages 2.5 m × 1.5 m × 1.5 m in size fixed in flowing waters with a current speed of 0.5 m/sec and a depth of 2 m. The cages are made of wooden planks and the side panels removable. The frame for fixing the side panels consists of bars 5 cm × 6 cm thick with battens 2.5 to 4.0 cm wide attached to them. The distance between the battens varies from 0.75 to 2.0 cm depending on the size of the trout. The lids of the fish cages, insofar as possible, block the light to minimum and protect the fish. They consist of two panels, each 1.50 m × 1.25 m in size. The frame of the lid consists of bars 5 to 6 cm thick to which a net with a mesh size of about 2 to 3 cm made of galvanized metal or stainless steel is fitted. The cages (Figure 110) are fixed on piles embedded in the bottom of the water body or in case of changes in the water level attached to metal or wooden pontoons. The bottom of the fish cage is made of planks and serves as a platform for feeding. A space 15 to 20 cm wide from each side of the cage is closed off by battens to allow the platform to be cleaned of food residue.

[14]Feeding is irregular in the months of December, January, and February.

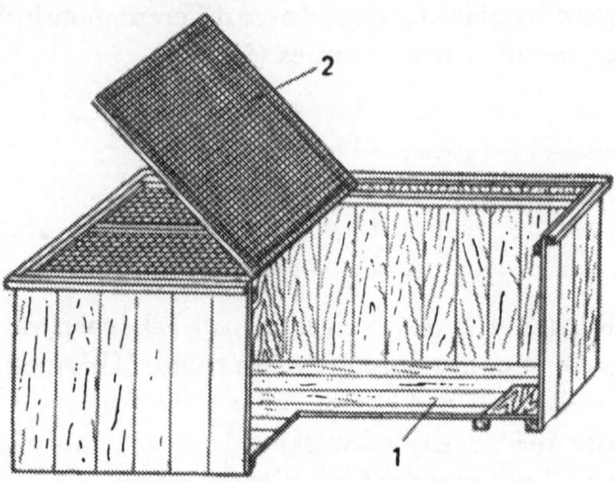

Figure 110. Fish cage in cross section.

1—bottom of cage in the form of a feeding platform; *2*—lid of cage divided into two halves.

The fish cages are arranged in a single row and so suspended that the upper part of the cage is above the water surface by 30 to 40 cm and the feeding platform in the water to a depth of about 1 m.

Marketable trout, replenishing stock, and brood fish have been raised in the cages just described. The density of trout stocking in cages depends on the age and size of the fish (number of fish/m^2 of cage area):

(a) Yearlings weighing 25 to 50 g stocked at the rate of 267 to 534 fish;

(b) Replenishing stock weighing 150 to 200 g stocked at the rate of 267 to 400 fish;

(c) Brood fish (males and females) weighing more than 600 g stocked at the rate of 80 to 133 fish.

Fish from cages toward the end of the rearing season (October) in the experiment conducted by the State Research Institute of Lake and River Fisheries (Farm "Ropsha", Leningrad district) yielded a net increase in weight of 74 kg per m^2. In the case of raising for five months (May to September) the yield of marketable fish with satisfactory stocking material (weight of yearlings about 50 g, and two-year fish 200 to 250 g) was 1.5 to 2.0 quintals for the given cage area.

Feed is extremely important in the cage culture of trout. The State Research Institute of Lake and River Fisheries has recommended the following rations for yearlings of rainbow trout cultured in cages (%):

A.	Fish mince	50	B.	Spleen	35
	Meat-bone meal	20		Meat-bone meal	40
	Fish meal	5		Phosphatides	10
	Phosphatides	10		Formula feed for fish	10
	Edible yeast	5		Edible yeast	5
	Formula feed for fish	10			

Under-yearlings of trout are fed in the first two weeks of stocking in ponds with spleen and 5% phosphatides. After this a small quantity of fish meal, edible yeast, etc. is added. When the trout are one month old they are fed the following mixtures (%):

Cattle spleen	70
Fish meal	10
Phosphatides	5
Edible yeast	5
Formula feed	5

Other rations and granulated feeds are also given; the composition of the latter has been detailed in Chapter XI. In the German Democratic Republic such feeds are used for carp as well as rainbow trout.

The approximate quantity of feed required over the entire period of fish culture is calculated by multiplying the productivity of trout by 4. Here the food coefficient is equal to 5, but multiplication done with 4 because the fish growth is taken as less due to their low initial weight. The requirements for individual components of the feed are calculated as follows: If a farm is planning to raise 50 tons of trout, the total quantity of feed required would be 200 tons (50×4). In using a ration consisting of fish waste the requirement for Baltic herring, herring, or small trash fish would be 160 tons (80%), edible yeast 20 tons (10%), formula feed 10 tons (5%), and phosphatides 10 tons (5%).

The daily ration for yearlings is 5% of the weight of the fish and for older age groups, 3%. The quantity of feed is modified after each control catch, which ought to be done twice a month.

The daily ration for trout of different weight, length, and food coefficients proposed by Ryle and Ray (1970) is given in Table 60.

Table 60. Daily ration for trout of different weight and length

Weight of fish, g	Length fish, cm	Daily ration of 1,000 fish (kg) with different food coefficients		
		1.3	1.4	1.5
0.503	3.55	0.047	0.050	0.054
1.071	4.57	0.077	0.083	0.089
4.978	7.62	0.216	0.232	0.249
10.066	9.65	0.346	0.370	0.396
30.608	13.97	0.720	0.779	0.833
50.33	16.51	1.005	1.083	1.115
101.34	20.83	1.604	1.726	1.853
152.52	23.87	2.111	2.274	2.437
184.15	25.40	2.392	2.577	2.758

To avoid pollution in the fish cages and to prevent the growth of algae on the side panels (which obstructs water exchange) the cages should be scrubbed regularly (once every 7 to 10 days) with a hard metal or plastic brush and the dead fish removed. During the rainy season, when the water temperature is high and a large quantity of organic matter tends to accumulate in the water body which deteriorates the oxygen regime, the fish should be carefully observed and, if necessary, some removed to reserve cages as a safeguard against mortalities.

Fish cages can be left in the water during winter even though the water ices over and need no special attention. If some fish are retained in them the cages are covered with two sheets of plywood or straw to protect the fish against freezing. Cleaning snow from the lids in order to supply feed in the cages is eliminated by affixing a wooden tube 1.25 m high and 25 cm in diameter to each cage through which food can be dropped. However, every cage (empty or with fish) should be cleared of snow, opened, and examined once or twice a month.

Experimental female brood fish maintained in fish cages and raised only on supplied feed of good quality laid high quality eggs which developed normally after fertilization. Egg loss was about 7% and the larvae, fry, and older fish also achieved normal development.

Cage rearing of under-yearlings of rainbow trout in the cooling reservoirs of the electric power station GRES (Moscow district) has also been successful. Fish weighing 7.20 to 8.07 g were stocked in October at the rate of 107 fish/m^2 and fed four times a day; by the end of April they had achieved a weight of 100 to 120 g with a survival rate of 93.7 to 96.5%.

Culture of Rainbow Trout in Fish Cages in Stationary Water Bodies. In addition to the culture of rainbow trout in cages suspended in flowing waters, studies have recently been conducted by the All-Union Research Institute of Pond Fisheries (Mikheev, Meisner, and Mikheev, 1970) on the culture of these fish in floating cages in stationary waters (lakes and artificial reservoirs). The cages are suspended in the upper lighted layers of water which have a good oxygen regime and are rich in phytoplankton, zooplankton, and other food organisms. Creeks protected from winds, with an area of 50 to 200 hectares, a consistent depth of 5 to 6 m, and a poor development of higher aquatic plants are also suitable. Creeks fed by rivulets from marshy areas or areas affected by industrial effluents are not suitable for cage culture.

The temperature at which caged rainbow trout feed and grow best in such waters is 16 to 20°C; they may live and grow for quite some time at 22°C and can withstand a brief temperature rise up to 28 and even 30°C. The light regime is very important. Strong sunlight cannot be tolerated by trout during the summer period and during winter they suffer from a shortage of light under the ice cover. Atmospheric air must be available to caged fish, especially during the winter period.

Depending on the nature of the culture, fish cages are classified as rearing, wintering, and foraging.

The production processes for culturing rainbow trout in floating cages installed in stationary waters (lakes and artificial reservoirs) usually commence with the preparation of brood fish. Their readiness for spawning under the water conditions in the central belt of the USSR (Pyalovskoe reservoir) is evident at the end of April–beginning of May. Sexual maturation in cage conditions is attained at the age of 2+ in males and more often at 3+in females. Egg incubation, according to the results obtained by fish cage culture at the Pyalovskoe reservoir, starts at the end of April–beginning of May when the water temperature is 5 to 8°C, and is completed toward May 20 to 25 when the water temperature is 18 to 20°C. The yolk sac of trout larvae is absorbed at a temperature of 11 to 22°C.

A high oxygen content in the water of reservoirs in early spring greatly influences the favorable progress of egg incubation and development of prolarvae and larvae.

The total consumption of heat during egg incubation is approximately 300 degree-days, and for the absorption of the yolk sac 100 degree-days.

Care of the eggs during the incubation period and at the time of large-scale hatching of larvae is generally similar to that described in the foregoing section.

When the free-swimming larvae start moving toward the lighted part of the apparatus in shoals, feeding should commence and the fish be transferred from the apparatus to the tanks in the rearing section and maintained there for 20 to 30 days. These tanks consist of concrete squares with rounded corners and a floor that slopes slightly toward the center. The size of the tanks is 4 m×4 m with a depth of 0.5 m.

The cages described above are stocked with 70,000 trout larvae. In this process shade should be created and a circular movement of the water ensured.

In the first 7 to 10 days the fry are fed zooplankton to which fish mince prepared from boiled fish is added, together with omelet prepared from egg powder and dry milk. The food is distributed on floating feeding tables.

For feeding fry in the early stages of development with natural foods live food organisms are reared. For this, special so-called crustacean tanks are installed—50 m long, 6 m wide, and 0.6 to 0.7 m deep—which are drainable.

Initially the fry are fed three to four times a day and the mince placed directly in the tanks. When the fry have adapted to the feeding tables, the mince is placed on the tables and, depending on its rate of consumption, supplied two to three times a day. The feeding tables are thoroughly washed after each feeding. To obtain a favorable oxygen regime in the tanks a flow of water is created and the tanks are regularly cleaned of excreta, feed residue, debris, etc. To prevent diplostomatosis the water is supplied through a gravel filter. During the day the tanks should be well lighted but half the area shaded from bright sunlight.

Toward July fry grow in weight to 0.5 to 1.0 g; they are then transferred to floating rearing cages (Figure 111) installed in a lake or reservoir. The cages are constructed of capron net with a mesh size of 3.6 to 4.0 mm. Fry which have not achieved the weight of 0.5 to 1.0 g are left in the tanks for further growth. Tanks emptied of fish are disinfected with chlorinated lime and washed with water.

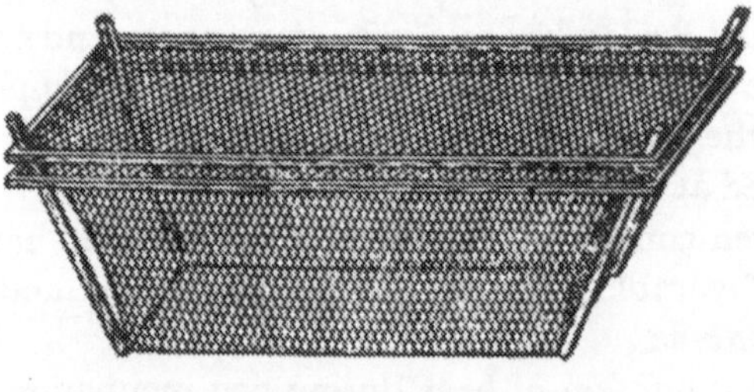

Figure 111. Cage for rearing larvae and fry.

The rearing cages are placed in less overgrown areas in creeks with a predominant depth of 4 to 6 m and protected from wind. To prevent diplostomatosis the cages are placed at the aforementioned depth at a distance of 80 to 100 m from the coastal overgrowth of plants, since the mollusk *Limnaea stagnalis,* an intermediate host of the parasite *Diplostomum spathaceum,* lives in such overgrowth. The rearing cages should have a volume of 50 m^3 with a length of 6 m and a depth of 3 m.

The density of stocking should be 800 fry/m^3 or 40,000 fry per cage. The fry are maintained in the rearing cages up to the end of the first fortnight of August, by which time they have attained a weight of 4 to 5 g. They are then transferred to fish cages made of capron net with a mesh size of 6.5 mm, and from these cages finally transferred to the wintering ones.

Fry held in rearing cages are initially fed the same feed used for fry in tanks (zooplankton, fish mince). To obtain a concentration of zooplankton inside the cages three electric lamps of 150 to 200 W each are mounted in each rearing cage. The lights are not switched on when the winds are strong because zooplanktonic organisms cannot overcome the waves and hence large concentrations will not occur under the lights.

In autumn live crustaceans caught in plankton nets towed behind boats in reservoirs are added in raw form to the feed mixture. Perch, roach, ruff, and other fishes must be boiled before their addition to the feed. Experiments have shown that feeding trout raw fish leads to a deficiency of vitamin B_1 and fish mortality in winter. This is due to the presence of the enzyme thiaminase in raw fresh-water fish which destroys thiamine in the food (B_1). Thiaminase is rendered inactive by boiling the fish up to 100°C.

In the case of rearing in floating cages the fry gradually adapt to the consumption of feed mixtures in which the components of animal origin should constitute not less than 90% of the daily ration (Table 61).

Table 61. Feed mixtures for under-yearlings in floating cages (from Mikheev, Meisner, and Mikheev, 1970)

Components	%	
	Up to August 15	After August 15
Fresh fish (boiled)	90	75
Crustaceans in raw form	2	—
Blood meal	1	—
Bran	—	18
Edible yeast	5	5
Casein	1	1
Fish oil with vitamins A and D	1	1
Total	100	100

To the feed mixtures given in Table 61 green leaves of sorrel (50 g/kg of feed) should also be added. In autumn and winter, when the vitamin content of plants decreases, multivitamins (A, B_1, B_2, C) as well as vitamin concentrates are added to the feed mixtures according to the following formula (mg/100 g of feed) (from B.I. Cherfas):

B_1 (thiamine).................	0.1– 1.0
B_2 (riboflavin)................	0.5– 1.5
C (ascorbic acid).............	25.0–50.0
B_3 (pantothenic acid)..........	1.0– 2.0
H (biotin).....................	0.005–0.025[15]

In using sorrel one must remember that the vitamins present in it are easily destroyed. Vitamin C is destroyed if the prepared feed is stored more than two hours. Chopped sorrel should not be exposed to air for a long time as this destroys its carotene. Hence freshly harvested sorrel should be passed through the meat mincer only just before its addition to the feed.

When the trout fry have transferred to feeding on granulated feed, fish mince is withdrawn from their diet.

[15]According to British pisciculturists the daily consumption of biotin by carp is 0.02 to 0.03 mg/kg of body weight of fish or at the rate of 0.1 mg/100 g of feed.

Casein glue, which binds the feed, is very important in the mixture because it increases the longevity of the pellets. Data also reveal that casein prevents fat degeneration in the liver.

The daily food norm for trout fry depends on their age and the water temperature, and ranges from 2 to 12% of the body weight. In August–first half of September feed is supplied at the rate of 10 to 12% of the body weight; in the second half of September at 8%; and in October, when the water temperature falls to 6°C and below, at 2%. Food is supplied three to four times a day until late autumn when feeding is done once a day.

In the experiments conducted by Mikheev, Meisner, and Mikheev (1970) the weight of under-yearlings toward autumn in floating cages fluctuated in different years from 14.0 (1968) to 36.0 g (1962). Fry growth slows down at the height of the summer period and becomes intensive in the second half of August, September, and the first half of October.

Under-yearlings reared in floating cages up to late autumn should attain an average weight of 15 g or slightly more. Before ice formation they should be transferred from the summer cages to the winter semisubmersible cages installed in reservoirs and lakes with an oxygen regime not below 5 ml O_2/l (Figure 112). Such cages are used for fish which must be fed during winter, including rainbow trout. As can be seen from Figure 112 these enclosures are

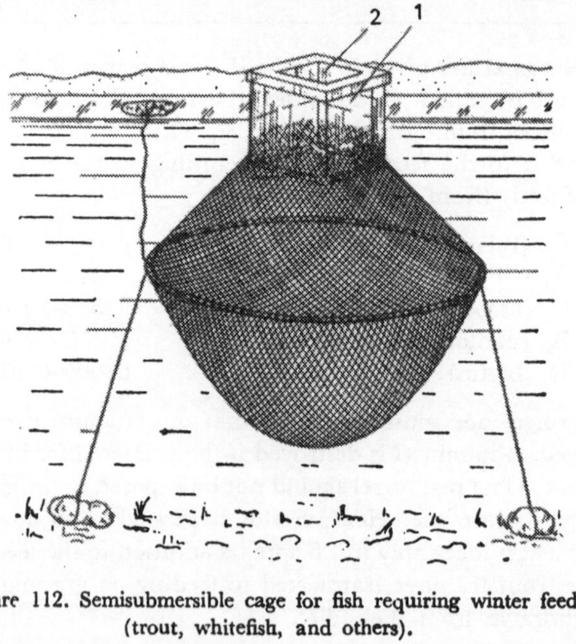

Figure 112. Semisubmersible cage for fish requiring winter feeding (trout, whitefish, and others).

1—funnel 1 m × 1 m, height 1.2 m, from which cage suspended; in winter funnel freezes in ice and fish are fed through an opening in it; *2*—abatjour of plexiglass for penetration of light in cage.

submerged in water up to the abatjour in which a funnel with an opening is placed for delivering feed in winter. Up to 50,000 under-yearlings of trout with an average weight of 10 to 15 g can be held in a cage of this type with these dimensions: 4 m × 4 m × 1 m.

Because of the low temperatures in the upper water layers in reservoirs and lakes, trout feed poorly during winter. Therefore, their feeding frequency is reduced to twice a week, using the same feed mixture as in summer, but increasing the vitamin supplement. A single feed delivery should approximate 1% of the body weight.

Sufficient care must be taken of cages housing hibernating trout. The glass and sides of the abatjour must be scraped clean of snow and ice every day to permit the passage of light, heat, and air—all three of which are needed by trout during the winter months.

In spring, before the ice thaws, the hibernating fish cages are released under the ice cover to prevent their being pressed by ice. However, the period of submersion of the cages in water under ice should be short and as quickly as possible after the ice has melted unloading of the trout should commence.

Submersible cages (Figure 113) are used for fish which do not need winter feeding (carp, catfish, etc.).

Floating enclosures are used for summer rearing of marketable trout and maintenance of replenishing stock and brood fish (Figure 114). They are constructed from capron net with a mesh size of 6.5 mm (twine 34/4), have

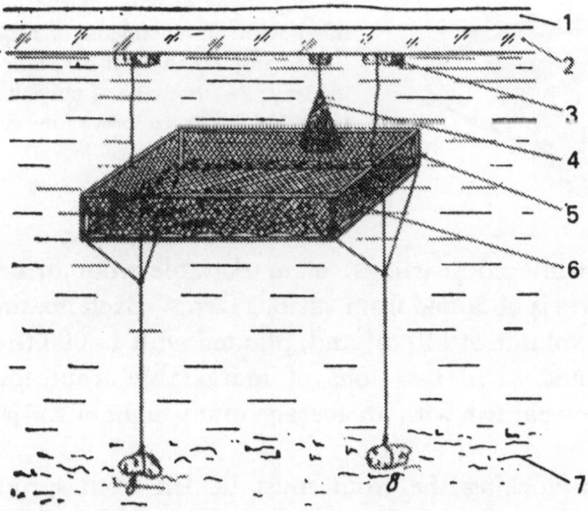

Figure 113. Installation of hibernating submersible cages for fish which do not require winter feeding (depth 1.0 to 1.2 m).

1—snow; *2*—ice; *3*—floats; *4*—sleeve; *5*—frame of box; *6*—net; *7*—bottom; *8*—sinkers.

a capacity of 140 m^3, installed at places with a depth of 6 to 8 m, and protected from wind. To prevent diplostomatosis they are situated at a distance of not less than 80 to 100 m from the coastal overgrowth of plants (outside the zone of mollusks). The frame of the floating enclosures (feeding, rearing, and hibernating) consists of airfilled polyethylene pipes or wooden battens to which the net is attached on the inner side (Figure 115).

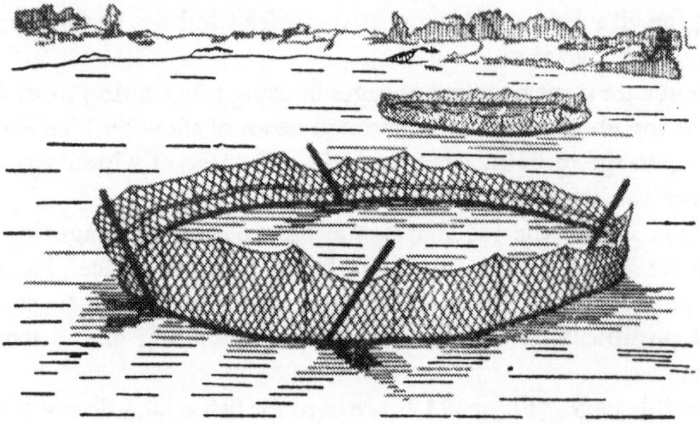

Figure 114. Floating enclosure or pool (volume 100 m^3, diameter 10 m, and depth up to 4 m).

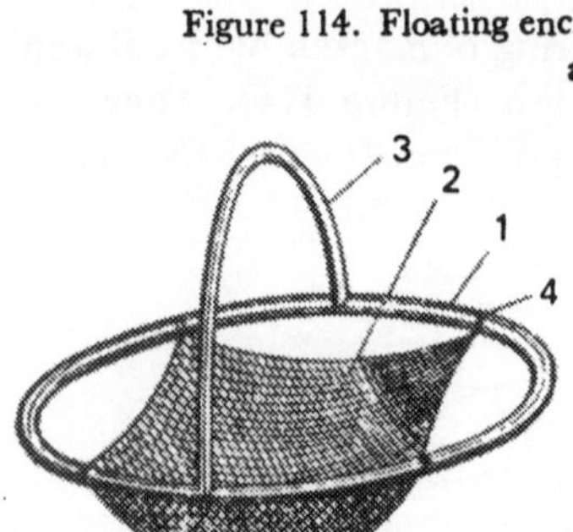

Figure 115. Floating tray for feeding fish in the summer.

1—ring made of polyethylene pipe (d=30 mm), filled with air (diameter of ring 1.5 to 2.0 m); *2*—rectangular sheet of net cloth No. 7, the feeding tray; *3*—arc made of polyethylene pipe(d=30 mm), height 80 cm, preventing dipping of box lid up to water; *4*—place for attaching corners of sheet to ring.

The stock (wintered yearlings) of marketable trout to be raised in the feeding enclosures is obtained from various farms. Each floating type feeding enclosure has a volume of 140 m^3 and, planted with 15,000 trout yearlings in the spring, should yield two tons of marketable trout toward autumn, consisting of two-year fish with an average unit weight of 150 g and a survival rate of 90%.

Under such stocking the trout must be fed from spring the mixture recommended for under-yearlings after August 15 (see Table 61). Subsequently the mixture described below (in %) is given:

Fresh fish for feed purposes (boiled) 50
Fresh zebra mussels (boiled), (or the quantity of fish in the mixure

increased to 70%)	20
Rye flour	15
Bran	10
Edible yeast	5

To make this feed the fishes are boiled whole (with scales, internal organs, and bones intact) and the zebra mussels boiled in their shells without adding water, which prevents losses in nutrients. Fishes and mussels are then minced and mixed with the dry feed mixture. The thickness of the pellets made from the final mixture is adjusted to accord with trout growth.

Sorrel leaves are added to the above mixture at the rate of 50 g per kg of feed. The daily food supply to two-year trout is 5 to 8% or, in the case of intensive food consumption, increased to 10% of the body weight. Food is given twice a day. The optimum temperature is 15 to 20°C. The consumption of feed in the course of raising two-year fish is 2.7 to 3.4 kg per kg of growth rate of trout, including the fish used for feed purposes (1.6 to 2.1 kg per kg of growth).

The large concentration of trout in the enclosures attracts various kinds of enemies. It is essential that the enclosures be protected from gulls in the first half of summer and from mink in the second half.

The norms of output of rainbow trout at different stages of raising in cages and enclosures are taken as follows:

Fertilized eggs	100%
Larvae after transfer to mixed feeding	80% of eggs
One-month-old fry	85% of larvae
2.5-month-old juveniles	70% of one-month-old fry
Fingerlings or under-yearlings	90% of 2.5 month-old juveniles
Yearlings	85% of under-yearlings
Two-year fish	90% of yearlings

Replenishing stocks and brood fish can be raised in floating enclosures installed in lakes and reservoirs. Under these conditions the males attain sexual maturity at the age of 2+, and the females more often at the age of 3+. Feed prepared from fresh small fish in the form of pellets is currently recommended. However, the feeding of replenishing stock and brood fish requires further study.

Trout cage culture must find a place in the Soviet fish industry, especially in the reservoirs and lakes of the central and northwestern parts of the country. The simplest type of cage culture is one-year raising of marketable trout. For this it is essential to have a network of cage-type nurseries to meet the requirements of foraging enclosures.

As revealed by studies conducted by the All-Union Research Institute of

Pond Fisheries, cage-type hatcheries appropriately installed throughout the country would yield two to five million trout yearlings per annum.

References

Bagenal, T.B. 1969. The relationship between food supply and fecundity in brown trout (*Salmo trutta* L.), *Fish, Biol.*, no. 2.

Birznek, O.A. and L.V. Kuzichkina. 1966. Ispol'zovania szhatogo vozodukha dlya vzyatiya ikry u foreli (Use of compressed air for egg extraction in trout). Sbornik *Nauchno-tekhn. Inform. VNIRO*. Moscow, no. 11.

Borovik, E.A. 1969. Raduzhnaya forel' (The Rainbow Trout). Minsk.

Brizinova, P.N. 1969. Izmenenie soderzhaniya zhira u raduzhnoi foreli v zavisimosti ot ratsiona (Change in the fat content of rainbow trout depending on rations). *Izv. VNIORKh.* Leningrad, vol. 68.

Brizinova, P.N. and S.V. Strel'tsova. 1967. Vliyanie fosfatidov i karbamida na usvoenie belkovogo korma raduzhoi forel'yu (Effect of phosphatides and carbamides on the assimilation of proteinic food by rainbow trout). Sbornik *Obmen Veshchestv i Biokhimiya Ryb*. Nauka, Moscow.

Brizinova, P.N. and S.V. Strel'tsova. 1969. Intensivnost' perevarivaniya belkovogo komponenta korma raduzhnoi forel'yu (Intensity of digestion of protein component of food by rainbow trout). *Izv. VNIORKh.* Leningrad, vol. 68.

Chittino, P. 1968. Piscicultura e ittiopatologia, *Riv. Zootech.*, vol. 41, no. 4.

Davis, H.S. 1956. Culture and Diseases of Game Fishes. University of California Press, Berkeley and Los Angeles.

Deufel, S. 1965. Pigmentierungsversuche mit canthaxanthin bei regenbogenforellen, *Arch. Fischereiwiss*, no. 2.

Galkina, Z.I. 1969. Vliyanie razmerov i intensivnosti okraski ikrinok na e'mbrional'noe razvitie i rost molodi raduzhnoi foreli (Effect of size and color intensity of eggs on embryonic development and growth of fry of rainbow trout). *Izv. GosNIORKh.* Leningrad, vol. 68.

Gambaryan, M. 1966. Udobreniya forel'nykh prudov (Manuring of trout ponds). *Rybovodstvo i Rybolovstvo*, no. 3.

Gorovaya, S.L. 1968. K voprosu o biokhimichekom sostave tela evropeiskogo ugrya i raduzhnoi foreli (On the problem of biochemical composition of European eel and rainbow trout). *Tezisy III Zool. Konf. BSSR*. Minsk.

Gracheva, M.N. 1953. Sovremennoe sostoyanie i perspektivy razvitiya forelovogo khozyaistva v SSSR (The present state and prospects for the development of trout farming in the USSR). *Trudy Soveshch. po Voprosam Prudovogo Rybovodstva*. Izd. AN SSSR, Moscow.

Gulidov, M.V. 1966. Preimushchestva vozvratnoi podachi vody v inkubatsionnye apparaty pri iskusstvennom razvedeniya ryb (Advantages of reverse water supply into incubation apparatus during artificial fish rearing). Sbornik *Nauchno-takhn. Inform. VNIRO*. Moscow, no. 12.

Iriskina, T.A. 1968. Karotinoidy raduzhnoi foreli (Carotenoids in rainbow trout). Avtoref. Kand. Diss., Leningrad.

Kamburova, St. 1949—1950. Kakvidite kato khrana pri izkustvenoto rybov"-dstvo god na selsko-stopanskata, Akademiya "Georgi Dmitsov". Sofia, vol. 2.

Knyazeva, L.M. 1967. Ustoichivost' raduzhnoi foreli i karpa k karbamidu (Resistance of rainbow trout and carp to carbamide). Sbornik *Obmen Veshchestv i Biokhimiya Ryb*. Nauka, Moscow.

Lavrovskii, V.V. 1960. Novyi pribor dlya viborki mertvoi ikry (A new apparatus for sorting out dead eggs). *Rybnoe Khoz-vo,* no. 10.

Malikova, E.M. 1956. Pishchevaya tsennost' nekotorykh bespozvonochnykh kak korma dlya ryb (The food value of some invertebrates for fish). *Biokhimiya,* vol. 21, no. 2.

Malikova, E.M., N.I. Kotova and N.S. Reznikova. 1969. Vyrashchivanie molodi Baltiiskogo lososya na rybnykh zavodakh (Raising young Baltic salmon at fish farms). Sbornik *Biotekhnika i Prigotovlenie Iskusstvennogo Korma—KRT*. Riga.

Mann, H. 1960. Gonadengeneicht und eizahlen bei regenbogenforellen, *Inform. Fischwirtsch,* no. 7.

Mikheev, P.V., E.V. Meisner and V.P. Mikheev. 1970. Sadkovoe rybnoe khozyaistvo na vodokhranilishchakh (Cage Rearing of Fish in Reservoirs). Pishchepromizdat, Moscow.

Mikheev, P.V., E.V. Meisner and V.P. Mikheev. 1970. Instruktsiya "Forelevye sadkovye khozyaistva v vodokhramilishchakh i ozerakh" (Instructions on "Trout Rearing in Fish Cages in Reservoirs and Lakes"). Izd. Minrybkhoz SSSR i VNIIPRKh, Moscow.

Privol'nev, T.I. 1968. Vyrashchivanie raduzhnoi foreli na rastitel'nykh kormakh (Raising rainbow trout on plant feeds). Sbornik *Prudovoe Rybnoe Khozyaistvo i Razvitie Ugrya.* Riga, vol. 1, pt. 2.

Privol'nev, T.I. 1968. Vyrashchivanie raduzhnoi foreli v sadkakh (Raising rainbow trout in fish cages). *GosNIORKh, Rybokhozyaistvennoe Izuchenie Vnutrennikh Vodoemov,* no. 1.

Privol'nev, T.I. 1969. E'kologo-fiziologicheskie i rybokhozyaistvennye osobennosti raduzhnoi foreli (*Salmo irideus* Gibb.) [Ecological and physiological culture peculiarities of rainbow trout (*Salmo irideus* Gibb.)]. *Izv. GosNIORKh.* Leningrad, vol. 68.

Privol'nev, T.I. 1970. Instruktsiya po sadkovoma vyrashchivaniyu raduzhnoi foreli (Instructions on Cage Raising of Rainbow Trout). Izd. GosNIORKh. Leningrad.

Privol'nev, T.I., S.V. Strel'tsova, P.N. Brizinova, I.N. Ostroumova and N.V. Koroleeva. 1964. Predokhranenie ot lipoidnoi degeneratsii pecheni raduzhnoi foreli vvedeniem v ratsion fosfatidov (Protection of the liver of rainbow trout from lipid degeneration by adding phosphatides to the rations). *Doklady Akad. Nauk SSSR,* vol. 156, no. 5.

Ryle, E.A. and O.L. Ray. 1970. Use of population numbers to estimate feed allowances for trout, *Progress. Fish. Cult.*, vol. 32, no. 2.

Sadlaev, K.A. 1962. Forelevoe rybovodnoe khozyaistvo (Trout Farming). Pishchepromizdat, Moscow.

Scott, D.P. 1962. Effect of food quantity on fecundity of rainbow trout (*Salmo gairdneri*), *J. Fish. Res. Board, Canada,* vol. 19, no. 4.

Shabaline, A.A. 1968. Deistvie khloristogo kobalta na fiziologicheskie pokazateli raduzhnoi foreli (Effect of cobalt chloride on the physiological parameters of rainbow trout). *Vopr. Ikhtiol.*, 8, 5, 52.

Skatkin, P.N. 1962. Biologicheskie osnovy iskusst vennogo ryborazvedeniya (Biological Basis of Artificial Fish Rearing). Izd-vo Akad. Nauk SSSR, Moscow.

Steffens, W. 1965. Die vermendung von frockenfutter in der forellentlichwirtschaft, *Dtsch. Fisch. Ztg.*, vol. 12, no. 1.

Steffens, W. 1966. Die zweckmässige grosse der weiblichen leichfische bei regenbogenforellen, *Dtsch. Fisch. Ztg.*, no. 11.

Steffens, W. 1968. Fütterungsversuche bei regenbogenforellen (*Salmo gairdneri*). Ergebnisse der verfütterung von trockenfuttermitteln in der Jahren 1963 und 1964, *Z. Fischerei,* vol. 16.

Steffens, W. 1968. Zuwachs und futterverwertung von regenbogenforellen bei verwendung trockenmischfutter, *Dtsch. Fisch. Ztg.*, no. 8, p. 15.

Timoshina, L.A. 1967. Aminokisloty embrionov raduzhnoi foreli (Amino acids of rainbow trout embryos). *Vopr. Ikhtiol.*, 7, 4, 45.

Titarev, E. 1967. Raduzhnaya forel' v vodoeme-okhladitele GRE'S (Rainbow trout in the cooling tanks of a state regional power plant). *Rybovodstvo i Rybolovstvo,* no. 2.

Tsirkova, M.K. 1966. Ispol'zovanie muki iz shelkopryada pri vyrashchivanii lososevykh (Use of silkworm meal in salmon culture). *Rybnoe Khoz-vo,* no. 4.

Yarzhambek, A.A. 1966. Karotinoidy lososevkh ryb i ikh svyaz' s vosproizvodstvom (Carotenoids of salmon and their relationship with fecundity). Avtoref. Kand. Diss., Moscow.

Chapter XVI

FISH BREEDING IN POND PISCICULTURE

The breeding of fish undertaken for pond culture is extremely important in the further development of pond fisheries. The methods of breeding in pond pisciculture are, in principle, identical to those in other fields of animal husbandry. The breeding of fish reared in ponds has its own specificity, however, in terms of the biological peculiarities of the fish and their living conditions.

Breeding in pond fisheries involves a complex of organizational and fish-rearing measures directed toward the improvement of the hereditary qualities of the fish and an increment in the productivity of the ponds. Such measures include technically correct rearing and ideal maintenance of the fish, provision of good quality nutrition to the fish at all stages of development, and the implementation of correct methods of breeding and selection. Since these measures are interrelated failure in one reduces the efficacy of breeding work.

Before initiating work on developing new breeds, it is important to determine which new or improved traits in the older breed are desirable in terms of economic significance, which particular geographic conditions the breed under development is required for, and in what way it should differ from the breed already existing in the given zone. There are no breeds of fish which are economically of equal importance in all climatic zones. The ultimate objective of breeding work is to develop such breeds and hybrids of fish for culturing as will increase output and exhibit improved qualities with a more economical consumption of feed and less labor and expenditure. New breeds or hybrids should yield at least 10% more marketable fish from the same pond area compared to the form already under culture at the fish farms in a given zone.

A new breed is checked by state breeding testing centers in three experimental ponds versus three control ponds.

The correct feeding of brood fish and replenishment stock cannot be overemphasized. It is futile to carry out breeding work under poor nutrition ("a breed develops through the mouth"). Good nutrition plays a major role in the physiological condition of the brood fish (especially females) which, in turn, exerts considerable influence on the quality of the sex products. It is possible to direct the growth and development of the progeny in the required direction by selecting the appropriate type of feed for the brood fish. For example, carp brood fish given supplementary feed from the early stages of postembryonic development produced a more viable progeny compared to fish maintained only on natural foods. The under-yearlings from spawners fed artificial feeds weighed 36.7% more than progeny reared only on natural foods. At the age of two years their growth superseded that of under-yearlings obtained from spawners fed only on natural foods by 100%.

Charles Darwin states in *Variability of Animals and Plants under Domestication* that proper feeding over several generations has a direct influence on the growth of a breed. He mentions later that in many domestic animals and one breed of carp a tendency toward a marked shortening in the facial bones could be attributed to the type of food consumed.

The following breeds are cultured in carp farms: scale, mirror with scales scattered over the body, mirror with scales arranged linearly in the middle of the body along its lateral line, naked or leather (without scales), Ukrainian frame carp, and Ukrainian scale carp. These breeds differ from one another in external appearance (Figure 116), hereditary characters, and economic significance. The more tolerant and winter-hardy scale carp and mirror carp with scattered scales are more widely preferred for carp culture. Leather carp is less tolerant and recommended for the southern regions of the country. Mirror carp with a linear arrangement of scales is not so common and like leather carp is less productive than scale carp and mirror carp with scattered scales. Compared to the latter leather carp weighs less (up to 27%), is shorter (up to 10%), and has a lower coefficient of feeding. However, leather carp is more resistant to German measles. Leather and linear-scaled carp are less fecund than scale carp and mirror carp with scattered scales. It is not possible to obtain a homogeneous progeny from them. For example, leather, mirror with scattered scales, and mirror with linear scales produce progeny with scales, with scattered scales, and without scales. Carp with linearly arranged scales are generally not undertaken in pond pisciculture for this reason.

Hence the main breeds for carp culture are scale and mirror with scattered scales and (for the eastern and partly western parts of Ukrainian SSR) the Ukrainian frame carp and scale carp.

In the northwestern part of the RSFRS (Novgorod, Pskov, and Leningrad districts), Estonian SSR, and Siberia the northern (Ropshinsk) carp is widespread, which is a hybrid of carp and Amur sazan and represents the fourth to fifth generation of this hybrid.

The following definition of a breed, given by Kislovskii (1951), is commonly accepted in animal husbandry and is applicable to pond pisciculture: "A breed is a large group of animals in which, due to fairly long systematically directed efforts, a specific generality of type and common requirements for survival develop. When the conditions satisfy the requirements of survival, the breed is capable not only of conserving its specificity, type, and productivity in pure breeding, but also to progress faster and exert beneficial effects when crossed with other breeds under appropriate conditions."

In pond pisciculture breeding work is more rewarding as fish are more prolific compared to most fertile farm animals. Hence more material is available from which, even under strict rejection, it is possible to select a sufficient quantity of specimens which are better in quality. In this process the economically valuable properties of not only the breed as a whole are studied, but also those individual peculiarities which play an important role in the formation of a breed. The selection of each species and breed is based on its biological peculiarities, economic value, and conditions of rearing.

Breeding and Selection of Brood Fish. Selection is the most important step in breeding work. The best specimens in terms of economically valuable characters are identified as a pure line and selected on the basis of very rigid criteria. Fish with even minute defects are rejected.

Selection for the replenishing stock used to be done with fry. Now it is recommended that selection be done with fingerlings during their transplantation from rearing ponds to wintering ponds. This ensures a better qualitative selection of fry after their first vegetative period. The selected fingerlings are released in separate wintering ponds which are highly productive in natural foods. Selection of yearlings, two-year fish, and fish of older age groups is done during each transfer from one category of pond to another; unsuitable specimens are rejected. For mating, spawners are selected which have a similar scale pattern, external appearance, and growth rate. It is

Table 62. Some body parameters of more common types of carp

Carp breed	Sex	Fulton's condition factor	Index of body height, l/H	Index of body thickness, $\frac{B \times 100\%}{l}$
Scale carp	Females	3.0–3.4	2.8–2.6	18–23
	Males	2.7–3.1	2.9–2.7	17–22
Mirror carp with scattered scales	Females	3.0–3.4	2.8–2.6	18–23
	Males	2.1–3.1	2.9–3.1	17–22
Ukrainian frame and scale carps	Females	3.1–3.6	2.7–2.2	
	Males	3.0–3.5	2.8–2.3	
Ropshinsk carp, strain B[1]	Females	2.4–2.8	3.3–3.1	15–18
	Males	2.2–2.6		
Ropshinsk carp, strain M[1]	Females	2.5–3.0	3.2–2.9	16–20
	Males	2.3–2.7		

[1]According to the findings of V.S. Kirpichnikov strain "B" has a greater proportion of inherited characters of the Amur sazan (about 75%); it is characterized by a slightly slower growth, but higher winter-hardiness compared to strain "M."

assumed that the type of scale pattern is related to the peculiarities of hormonal activity of carp under the influence of the external environment. The sex of the spawners must be carefully ascertained. Other indexes in the process of selection include: external appearance, weight, state of health, activity at time of spawning, degree of fecundity and, presence/absence of deformities.

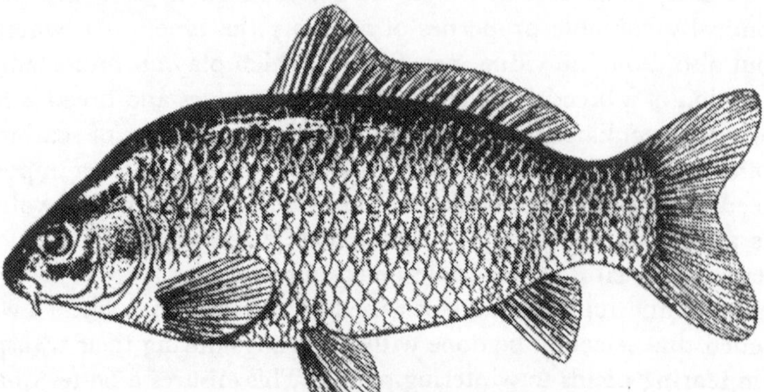

A

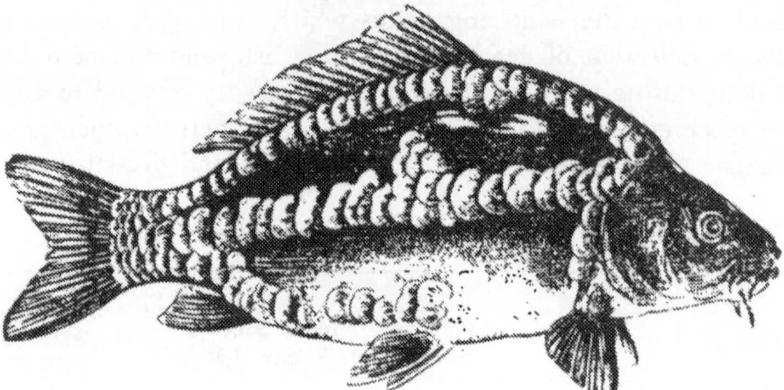

B

C

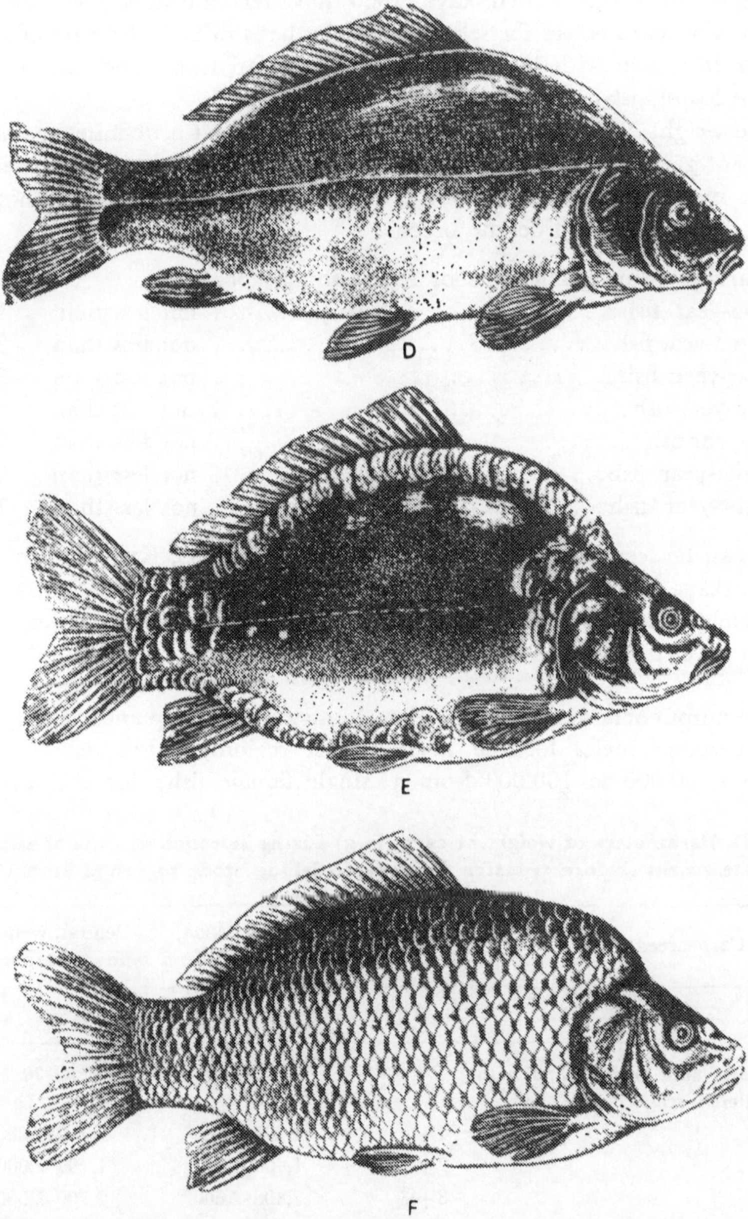

Figure 116. Common carp.

A—scale carp; *B*—Mirror carp with scales scattered all over the body; *C*—Mirror carp, with linear arrangement of scales in center of body (along lateral line); *D*—naked or leather carp (without scales); *E*—Ukrainian frame carp; *F*—Ukrainian scale carp.

Kirpichnikov and Golovinskaya (1966) have recommended the following approximate parameters for selection on the basis of body features of scale carp, mirror carp with scattered scales, Ukrainian frame and scale carps, and the Ropshinsk carp (Table 62).

The weight of brood fish and replenishing stock at a definite age is an important gauge of their quality. The following published norms of weight of carp in spring (in kg) are used irrespective of zone, sex, or method of raising brood fish of different breeds:

Yearlings (during selection for replenishing stock)..........	0.05
Two-year fish................................not less than	1
Three-year fish..............................not less than	2
Four-year fish...............................not less than	3
Five-year fish...............................not less than	4
Six-year fish................................not less than	5
Seven-year fish..............................not less than	6
Eight-year fish..............................not less than	7

As can be seen from the data presented in Table 63, Kirpichnikov and Golovinskaya (1966) have approached this problem in a different manner, one which makes it possible to differentiate the weight index more correctly not only over different years but also for different zones, breeds, and sex of carp.

The output of larvae, weighing 10 to 14 mg, of scale carp and mirror carp with scattered scales for the northwestern regions, Baltic regions, and Siberia is 50,000 to 100,000 from a single female fish; for the central,

Table 63. Parameters of weight of carp (in g) during selection at time of autumn assessment (before transfer from replenishing stock to brood stock)

Carp breed	Age group and sex	Northern region, Baltic region, Siberia	Central, western, and southern regions
1	2	3	4
Scale carp and mirror carp with scattered scales	0+	30–50	50–70
	1+	600–800	750–1,000
	2+	1,400–1,700	1,700–2,000
	3+♀	2,400–2,600	2,700–3,000
	3+♂	2,200–2,400	2,500–2,700
	4+♀	3,000–3,300	3,300–3,800
	4+♂	2,800–3,000	—
	5+♀	3,500–3,800	—
	5+♂	—	—

1	2	3	4
Ukrainian frame carp and scale carp	0+		60–100
	1+		1,000–1,500
	2+♀		2,900–3,400
	2+♂		2,700–3,000
	3+♀		3,800–4,200
	3+♂		3,500–3,800
	4+♀		4,800–5,200
	4+♂		4,500–4,800
Ropshinsk carp		*Strain B*	*Strain M*
	0+	30–50	30–50
	1+	500–700	600–800
	2+♀	1,400–1,600	1,500–1,700
	2+♂	1,300–1,500	1,400–1,600
	3+♀	2,200–2,400	2,400–2,600
	3+♂	2,000–2,200	2,200–2,400
	4+♀	2,700–3,000	3,000–3,300
	4+♂	2,400–2,700	2,700–3,000
	5+♀	—	3,500–3,800
	5+♂	—	—

southern, and western regions 100,000 to 200,000 larvae. In the case of Ukrainian carp the output is 150,000 to 250,000 larvae and for Ropshinsk carp 75,000 to 150,000 larvae (the number of larvae from larger females reaches 175,000).

Weight is also an important parameter in rearing brood stock of phytophagous fish. The parameters of recommended weights are given in Table 64 for the conditions of northern Caucasus, southern Ukrainian SSR, Moldavia, Trans-Caucasus, and Central Asia.

Table 64. Weight parameters of brood stock

Age group	Silver carp, kg	Grass carp and bighead, kg	Remarks
Under-yearlings	0.05	0.1	Higher weight indexes are not advantageous since they complicate the work of obtaining progeny, especially in the case of very large progenitors
Two-year fish	1.0	1.5	
Three-year fish	2.0	3.0	
Four-year fish	3.0	5.0	
Five-year fish	4.0	7.0	

In the course of rearing brood fish of phytophagous fish the same methods of breeding are followed as in the case of carp culture. Brood fish are reared in the usual carp ponds. The bottom of the various categories of ponds should be suitable, ensuring complete drying, and the supply and discharge of water independent.

Shortcomings in the external appearance of carp have a negative influence on their growth and development; hence specimens with deformities must be rejected. It is highly probable that deformations of the back and abnormal development of the fins are hereditary in character; they are mostly found in the case of relatively prolonged matings, especially in leather carp. Genetic characters such as irregular development of the fins and a curved spinal column inhibit fish movement, which is reflected in growth. In very severe conditions the body of under-yearlings is shortened, as in the case of tadpoles, swimming poor, weight toward autumn 35.3% less than normal, and the ribs wavy, twisted, etc.

The underdevelopment or total absence of gill opercula create difficulties in respiration and lead to branchial diseases. An oblique snout hampers food intake. A convex spine and bulging head significantly retard growth.

Fish for the replenishing stock are selected from those ponds wherein the conditions of the external environment have met the requirements of a given species and breed.

Selection and breeding are links in a single process directed toward the systematic improvement of the quality of individual stock and breed. To improve existing breeds and create desirable new ones breeding work must be done properly. Brood fish are picked on the principle of "the better among the best". Here consideration is given to health, external characters, fleshiness, size of head, height and width of back, scales, and other economically valuable characters for the purpose of ensuring that the better qualities of the parents are transmitted to the progeny. The same requirements hold when rejected spawners are replaced by new ones.

Breeding work in carp culture involves another responsibility: a sufficient quantity of brood fish and replenishing stock, which are not closely related, must be included in the composition of stocks at commercial farms. Inbreeding, a common practice in fish farms, has a harmful effect on the viability and productivity of fish; the closer the relationship among the fish, the stronger this effect is manifested.

Studies carried out by the Department of Pond Fisheries at the Timiryazev Agricultural Academy (Dubnik, 1968) reveal that the harmful effect of inbred fish becomes manifest from the very early stages of development of carp. The embryos obtained from continuous inbreeding are characterized by a low adaptability to the surrounding environment, especially to unfavorable conditions. Their survival is two to four times less than carp embryos obtained from distant matings. Pulina (1968) has also shown that the first inbred progeny of under-yearlings (with a coefficient of inbreeding 25%) were under the average weight by 22.3% and productivity by 25.4%.

The following measures are recommended to prevent inbreeding: brood fish should be periodically exchanged (after eight to ten years) between fish farms within the limits of a given zone; two-line mating should be followed, i.e., marketable fish should consist of the first generation from mating two lines; proper aftercare and maintenance of the replenishing stock and brood fish must be given; and selection of the replenishing and brood fish stock must be done carefully.

To ensure the required quantity of carp brood fish at a farm, the proportion for each age group selected should comprise 24 under-yearlings, 12 two-year fish, 4 three-year fish, and 3 four-year fish. For rainbow trout and largemouth bass a similar number of fish in different age groups are raised. For Ukrainian carp for the selection of a single set of spawners it is recommended that the replenishing stock consist of 100 yearlings, 25 two-year fish, and 12 three-year fish for the breeding stock. As a result of rejection it is essential to ensure replacements by one-fifth of the brood fish stock. Thus, the replenishing stock has to be properly planned. This makes an annual recruitment of the brood fish stock possible which, in turn, ensures fulfilling the planned targets of the fish farm.

According to the findings of Kirpichnikov and Golovinskaya (1966) carp can be used as spawners in the following age groups:

(a) Scale carp and mirror carp with scattered scales (the main objects of rearing in the central and northern regions of the USSR): females at 6 to 12 years, males at 4 (3) to 11 years;

(b) Ukrainian frame carp and scale carp; females at 4 to 10 years, males at 3 to 10 years;

(c) Ropshinsk carp: females at 6 (5) to 12 (14) years, males at 4 (3) to 10 (12) years.

As revealed by studies carried out in the central belt of the European part of the USSR, the progeny from matings of spawners of early age and at an age higher than nine years is qualitatively inferior (see below "Selection of brood fish by age").

In the course of rearing breeding stock, brood fish are carefully selected at all stages. It is essential to have two to three hatching ponds with an area of 0.1 hectare for checking their quality at each farm in which one set is placed. An increase in the productivity from each set of carp spawners by continuous improvement in their quality makes it possible to reexamine the existing norms in the direction of their significant improvement.

The parameters of productivity of a single set of carp spawners under the existing and possible norms of improvement in breeding work are presented in Table 65.

By properly observing the technique of managing a carp farm and with an efficient level of breeding work, the existing fish farms of collective and state enterprises can increase the output from a single set as follows: larvae—up to

Table 65. Productivity per set of carp spawners in the central and adjacent climatic zones

Index of calculation	Productivity in accordance with provisional norms approved by Ministry of Fisheries of the USSR	Potential productivity under enhanced fish rearing norms (according to Golovinskaya, 1962)
Output of larvae of 4 to 8 days from hatching ponds from single female fish	80,000	400,000
Output of under-yearlings	56,000 (70%)	320,000 (80%)
Output of yearlings	36,400 (65%)	288,000 (90%)
Output of 2-year fish	About 31,000 (85%)	274,000 (95%)
Average weight of 2-year fish	400 g	500 g
Total weight of marketable fish obtained from set over one season	124 q	1,370 q

5 times, under-yearlings—more than 5.7 times, yearlings—almost 8 times, and marketable (table fish) carp—more than 11 times.

For large-scale selection or to obtain 20 sets of carp spawners 100,000 under-yearlings are kept in the replenishing stock out of the progeny from the better parents. Selection in subsequent years proceeds approximately in the following manner:

Age of fish	Number of fish rejected	Number of brood fish selected
Under-yearlings	98,000	2,000
2-year fish	1,500	500
3-year fish	250	250
4-year fish	190	60

These repeated cycles of mass selection are carried out among phytophagous fishes: the first from yearlings in which 50% are kept, the second from two-year fish in which 10% are kept, and the third from the first maturing fish in which 25% are kept. To avoid inbreeding two-line mating and heterosis should be practiced.

Sevost'anova (1969) studied selection from yearlings and two-year fish of Danish, Czechoslovakian, and Gostilitsk (local) trout, and found that the brood fish of the Danish trout are heavier than the other groups, with the difference in the fourth year reaching 0.5 to 0.7 kg. The females of the Danish trout are characterized by higher fecundity and larger number and size of eggs. However, under conditions of intensive feeding Danish trout is more inclined to fattening and cirrhosis of the liver. The Gostilitsk trout lags far behind the Czechoslovakian in maturation rate and fecundity, while a mixture of these fishes is closer to the Czechoslovakian trout in these parameters and some of the local females have an even higher fecundity than the Czechoslovakian trout. One of the most effective methods of selecting brood fish of rainbow trout is to separate those individuals from the stock which give a minimum loss of eggs.

Selection of Brood Fish by Age. To bring about improvements in existing breeds selection on the basis of age of the brood fish is important. Changes which take place in a living organisms with increasing age undoubtedly influence its weight, reproductive function, and transmission of hereditary traits. With advancement in age qualitative changes take place in the blood, tissues, body cells, and organs. For example, the relative fecundity (number of eggs/kg of body weight) in a fish increases up to a certain age, after which, in spite of continuous absolute growth, it decreases. As the female advances in age the egg size (diameter and weight) also changes. It is known that different types of progenies are obtained from eggs of different sizes, which is attributed to the physiological condition of the fish. A poor quality progeny is generally obtained from eggs of very young and very old females.

As revealed by studies conducted by the Fisheries Laboratory of Timiryazev Agricultural Academy (1967) over a number of years, the duration and nature of spawning of carp spawners at different ages are not similar in mating fish of the same age. Fish spawning for the first time spawn poorly—for a period of 3 hr 10 min; at the age of six to seven years the selfsame pair spawned for a period of 5.5 to 7.0 hr, and at the age of nine years for 10 hr but less vigorously. Another pair at the age of eight years spawned vigorously for 9 hr and at the age of twelve years for 11 hr but less vigorously and somewhat intermittently.

A progeny differing in quality is obtained from mating spawners of the same age. The percentage of fertilization in the case of fish spawning for the first time is the lowest; the extent of losses of fertilized eggs during incubation is minimum from spawners 6 to 8 years old and more from fish 9 to 12 years old. Between the age of 3 to 7 years the quality of the milt improves due to an increase in the number of viable spermatozoa. In the case of early maturation of the male (at age 3+) the number of nonviable spermatozoa in the milt may be considerably higher compared to males which mature at the age of 4+. The progeny obtained from spawners of different ages differ in the accumulation of subcutaneous fat and thickening of the myones. A more rapid accumulation of subcutaneous fat and a greater thickness of the myones were observed in under-yearlings from carp spawners of middle age (6+, 7+, and 8+) than in under-yearlings from older (10+, 11+, 12+) and younger (5+) spawners.

The chemical composition of the eggs, larvae, and fry from different age groups is not same. The eggs and fish obtained from middle-aged spawners contain larger quantities of fat and proteins than the eggs and fish from younger spawners. The progeny from older spawners occupy an intermediate position with regard to these parameters. Spawners 6 to 8 years old and sometimes even 9 years of age produce better progeny in terms of these parameters. The best parameters were recorded for under-yearlings from spawners 7+ and 8+ years old (the highest output of under-yearlings, i.e., 83 and 94% from the rearing ponds). These age groups should also be considered the most active with regard to sexual activity for the central belt of the European part of the USSR.

The losses of under-yearlings obtained from spawners of different ages during the winter period are also not similar. They are higher in progeny from spawners aged 4+ and even 5+, even higher than in spawners 10 years old. The minimum loss during winter was observed in the case of under-yearlings from middle-aged parents of 7+ and 8+ years. Growth of the progeny (to the attainment of sexual maturation) also indictaes the advantages of selecting spawners 8+ years old (Table 66).

Table 66. Growth of carp from spawners of different ages

Age of spawners	Weight of progeny, g		
	2-year	3-year	4-year
4+	510.0	2,046	3,500
8+	533.5	3,087	3,975
17+	375.7	1,730	2,225

The second generation of under-yearlings, raised from the first generation of ages 4+, 8+, and 17+, had different weights in autumn.

The correlation between size, age, and condition of females and size and viability of their eggs has been studied in rainbow trout, Baltic and White Sea salmon, and Atlantic salmon (Galkina, 1968) and the data are summarized below:

(a) As the female of rainbow trout advances to the age of 5 years and salmon up to 6 years, the average weight of the eggs increases; subsequently the egg weight remains almost the same.

(b) The viability of the eggs from young female trout spawning for the first time is lower than that of eggs from middle-aged females; however, such a correlation was not seen for the eggs of Black and White Sea and Atlantic salmon.

(c) The larvae from eggs of different size revealed little difference in length and weight (except for eggs weighing below 35 mg in trout and below 80 mg in Atlantic salmon), but differed notably in the quality of the yolk in the yolk sac.

(d) Eggs of rainbow trout weighing less than 35 mg (obtained from females spawning for the first time) and 80 mg from Atlantic salmon are not suitable and should be rejected.

According to the findings of N.P. Novozhenin (1970) the viability of the progeny of rainbow trout obtained from fish spawning for the first time is lower than from spawners of older-aged groups, and the output of fry is more than twice less compared to the output of five-year fish.

Mating spawners of different ages, frequently undertaken in commercial practices, does not produce a positive effect in every combination. Studies by the Department and Laboratory of Pond Fisheries of the Timiryazev Agricultural Academy reveal that in different age pairing of carp spawners, the nature of spawning, quality of the progeny, and output of marketable fish depend on the age of the fish paired (Figure 117).

For example, pairing first-time spawners with older fish and even with middle-aged fish yielded the following results: considerable delay sometimes in spawning (up to August); lower percentage of egg fertilization; higher

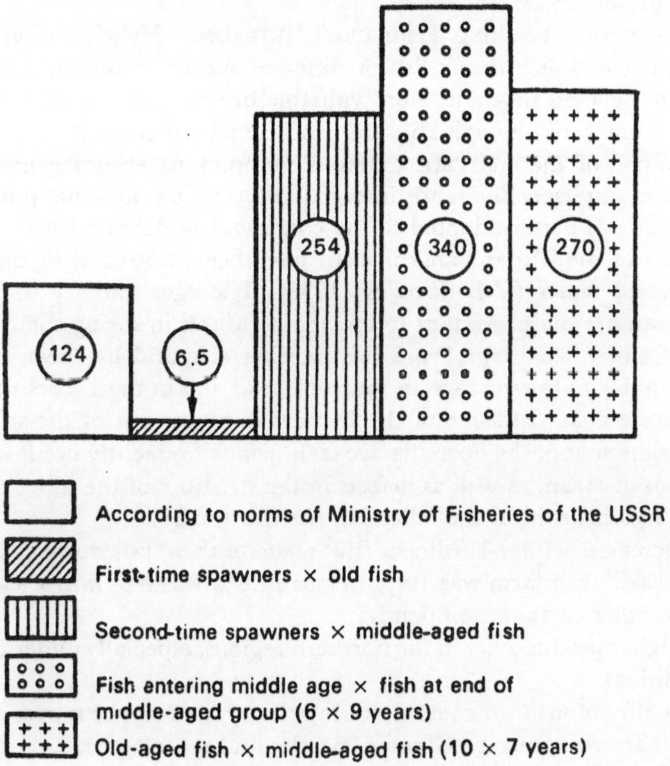

Figure 117. Total weight of marketable fish obtained from parents paired from different age groups of carp spawners (in quintals).

losses of eggs and larvae (sometimes reaching 100%); and very low output of fry from the hatching ponds (the lowest of the trials was 1,153 fry/kg of body weight of female fish). Good results were obtained from pairing middle-aged spawners with second-time spawners, spawners entering the middle-aged group (six years old), and spawners at the end of this age group (nine years old). The output of fry was highest in these pairings (up to 31,337 to 55,814 fish/kg of weight of female). Hence spawners from extreme age groups should not be paired since their progeny, especially in the early period of life, are characterized by reduced viability.

Progeny of the same age obtained from different age combinations of parents varied in quality in a number of piscicultural and biochemical parameters during their development in rearing, wintering, and foraging ponds: growth, chemical composition of the body, and winter-hardiness (in particular, waste and loss of nutritional substances during the period of winter starvation). Some groups differed notably in hematological parameters (hemoglobin content, number of erythrocytes and leucocytes per cc of blood, and intensity of blood formation).

HYBRIDIZATION IN POND FISHERIES. HETEROSIS. Hybridization in this case is understood as crossing fish of different species to obtain marketable fish, and to develop new and more valuable breeds and hybrid forms, i.e., ones which combine the valuable traits of the parent species.

A number of methods are followed in pond fisheries for interbreed, intraspecific, interspecific, and intergeneric crossing for the purpose of obtaining breeds better adapted to the conditions of definite zones.

Females of the Taparovanskii sazan have been crossed with male carp with scattered scales (V.I. Savel'ev, 1939). The eggs and fry of Taparovanskii sazan are more resistant to low temperatures in spring than the eggs and fry of cultivated carp. From this crossing a hybrid has been obtained which is similar to mirror carp in shape of head and in most cases similar to sazan in scale coat. In some of the hybrids displacement of the scales and underdevelopment of the opercula are seen, which frequently occur in young Taparovanskii sazan, as well as defects in the structure of the fins.

These hybrids exhibit the following positive traits:

(a) Increased winter-hardiness (the waste of these fish during winter at the "Nikol'skii" fish farm was 1.4% versus 24.8% waste of mirror carp, and also less in other categories of ponds);

(b) Higher productivity in the northern regions, especially under unfavorable conditions;

(c) Quality of meat in general not inferior to the meat of mirror carp;

(d) Higher resistance to German measles and saprolegniosis.

However, all these desirable traits are not preserved in the second generation, and back-crossing females of mirror carp with males of Taprovanskii sazan does not yield hybrids with the above-mentioned advantages.

Volga sazan has been crossed with various breeds of cultivated carp. The hybrids of Volga sazan and carp with scattered scales have scales similar to sazan but in many individuals the scales are irregular in arrangement. The lateral line changes in hybrids; it is either disrupted or totally absent but deformations are fewer than in carp. These hybrids have a number of economically valuable properties, for example, a higher tolerance toward certain factors of the external environment (oxygen deficiency, higher alkalinity) and greater resistance against diseases. In a pond polluted with sewage water all the carp died while 32.4% of the hybrids survived. In experimental foraging ponds the loss in hybrids was 4%, in carp with scattered scales 8%, and in leather carp 10%. The loss of carp in rearing ponds was 82.8% versus 9.7% for hybrids. In the first year of life, especially in the first two to three weeks, hybrids grow more rapidly than carp with scattered scales, and more rapidly still than leather carp. Hybrids also achieve a higher weight in foraging ponds in the second year of life. With a uniform weight of stocking material hybrids lag behind carp with scattered scales in terms of weight in the second year of life by 5 to 10%, and in the third year by 16.7%. However, the meat of hybrids is characterized by a slightly higher fat content compared to sazan (sazan—7.67%, hybrids—6.74%) and a much higher fat content compared to carp with scattered scales (3.57%).

All these desirable qualities of the Volga sazan-carp hybrids unfortunately are not preserved in the second generation.

The Amur sazan is characterized by higher winter-hardiness and a good growth rate in ponds. However, its growth slows down in the second year of life and slackens even more later on. It has other deficiencies such as a lower utilization of feed, slender body, and low meat content. However, because of its positive traits the Amur sazan is a very valuable object for crossing with carp and rearing hybrids of the first generation in ponds, especially in the northern regions of the country.

Work on the creation of new hybrid breeds of carp for these regions commenced in 1949. Mass-scale crossing and development of hybrids of the first generation between carp and Amur sazan have shown a higher winter-hardiness, rapid growth, higher viability, and better utilization of natural foods by the hybrids compared to carp. The cold resistance of the first and second generations of these hybrids was better in the first year of life; the lower temperature threshold for survival in the embryonic period of development was 7 to 8°C. The cold resistance of the eggs increased significantly when the spawners were kept at a temperature of 8 to 12°C before breeding.

A specially high sensitivity to cold was noticed at the stage prior to hatching of the larvae at a temperature of 10 to 12°C, leading to injury and even mortality. With further development they became more resistant to low temperatures. Two- and three-month-old fingerlings of hybrids weighing 5 to 15 g could withstand a temperature drop to 2.5°C and during hibernation close to 0°C. In general, the fingerlings of hybrids of Amur sazan and carp hibernate in wintering ponds with a water temperature falling to 0.1 to 0.2°C and even to 0°C for brief spells of time.

The hybrids of the first generation withstand hibernation better than carp. The most winter-hardy hybrids were obtained from females of Amur sazan crossed with carp males. In the Leningrad district hybrids under favorable conditions attain a 37% higher weight than carp and even exceptionally 110%. In an experiment in Latvian SSR all the fingerlings of the local carp perished (100% mortality) as a result of sanguinicoliasis, while the mortality among the hybrids was only 43.3%.

Since the Amur sazan commencing from the third year of life, and sometimes even from the second, slows down in growth, it is recommended that the following weight parameters (from Kirpichnikov and Golovinskaya, 1966) be used in selecting breeding stock for the purpose of crossing with carp with scattered scales.

Age group	Weight, g	
0+	30–50	
1+	450–600	
	♀	♂
2+	1,000–1,200	900–1,100
3+	1,600–1,800	1,400–1,600
4+	2,000–2,300	1,800–2,000
5+	2,300–2,500	—

The parameters of body structure recommended for five- to seven-year spawners of Amur sazan by the foregoing authors are given in Table 67.

Table 67. Parameters of body structure of Amur sazan

Sex	K	l/H	$\frac{B \times 100}{1}$ %	Other parameters
Female	2.4–2.8			Working fecundity of a single female
Male	2.2–2.6	3.6–3.2	15–18	weighing 2–3 kg, 75,000 to 125,000 larvae

In addition to intraspecific hybrids, remote hybrids—interspecific and intergeneric—are also bred. Experiments have shown that it is possible to cross sazan with grass carp, silver carp, and bighead (eggs of sazan and milt of phytophagous fishes) belonging to different subfamilies of the carp family, in spite of their notable differences in biological properties and peculiarities of reproduction (Makeeva, 1968).

In waters with a predominance of Crucian carp and in peat ponds it is more appropriate to use hybrids of carp and Crucian carp (*Carassius carassius*). Although the latter grows at a slower rate than carp, under proper fish management it can attain a weight of 800 g at the age of three years. In tolerance it is closer to goldfish and in ponds with an unfavorable hydrochemical regime and an abundance of small poorly grown Crucian carp of no food value, it is more appropriate to culture hybrids than either carp or goldfish. Furthermore, this hybrid is sterile (fertile specimens, although extremely rare, are nevertheless found to occur), which excludes the possibility of overpopulation of ponds with fry (often seen in the culture of goldfish). Hybrids obtained from crossing female mirror carp with male Crucian carp and female Crucian carp with male wild sazan, at the age of seven years are closer to carp in shape (Nikolyukin, 1952).

A small number of fry have been obtained as a result of fertilizing Crucian carp eggs with the spermatozoa of hybrids of Crucian carp and sazan, and a large number of progeny from crossing female Crucian carp-sazan with male mirror carp. The latter hybrid has a slightly higher growth rate than carp at the end of the second summer (hybrid—493 g, carp—477 g). To obtain fertile carp-Crucian carp hybrids as brood fish it is recommended that one use females obtained from crossing carp with veiltail, and triple Crucian carp hybrid males obtained from repeated crossings of lake goldfish with veiltail and goldfish (*Cyprinus auratus*) (Kuzema, Kucherenko, and Tomilenkov, 1967). To strengthen the qualities of carp, carp males may be used.

Work on crossing carp and tench has also started (Nikolyukin, 1962* and Viktorovskii, 1966). To obtain hybrids from this cross the eggs from the female Ropshinsk scale carp (strain B) were fertilized with a mixture of milt from two tench males. Viable hybrids with a good growth rate were obtained which are closer to carp in their taxonomic properties and, in some cases, do not even differ from carp. These hybrids are greenish in color and their body surface is covered with slime.

Hybrids have also been obtained from crossing silver carp and bighead. They are similar to silver carp in body color but differ from the parental species in size of head and structure of the branchial apparatus. The percentage of fertilization of the eggs of silver carp with the milt of its males is 92 to 99%, survival from the eggs to larvae 30%, and survival from larvae to fingerlings 21.0%. These figures in the case of hybrids obtained from females of silver carp and males of bighead are respectively: 90.0 to 93.5%, 43.6%, and 83.0% (V. Vinogradov and L. Erokhina).

Vinogradov and Erokhina also report that the survival of larvae from eggs of bighead fertilized with its own milt is 20 to 70%, and the survival from the larvae to fingerlings 20.5%. In the case of hybrids obtained from crossing females of bighead with male silver carp these figures are 48.1 to

*Publication date given as 1952 in bibliography—General Editor.

97.0 and 42.2% respectively. In behavior the hybrids are much more quiet than silver carp. They are easily raised in nondrainable ponds and reservoirs and can replace silver carp, the catching of which is difficult in such water bodies. Differences in the development of such hybrids are evident in the size of the eggs, embroys, and larvae, the number of caudal myotomes, and in the swelling properties of the eggs (Makeeva and Sukhanova, 1966). In coloration these hybrids are more pigmented than bighead.

Under the conditions of Turkmenian SSR the fertilization of eggs obtained from a cross between female grass carp and male of silver carp was 62 to 98%, and in a reciprocal cross 70 to 100%. Together with other traits the larvae of both these hybrids are characterized by pigmentation: In the first cross (female grass carp and male silver carp) they are closer to the pigmentation of silver carp, but at the end of the larval period more similar to grass carp with a more intensive pigmentation than in the mother parent. In the reciprocal cross the intensity and peculiarities of pigmentation are expressed less vividly than in common silver carp.

After crossing female grass carp with male bighead the percentage of fertilized eggs was 89 to 92%. The percentage of mortality during the incubation period was higher (25 to 33%) compared to nonhybrid eggs (10 to 18%). Differences exist in the number of myotomes: their number in the trunk is 27 to 29, which makes the hybrids closer to grass carp, and in the caudal part 17 to 18 as in the case of bighead. Other differences are evident in the pigmentation of the hybrid larvae—in the preanal fin fold as well as between the chords and the intestines—which are characteristic of the male of bighead. In general the pigmentation of these hybrids is more intensive. They develop slightly more than the "pure" species.

Studies have recently been initiated in the USA and other countries on hybridization between different salmonids.

In the USA work is in progress on hybridization of rainbow trout, brown trout, brook trout, and the common trout introduced from Europe. The most satisfactory results from different combinations of trout species have been obtained from crossing brook trout and brown trout (hatching of larvae 28%). These hybrids are fertile and under artificial fertilization have given good results. Egg survival reaches 70%. The hybrids grow better than brown trout, the primary object of trout farming in the USA, which is more sensitive to diseases than rainbow trout. The hybrids, as expected, are more resistant to diseases.

Taking into account that at the end of the second and sometimes even the first year of life, the males of brown trout become sexually mature, leading to a slowing down in growth, reduction in fattiness, and greater sensitivity to diseases, crosses have been made between brown trout and alpine char in order to obtain sterile hybrids with better growth and greater resistance to diseases. The hybrids obtained at the farm "Kotere" remained sterile for five to six years. Hybrid larvae of alpine char and brown trout hatched on the ninth day, while those of lake trout and brown trout hatched in three to four days.

The six-year-old hybrids at "Kotere" weigh more than 1 kg and externally resemble brown trout, except for more vivid yellowish-white stripes. Red spots do not appear on either side of the body. The fins, similar to alpine char, are long, ash-gray, and with a white margin (latter absent on pectoral fins; dorsal fin entirely covered with stripes). These hybrids grow well at a temperature of 2°C in winter and 12°C in summer. They are characterized by a higher resistance to diseases compared to brown trout; diseases of the thyroid gland and fins, so often occurring in brown trout, have not been seen in the hybrids. Under epidemic conditions of furunculosis brown trout suffers greater losses than the hybrids.

A hybrid of the lake trout has been obtained in Canada, which does not suffer from lamprey attacks. In the USA work is in progress on the hybridization of chun salmon and sea trout and between a number of other fish species.

Of the members of the salmon family hybridization between some species of whitefish (coregonids) is of particular interest. The hybrids of these fish, compared to the parent species, possess a number of economically important characters, notably rapid maturation and fairly high fecundity.

A hybrid obtained at Ural from crossing ripus and Lake Chudskoe whitefish has proved quite suitable for releasing in small water bodies (lakes and production ponds). Compared to the initial forms this hybrid is characterized by rapid growth (under-yearlings attain a weight of 80 to 150 g and in autumn two-year fish 300 to 500 g and three-year fish 600 to 900 g), and a tolerance and survival rate superior to the ripus by more than 20 times. The hybrid attains sexual maturity at the age of 18 months and maintains its capacity for reproduction. Under good food conditions it reaches marketable weight even at the age of seven to eight months, which makes it possible to utilize hybrids of the first generation not only in a two-year rotation, but also in one-year culture in lakes and foraging carp ponds.

A number of anomalies are evident in the sexual maturation of the Ural hybrids: hypertrophy of the male sex glands (in six-month-old under-yearlings the weight of hypertrophied gonads was 30 to 40 times more than normal), sterility of the male fish, and incomplete termination of the process of spermatogenesis. Yet in many males the gonads are almost normal in terms of weight, differing instead in structure—four to eight lobes versus the usual bilobate. The number of lobes sometimes reaches 15 to 20. However, only with external disturbance of the gonads, without any symptom of hypertrophy, the process of spermatogenesis proceeds normally and the males discharge dense milt containing motile and good quality spermatozoa.

According to Nesterenko (1957) hypertrophy and lobate male gonads are characteristic of the hybrids of whitefish (female) and ripus (male); sterility occurs in 8 to 12% of the males, while the rest remain normal. Hermaphroditism and a sharp deviation from the normal sex ratio sometimes occur in the hybrids of ripus (female) and whitefish (male). In hybrids of whitefish (female) and ripus (male) the sex ratio is close to 1:1, but in reciprocal hybrids—ripus (female) and whitefish (male)—this ratio shifts in favor of the female (6:1). These data reduce the value of such hybrids as brood fish but do not detract from their utilization in fish farms and breeding stations. Hermaphroditism can be controlled to some extent. The number of hermaphrodites was found to be greater in waters with a high mineral content. When mineralization was reduced to 0.07 to 1.10% hermaphroditism occurred only in isolated cases.

Hybridization of female Lake Ladoga whitefish and male ripus is not recommended: the males of such crosses were sterile due to the degeneration of their spermatozoa.

Under the conditions of Ukrainian SSR the hybrids of Lake Chudskoe whitefish and peled from Lake Endyr' (basin of Ob River) exhibit excellent piscicultural qualities. The under-yearlings of such hybrids reach a weight of 100 to 200 g over the vegetative growth period, two-year fish 250 to 300 g, three-year fish 450 g, and four-year fish 680 g, with a maximum of up to 950 g, i.e., their growth is better than that of the initial forms. The hybrids consume zooplankton during the summer, which is available at that time in sufficient quantities, and in spring and toward autumn also feed on benthos and sometimes only benthos. In this respect the nature of their feeding is closer to that of Lake Chudskoe whitefish.

The fish productivity of experimental ponds of carp and hybrids was 515 kg/ha for two-year carp and 462 kg/ha for hybrids.

Under the conditions of Ukrainian SSR male and female hybrids of Lake Chudskoe whitefish and peled in general reach sexual maturity at the age of 2+, but sometimes at the age of 1+. The initial forms attain sexual maturity much later (Lake Chudskoe whitefish at the age of 4+ and peled at 2+ or 3+). The progeny of the hybrids are viable. These hybrids exhibit the following peculiarities in pond culture:

(1) They differ from the initial forms and in many characters hold an intermediate position;

(2) They are more fertile than Lake Chudskoe whitefish; their larvae are larger than those of peled and smaller than those of Lake Chudskoe whitefish;

(3) They are more tolerant in relation to changes in water temperature, oxygen content, and diseases;

(4) They have a higher vitality rate, more rapid growth, earlier sexual maturation, and better taste and food qualities.

Fishes from the sturgeon family have rarely been undertaken for pond culture. N.I. Nikolyukin has demonstrated the economic viability of crossing beluga and sterlet, common sturgeon and sterlet, and beluga and common sturgeon.

The most promising hybrid, that obtained from crossing beluga and sterlet, combines the good qualities of rapid growth of beluga and earlier sexual maturation of sterlet. Male beluga attain sexual maturity not earlier than 12 years and females not earlier than 16 years, while their hybrid matures at the age of 3 to 4 years.[1] Even in ponds with a poor availability of natural foods hybrids attain the same weight toward the second year of life as sterlet in the fifth to sixth.

The hybrid obtained from crossing beluga and sterlet grows more rapidly than the hybrid obtained from crossing sterlet and beluga and is characterized by higher viability and a better growth rate. The eggs and larvae of the female beluga are larger than those of sterlet. It is more difficult to work with the smaller, more adhesive eggs of starlet, and its smaller, more delicate fry require much more attentive aftercare. The average unit weight of the under-yearlings of hybrids from beluga and sterlet with a stocking density of fry at the rate of 10,000 to 14,000 per hectare and an output of 40 to 60% is 71.9 g; the weight of two-year fish (with a stocking density of up to 600 fish per hectare and an output of 98 to 100%) 446 to 554 g; and the weight of three-year fish 900 to 1,000 g. Fish productivity from the culture of under-yearlings reaches 581 kg/ha.

The larval behavior of the hybrids of beluga and sterlet is closer to that of beluga, while the behavior of hybrids of sterlet and beluga is closer to that of sterlet. Both hybrids occupy an intermediate position however in external appearance and morphological characters, leaning toward a greater similarity to beluga.

Since the male sterlet produces a small quantity of milt 10 to 20 mature sterlet males should be allowed for each female of beluga and sterlet.

In crosses between female common sturgeon and male sterlet hybrids have been obtained which are characterized, even at ten months of age, by a higher weight than sterlet and a greater viability. The weight of under-yearlings varies from 25 to 77 g toward autumn, of two-year fish with a stocking density of 500 to 600 yearlings per hectare and an output of 90% about 500 g (fish productivity 180 to 230 kg/ha), and three-year fish up to 800 g.

Sturgeon hybrids grow well in the usual carp ponds—ponds without much silt deposition, having an area of 0.5 ha, an average depth of 1 to 2 m, shallow water parts, and devoid of a dense overgrowth of aquatic plants. Filamentous algae are particularly harmful to hybrid fry as they obstruct their search for

[1]According to G. Merla (*Dtsch. Fisch. Ztg.*, 1970, no. 19) hybrids obtained by crossing beluga and sterlet mature at the age of five to seven years (females) or three to five years (males).

benthic organisms and cause mortality due to fish entanglement in the filaments. The oxygen content in ponds should not fall below 5 mg/l and the water temperature should not exceed 28°C. With a reduction in oxygen content to 2 to 3 mg/l sturgeon consumes food poorly and its growth suffers. There should be sufficient quantity of benthos in the ponds (not below 8 g/m^2 on the average over the vegetative growth period). The ponds should be well prepared, small trash fish removed, and fertilizers applied.

To avoid competition for food sturgeon hybrids should not be raised with carp, even though combined raising with phytophagous fishes is highly desirable. However, the hibernation of sturgeon hybrids and carp is permissible. Hybrids of beluga and sterlet overwinter better than those of beluga and sturgeon (losses 50%).

The bibliography given at the end of this chapter on breeding or hybridization of fish is not exhaustive. It merely reflects the wide and variable possibilities available in pond fisheries for increasing fish productivity, developing new and valuable breeds, and improving existing forms.

Hybridization is important not only as a means of increasing fish types, but of improving the productivity of pond fisheries through the effect of heterosis.

Heterosis is understood here as an increase in the productivity of hybrids of the first generation over parental forms. The basis for this phenomenon is a complex of genetic and biological processes, whereby fish growth is augmented and other qualities altered. Charles Darwin long ago mentioned the advantages of combining several different sex elements and the harm accruing from combinations of identical types.

It is possible through heterosis to increase viability and fertility, advance maturation, and augment the productivity of progenies derived from crossing organisms which differ in hereditary traits or which have been raised under different conditions. Heterosis in pond culture is expressed in hybrids by a faster rate of growth, higher tolerance toward unfavorable environmental conditions, winter-hardiness, earlier sexual maturation, and higher productivity. Not only intraspecific, but also more distant hybrids—interspecific and intergeneric—have been found to be fertile, which makes it possible to practice heterosis not only in the first generation but in subsequent ones. Heterosis, manifested in full measure in the first generation, declines in subsequent ones. Hence hybrid individuals should not be maintained as brood stock.

Heterosis has acquired maximum significance in carp culture. Heterosis for improving growth rate and viability is distinctly expressed in the hybrids of carp and sazan, especially of Amur sazan. In Ukrainian SSR the Ropshinsk hybrids serve as material for crossing with Ukrainian carp and a strong heterosis for growth has been recorded in the hybrids derived from such crosses. The hybrids produced by mating female Ukrainian scale carp with male Ropshinsk carp exceed the parental forms by 25% in growth; the nutritional value of these hybrids is comparable to that of the Ropshinsk carp and better than that of the Ukrainian scale carp. The hybrids obtained from the female Ropshinsk Ukrainian hybrid and male Ukrainian scale carp attain a higher weight than the hybrids obtained by crossing the female Ukrainian scale carp with the male Ropshinsk Ukrainian carp. Thus, brood fish from the first hybrid generation are not comparable in terms of vigor of transmission of hereditary traits to their progeny; a terminating maternal effect has been reported in a number of experiments.

The winter-hardiness of hybrid under-yearlings of the second generation (progeny from the female Ropshinsk Ukrainian hybrid and the male Ukrainian scale carp) was greater than that of the pure line stock.

Heterosis is also manifested in the hybrids of salmonid species crossed with whitefishes.

The effect of heterosis on fish productivity is seen when largescale commercial cross-breeding programs are undertaken for the exclusive purpose of producing cheaper, better quality marketable fish, which usually constitute hybrids of the first generation. For a better manifestation of heterosis in fish, as in all farm animals, the conditions of feeding and maintenance must correspond directly to the manifestation of the inherited properties of the improved breed.

One of the common objectives of breeding work in carp culture is to develop economically valuable breeds of carp adaptable to different zones and different conditions of nutrition and maintenance (ponds, floodland lakes, rice fields, peat quarries, irrigation networks, and saline waters).

A group of pisciculturists, under the direction of A.I. Kuzema, have recently developed the Ukrainian scale carp and Ukrainian frame carp. Work is in progress under the leadership of V.S. Kirpichnikov on the development of new carp breeds for the northwestern regions of the European part of the USSR (Novgorod, Pskov, and Leningrad districts). Breeding work has also been started in Estonia and Siberia, including crossings with the Ukrainian carp forms. The latter have been named the "Northern (Ropshinsk)" carp. The possibility of a Siberian breed of carp is being investigated. The "local carp" may be of special value in breeding work as it is from the Aleiskii region and has adapted over several generations to local conditions. Work is also in progress on developing a Belorussian breed of carp. Studies are underway at the All-Union Research Institute of Pond Fisheries under the guidance of K.A. Golovinskaya for developing a breed of "middle origin" carp for the central part of the USSR.

A few new breeds of carp have been developed in Hungary: "bikharugraisk" for the western districts, and "hartobad" for the eastern, which is adapted to water bodies with saline soils. In southern China a fast-growing breed of carp has been developed, which is particularly suited for raising in rice fields. New breeds of carp are also being developed in other countries.

In addition to breeding carp, in the USSR and a number of other countries (the USA, the German Democratic Republic, etc.) attention has focused recently on the breeding of trout. Studies have been conducted at the Ropshinsk farm (near Leningrad) to improve the hereditary characters of rainbow trout. In the USA rainbow trout stocks have been improved over many years by the Fisheries School of the University of Washington. In the first ten years of study a fast-growing, early-maturing fish was developed through mass selection. Work on mass selection was subsequently replaced by work involving pure line selection in which fish of the largest size, earliest maturation, and highest fecundity were selected, having a convenient time of spawning. The initial four-year female trout (1932) had a body weight of 560 g and a fecundity of 800 to 1,200 eggs. The three-year trout in the progeny raised in 1952 had an average weight of 2,100 g and produced 4,500 eggs. The 1952 stock is distinguished by rapid growth, earlier sexual maturation, adaptation to higher temperatures, resistance to diseases, and high fecundity. In California a breed of trout has been developed through selection and appropriate maintenance which is characterized by high fecundity, good egg quality, and a rapid growth rate in the first year of life.

Embody and Heiford (1925) worked on the breeding of brown trout to obtain a form resistant to furunculosis. After five generations two stocks were obtained which are distinguished by resistance to this disease, higher growth rate, and increased fecundity. Since 1952 continuous work has been done to develop rainbow trout stocks resistant to furunculosis and ulcerous diseases (Wolf, 1954). Stocks were artificially infected with the bacteria causing furunculosis and the less resistant individuals rejected. The progeny of these stocks were given the same treatment, the less resistant individuals rejected, and the process repeated. The most resistant lines were then crossed, and the growth rate, fecundity, and color and survival of the eggs recorded during selection. Mortality due to furunculosis was around 57% in the control group, but did not exceed 12% in the experimental stocks.

Estimation and Assessment of Brood Fish and Replenishing Juvenile Stocks. Progress in breeding work in pond pisciculture requires careful planning and a systematic study of the brood and replenishing juvenile stocks and the breeding qualities of fish. An estimation and assessment of these stocks occupy a very important place in planning since the question of rejection or selection of brood fish and replenishing stocks is decided on this basis. A qualitative assessment is made individually every spring of a complex of characters (productivity, constitution, body structure, origin, quality of offspring, etc.) at the time of catches from the wintering ponds and fish transference to the summer ponds. Brood fish and replenishing stocks received from other farms must be accompanied by the necessary documents, giving details of the fish in previous years. If such documents are not attached,

Table 68. Certificate

List of brood fish and juvenile replenishing stock of fish farm——————dated——————, 19——

with the participation of——————

As a result of the estimation, the following qualitative and quantitative composition was observed————————————

(brood fish, replenishing stock)

S. No.	Tag No.	Scale condition	Age	Sex	Weight, g	External appearance				Relative parameters		
						Body length excluding caudal fin, cm	Length of head (maximum), cm	Body height (maximum), cm	Body thickness (maximum), cm	Relative length of head to length of body	Trunk height (ratio of body length to height)	Body thickness (ratio of thickness to body length), %

Signature

Table 69. Permit for brood fish No.——————

Fish breeding farm (collective, state)——————Region——————District, Republic——————Scale condition——————

Obtained in 19——————from female No.——————and male No.——————. Sex——————

Year	Date of inspection	Age, year	Weight, g	External appearance of fish				Relative parameters			Remarks
				Body length without caudal fin, cm	Length of head, cm	Body height, cm	Body thickness, cm	Ratio of head length to body length	Trunk height (ratio of body length to body height)	Body width (ratio of thickness to length)	
											External defects and possible causes for diseases are indicated

Progeny from spawning												
				Fecundity: output of fry, in thousands		Productivity on basis of under-yearlings[1]						
Year	No. of female or male	No. of pond	Rearing period of fry, days	Total	Per kg of body weight of female fish	No. of pond	No. of fish released, in thousands	No. of fish caught, in thousands	Average weight, g	Survival, % of fish stocked	Total weight (productivity), kg/ha

	Winter hardiness[2]				Productivity of 2-year fish					
Year	No. of pond	No. of fish stocked, in thousands	No. of fish caught, in thousands	Survival, %	No. of pond	No. of fish released, in thousands	No. of fish caught, in thousands	Survival, %	Average weight, g	Total weight, kg/ha

[1]The productivity of under-yearlings is estimated only from a particular female parent. If these are raised to marketable weight, the respective graphs are based on the values of quantitative output of fry from a given female fish.

[2]Only data on the progeny obtained from a pair or, exceptionally, from a redd of spawners, are used for drawing the curve of fecundity and winter-hardiness.

estimation and assessment are done upon receipt of the fish. All the brood fish (males and females) are subjected to inspection and an average random sample of 30 to 50 fish taken from each age group for assessment of the replenishing stock. In addition to the parameters of the average sample, the weight and composition of the replenishing stock are taken into account on the basis of scales, and total quantity and weight (average, maximum, and minimum) of young fish from each age group and breed. On the basis of annual estimation and assessment of brood fish and replenishing stock a certificate (Table 68) and permit for brood fish are prepared (Table 69).

Changes in the quality of the brood fish and replenishing stock throughout the year—number of rejected fish and fish transferred from the replenishing stock to the active stock—are indicated in the certificates.

Assessment of the replenishing stock is done in different age groups, while brood fish are assessed sex-wise. Changes occurring during the year and reasons for losses and rejection are indicated in the certificate.

The results of estimation and evaluation provide data on the quantity of stock raised and make it possible to follow systematically the quality of the breeding stock, timely recording and rectification of defects, improving maintenance, care, and nutrition, and conducting selection.

All documents pertaining to estimation and evaluation of brood fish and replenishing juvenile stock are kept in the records of the collective or state farm and copies of these sent to the higher office.

Rules for Determining and Recording Individual Parameters. To record their weight brood fish and replenishing recruits are placed in a special crib, the weight of which is already known (Figure 118). The crib is placed on the balance and the weight of the fish taken when it lies still. The weight of the crib, checked beforehand, is rechecked periodically during the course of the work. The precision of weighing brood fish and four-year fish is 100 g and that of two-year and three-year fish up to 50 g. Fish measurements are taken on a measuring board (Figure 119) with the help of the usual set square with a precision of 0.1 cm. The dimension of the measuring board should approximate the size of the brood fish and replenishing stock (about 80 cm long, 25 cm wide, and 20 cm high). The measuring board has a smooth surface, fitted with small side panels; the table tops are covered with straw and tarpaulin. The manner of recording length, maximum height, and thickness is shown in Figure 119.

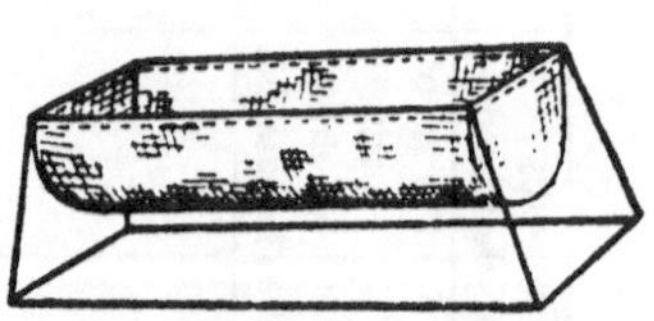

Figure 118.
Crib for weighing brood fish.

To summarize the data recorded, the indexes of body height (ratio of body length to maximum height) and body width (ratio of body thickness to body

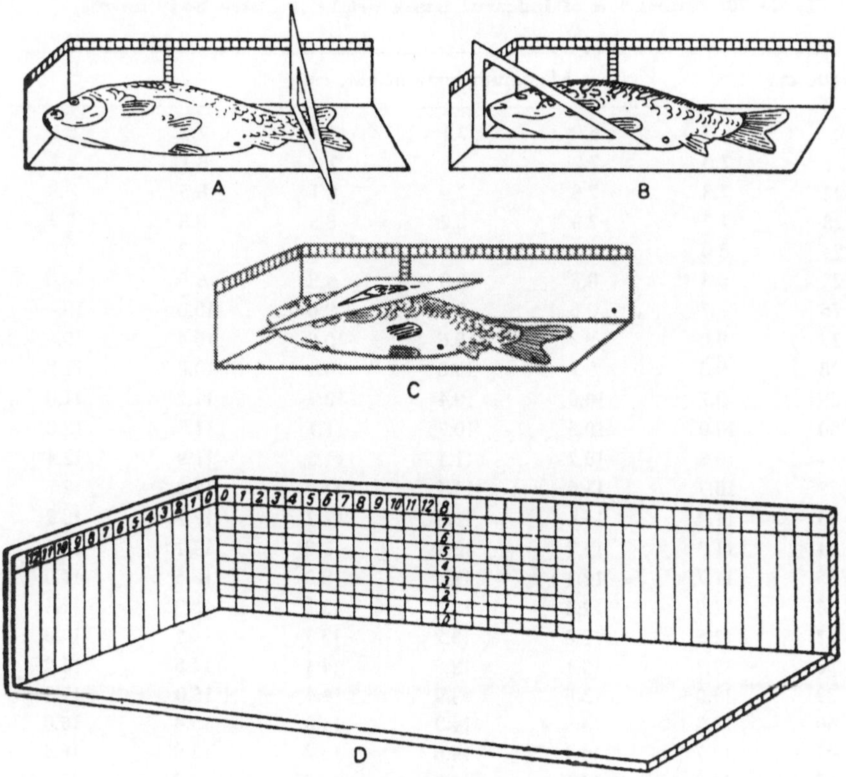

Figure 119. Measuring board.
A—measuring body length; *B*—measuring bodv height; *C*—measuring body thickness; *D*—general view of measuring board.

length) are determined. The index of relative body height in a normally developed cultivated carp in close to 2.5 to 2.7, and the index of relative body thickness 20 to 22%. These indexes have been determined for different lengths of fish (Tables 70 and 71).

The sex of the fish is recorded only when it is clearly discernible; otherwise, "sex not discernible" should be written. The same holds true for the age of brood fish and replenishing stock when, due to mixed grouping, the age is not precisely known. External symptoms of disease, scars from healed wounds and ulcers, and deformities are also recorded in the respective columns. Particular attention is paid to the correctness of the lateral line. The properties of spawners must be conscientiously detailed.

Types of Breeding Work in Pond Pisciculture. Breeding work should be carried out at specialized breeding farms, fish hatcheries, state and collective fish hatcheries, and full-system fish farms in which seed fish and marketable

Table 70. Estimation of index of trunk height (relative body height)

Length, cm	Maximum body height, cm					
20	6.7	6.9	7.1	7.4	7.7	8.0
21	7.0	7.2	7.5	7.8	8.1	8.4
22	7.3	7.6	7.9	8.1	8.5	8.8
23	7.7	7.9	8.2	8.5	8.8	9.2
24	8.0	8.3	8.6	8.9	9.2	9.6
25	8.3	8.6	8.9	9.3	9.6	10.0
26	8.7	9.0	9.3	9.6	10.0	10.4
27	9.0	9.3	9.6	10.0	10.4	10.8
28	9.3	9.7	10.0	10.4	10.8	11.2
29	9.7	10.0	10.4	10.7	11.2	11.6
30	10.0	10.3	10.7	11.1	11.5	12.0
31	10.3	10.7	11.1	11.5	11.9	12.4
32	10.7	11.0	11.4	11.9	12.3	12.8
33	11.0	11.4	11.8	12.2	12.7	13.2
34	11.3	11.7	12.1	12.6	13.1	13.6
35	11.7	12.1	12.5	13.0	13.5	14.0
36	12.0	12.4	12.9	13.3	13.9	14.4
37	12.3	12.8	13.2	13.7	14.2	1[illegible].8
38	12.7	13.1	13.5	14.1	14.6	15.2
39	13.0	13.5	13.9	14.4	15.0	15.6
40	13.3	13.8	14.3	14.8	15.4	16.0
41	13.7	14.1	14.6	15.2	15.8	16.4
42	14.0	14.5	15.0	15.6	16.2	16.8
43	14.3	14.8	15.4	15.9	16.5	17.2
44	14.7	15.2	15.7	16.3	16.9	17.6
45	15.0	15.5	16.1	16.7	17.3	18.0
46	15.3	15.9	16.4	17.0	17.7	18.4
47	15.7	16.2	16.8	17.4	18.1	18.8
48	16.0	16.6	17.1	17.8	18.5	19.2
49	16.3	16.9	17.5	18.2	18.9	19.6
50	16.7	17.2	17.9	18.5	19.2	20.0
51	17.0	17.6	18.2	18.9	19.6	20.4
52	17.3	17.9	18.6	19.3	20.0	20.8
53	17.7	18.3	18.9	19.6	20.4	21.2
54	18.0	18.6	19.3	20.0	20.8	21.6
55	18.3	19.0	19.6	20.4	21.2	22.0
56	18.7	19.3	20.0	20.7	21.5	22.4
57	19.0	19.7	20.4	21.1	21.9	22.8
58	19.3	20.0	20.7	21.5	22.3	23.2
59	19.7	20.3	21.1	21.9	22.7	23.6
60	20.0	20.7	21.4	22.2	23.1	24.0
61	20.3	21.0	21.8	22.6	23.5	24.4
62	20.7	21.4	22.1	23.0	23.9	24.8

Length, cm	Maximum body height, cm					
63	21.0	21.7	22.5	23.3	24.2	25.2
64	21.3	22.1	22.9	23.7	24.6	25.6
65	21.7	22.4	23.2	24.1	25.0	26.0
66	22.0	22.8	23.6	24.4	25.4	26.4
67	22.3	23.1	23.9	24.8	25.8	26.8
68	22.7	23.5	24.3	25.2	26.2	27.2
69	23.0	23.8	24.6	25.6	26.5	27.6
70	23.3	24.1	25.0	25.9	26.9	28.0
Index of trunk height	3.0	2.9	2.8	2.7	2.6	2.5

Table 71. Estimation of index of body thickness (relative body thickness)

Length, cm	Maximum body thickness, cm					
20	3.4	3.6	3.8	4.0	4.2	4.4
21	3.6	3.8	4.0	4.2	4.4	4.6
22	3.7	4.0	4.2	4.4	4.6	4.8
23	3.9	4.1	4.4	4.6	4.8	5.1
24	4.1	4.3	4.6	4.8	5.0	5.3
25	4.3	4.5	4.8	5.0	5.3	5.5
26	4.4	4.7	4.9	5.2	5.5	5.7
27	4.6	4.9	5.1	5.4	5.7	5.9
28	4.7	5.0	5.3	5.6	5.9	6.2
29	4.9	5.2	5.5	5.8	6.1	6.4
30	5.1	5.4	5.7	6.0	6.3	6.6
31	5.3	5.6	5.9	6.2	6.5	6.8
32	5.4	5.8	6.1	6.4	6.7	7.0
33	5.6	5.9	6.3	6.6	6.9	7.3
34	5.8	6.1	6.5	6.8	7.1	7.5
35	6.0	6.3	6.7	7.0	7.4	7.7
36	6.1	6.5	6.8	7.2	7.6	7.9
37	6.3	6.7	7.0	7.4	7.8	8.1
38	6.5	6.8	7.2	7.6	8.0	8.4
39	6.6	7.0	7.4	7.8	8.2	8.6
40	6.8	7.2	7.6	8.0	8.4	8.8
41	7.0	7.4	7.8	8.2	8.6	9.0
42	7.1	7.6	8.0	8.4	8.8	9.2
43	7.3	7.7	8.2	8.6	9.0	9.5
44	7.5	7.9	8.4	8.8	9.2	9.7
45	7.7	8.1	8.6	9.0	9.5	9.9
46	7.8	8.3	8.7	9.2	9.7	10.1

(Contd.)

Length, cm	Maximum body thickness, cm					
47	8.0	8.5	8.9	9.4	9.9	10.3
48	8.2	8.6	9.1	9.6	10.1	10.6
49	8.3	8.8	9.3	9.8	10.3	10.8
50	8.5	9.0	9.5	10.0	10.5	11.0
51	8.7	9.2	9.7	10.2	10.7	11.2
52	8.8	9.4	9.9	10.4	10.9	11.4
53	9.0	9.5	10.1	10.6	11.1	11.7
54	9.2	9.7	10.3	10.8	11.3	11.9
55	9.4	9.9	10.5	11.0	11.6	12.1
56	9.5	10.1	10.6	11.2	11.8	12.3
57	9.7	10.3	10.8	11.4	12.0	12.5
58	9.9	10.4	11.0	11.6	12.2	12.8
59	10.0	10.6	11.2	11.8	12.4	13.0
60	10.2	10.8	11.4	12.0	12.6	13.2
61	10.4	11.0	11.6	12.2	12.8	13.4
62	10.5	11.2	11.8	12.4	13.0	13.6
63	10.7	11.3	12.0	12.6	13.2	13.9
64	10.9	11.5	12.2	12.8	13.4	14.1
65	11.1	11.7	12.4	13.0	13.7	14.3
66	11.2	11.9	12.5	13.2	13.9	14.5
67	11.4	12.1	12.7	13.4	14.1	14.7
68	11.6	12.2	12.9	13.6	14.3	15.0
69	11.7	12.4	13.1	13.8	14.5	15.2
70	11.9	12.6	13.3	14.0	14.7	15.4
Index of body thickness, %	17	18	19	20	21	22

fish are reared by pond culture. At one-year production farms, which receive stocking material and raise only marketable fish for one year, breeding work is not done. However, the objectives of each type of farm involved in breeding work differ.

Specialized breeding farms have been established in various geographic zones, with different climatic conditions, primarily to develop new breeds and lines with outstanding productivity and to radically improve existing breeds. The progress of a breed depends on the development of lines. To solve the problems confronting a specialized breeding farm, relative or connective breeding on a definite line is permitted; with the skillful utilization of this method existing breeds can be radically improved and new breeds developed. The objective of relative breeding is the improvement and fixation of the valuable traits of the parents. The pair of related fish and their progeny must be raised in good conditions. Work is conducted in specialized breeding farms under the aegis of research institutes.

As experience in animal husbandry has revealed, unskilled use of closely related matings can lead to decline in fertility or even to total sterility, deterioration in metabolism, and lowered resistance to harmful organisms. The constitution of the animal becomes more delicate, the size of the animal smaller, and deformities appear. These undesirable consequences occur in pond pisciculture also, decreasing productivity at all stages of fish culture.

The work of specialized breeding farms is expedited by state fish breeding farms. The main work of the latter is to raise the highly productive pure fish breeds obtained from the former, and distribute them (as an improved stock of brood fish) to fish hatcheries, full-system fish farms, and collective and state farms. Pure line breeding is not done at these farms.

Mass-scale breeding work must be regularly carried out at different state and collective fish hatcheries, full-system fish farms, and collective and state enterprises. The main objective of these farms is not only production but also improvement of the productive qualities and breeding values of brood fish through correct maintenance and care. Improvement in quality and increase in productivity is achieved in full-system fish farms by carefully observing the conditions of culture and maintenance of the replenishing stock and brood fish, i.e., provision of balanced and proper diet, correct and systematic selection and rejection of less productive individuals, elimination of inbreeding through a timely exchange of males with other farms, and so on. In the case of mass-scale selection work pure line breeding is not essential and related matings are not practiced.

All the forms of breeding work are closely interrelated and directed toward the main objective—improving the stocks of commercial farms and ensuring a higher output of fish. Productivity, constitutional strength, viability, and transmission of valuable traits to the progeny are primary in achieving the main objective.

State prizes are awarded annually to encourage fish breeders to develop fast-growing and highly productive fish breeds and to improve present culturable breeds.

With regard to trout, whitefish, and phytophagous fish, specialized work can be combined at breeding farms of a higher order and state fish breeding farms.

The following are the most important objective of breeding work in the USSR:

(1) Organization of breeding work and guidelines from the Ministry of Agriculture and Fisheries of the USSR and construction of specialized breeding farms and state fish breeding farms.

(2) Development of research work on breeding of fish, fish culture at farms, genetic studies of pond fishes, theory of selection of fish, study of

heterosis under commercial matings, and further study on the problems of gynogenesis and induction of artificial mutations.

(3) Ensure a thorough control of mating and hybridization of fish and their concentrations in specialized breeding nurseries.

(4) Prohibit crossing of carp with Amur sazan, which has become a mass-scale phenomenon without the special permission of the Ministry of Agriculture of the USSR.

(5) Prohibit the mass-scale release of interspecific hybrids of fish into natural waters, especially the hybrids of salmon and sturgeon, since this may lead to considerable harm in the reproduction of these valuable fish species as food objects. The release of such hybrids in a small quantity must be controlled by research organizations through special permission.

References

Dambrovskii, V.K., G.A. Prokhorchik and A.I. Chutseva. 1970. Formirovanie Belorusskoi porody karpa (Formation of the Belorussian carp breed). *Trudy Belorussk. Nauchno-issled. In-ta Ryb. Khoz-va,* vol. 7.

Danaldson, L. and P. Olson. 1955 (1957). Development of rainbow trout brood stock by selective breeding, *Trans. Amer. Fish. Soc.,* vol. 85.

Darwin, Ch. 1941. Izmenenie zhivotnykh i rastenii v domashnem sostoyanii (Changes in Animals and Plants under Domestication). OGIZ, Moscow.

Dudnik, Yu. 1968. Ob otritsatel'nom vliyanii inbridinga (The harmful effects of inbreeding). *Rybovodstvo i Rybolovstvo,* no. 4.

Dudnik, Yu. 1968. Vliyanie inbridinga na nekotorye storony rannego razvitiya karpa (Effect of inbreeding on certain aspects of early development of carp). *Sbornik Nauchno-tekhn. Inform., VNIRO.* Moscow, no. 9.

Engliger, N.F. 1964. Selective breeding of trout for resistance to furunculosis, *Fish and Game,* vol. 7, no. 2.

Galkina, Z.I. 1968. Vliyanie razmerov i vozrasta samok lososevykh na potomstvo (Effect of size and age of female salmon on the offspring). Avtoref. Kand. Diss., Leningrad.

Golovinskaya, K.A. 1962. Plemennoe delo v prudovom rybovodstve (Breeding work in pond fisheries). *Rybovodstvo i Rybolovstvo,* no. 3.

Golovinskaya, K.A. et al. 1969. Formirovanie karpovogo plemennogo fonda rybplemkhoza "Yakot' " i nekotorye osobennosti sostovlyayushchikh ego otvodok (Formation of carp breeding stock at the fish breeding farm "Yakot" and some peculiarities of its constituent strains). Sbornik *Voprosy Prudovogo Rybovodstva.* Izd. VNIIPRKh, Moscow, vol. 16.

Gorodetskaya, A.S. 1959. O vliyanii organizma materi na kachestvo potomstva (Influence of the mother on the quality of the progeny). *Trudy Voronezh Zoovet. In-ta,* vol. 16.

Grinevich, S.I. 1965. Ukrainskie karpy (Ukrainian Carp). Urozhai, Kiev.

Kirpichnikov, V.S. 1954. Sozdanie novoi severnoi gibridnoi porody karpa (Creation of a new northern hybrid breed of carp). *Trudy Soveshchaniya po Rybovodstvu.* Izd-vo Akad. Nauk SSSR, Moscow.

Kirpichnikov, V.S. 1959. Geneticheskie metody selektsii ryb (Genetic methods of fish breeding). *Bull. Mosk. Ob-va Ispyt. Prirody, Odt. Biol.*, vol. 64, no. 1.

Kirpichnikov, V.S. 1959. Gibridizatsiya karpa s Amurskim sazanom primenitel'no k zapadnym oblastyam SSSR (Hybridization of carp with Amur sazan as applicable to the western regions of the USSR). *Trudy V Konf. po Izuch. Vnutr. Vodoemov Pribaltiki.* Minsk.

Kirpichnikov, V.S. and L.I. Lebedeva. 1953. Povyshenie zimoustoichivosti segoletkov karpa, Amurskogo sazana i ikh gibridov (Increasing the winter-hardiness of under-yearlings of carp, Amur sazan, and their hybrids). *Trudy Vsesoyuzn. Gidrobiologicheskogo Obshchestva,* no. 5.

Kirpichnikov, V.S. and K.A. Golovinskaya. 1966. Kharakteristika proizvoditelei osnovnykh porodnykh grupp karpa, razvodimykh v SSSR (Characteristics of brood fish of the main groups of carp breeds reared in the USSR). *Izv. GosNIORKh.* Leningrad, vol. 61.

Kirpichnikov, V.S., O.N. Bauer and N.A. Mosevich. 1956. Metodicheskie ukazaniya po provedeniya zimovki segoletkov karpov, gibridov i sazanov v rybkhozakh severnykh i severo-zapadnykh raionov SSSR (Methodical instructions on conducting hibernation of fingerlings of carp, hybrids, and sazan in fish farms of the northern and north-western regions of the USSR). *Izv. VNIORKh.* Leningrad.

Kislovskii, D.A. 1951. Razvedenie sel'skokhozyaistvennykh zhivotnykh (Breeding of Agricultural Animals). Sel'khozgiz, Moscow.

Korovina, V.M. 1961. Zavisimost' stoikosti zarodyshei ryb ot vozrasta proizvoditelei (The dependence of the hardiness of fish embryos on the age of the spawners). *Izv. VNIORKh.* Leningrad, vol. 51.

Kozhin, N.I. 1969. Zadachi genetiki i selektsii ryb v prudovom rybovodstve (Aims of genetics and fish breeding in pond fisheries). *Nauchnye Trudy. Rybovodstvo i Bolezni Ryb. Izd. VASKhNIL.* Moscow.

Kucherova, F.N. 1952. Upravlenie e'mbrional'nym razvitiem zhivotnykh putem vozdeisviya na materinskii organizm (Regulation of embryonic development of animals through action on the maternal organisms). *Usp. Sovr. Biol.*, 34, 3, 6.

Kuzema, A.I., A.P. Kucherenko and V.G. Tomilenko. 1967. Khozyaistvennaya kharakteristika vyrashchivaniya ropshinsko-ukrainskikh pomesnykh karpov (Economic aspects of raising Ropshinsk and Ukrainian local carp). Sbornik *Rybnoe Khoz-vo.* Urozhai, Kiev, no. 6.

Lavrovskii, V.V. 1964. Opyt po gibridizatsii nekotorykh vidov roda *Salvelinus* (Experiments on hybridization between some species of the genus *Salvelinus*). Sbornik *Nauchno-tekhn. Inform. VNIRO.* Moscow, no. 7.

Makeeva, A.P. 1968. Gibridizatsiya sazana s rastitel'noyadnymi rybami (Hybridization of sazan with phytophagous fish). *Vopr. Ikhtiol.*, 8, 2, 49.

Makeeva, A.P. and A.I. Sukhanova. 1966. Razvitie gibridov rastitel'-noyadnykh ryb (Development of hybrids of phytophagous fish). *Vopr. Ikhtiol.*, 6, 3, 40.

Martyshev, F.G. 1953. Vliyanie vozrasta proizvoditelei karpa na khozyaistvennye kachestva potomstva (Effect of age of carp spawners on the economic qualities of the progeny). *Nauchn. Konf. TSKhA.* Sbornik *Materialov.* Moscow.

Martyshev, F.G. 1961. Rost i razvitie potomstva karpa ot proizvoditelei raznogo vozrasta (Growth and development of carp progeny from spawners of different ages). *Doklady TSKhA,* vol. 69.

Martyshev, F.G. 1964. O vliyanii vozrasta proizvoditelei karpa na potomstvo (Effect of age of carp spawners on the progeny). *Doklady TSKhA,* no. 95.

Martyshev, F.G., I.M. Anisimova and Yu.A. Privezentsev. 1967. Vozrostnoi podbor v karpovodstva (Age Composition in Carp Farming). Kolos, Moscow.

Menyuk, N.S. 1967. Pitanie gibridov chudskogo siga s pelyad'yu v usloviyakh prudov sigovogo khozyaistva "Pushcha-Voditsa" (Feeding of hybrids of Lake Chudskoe whitefish and peled under the pond conditions of the coregonid farm "Pushcha-Voditsa"). Sbornik *Rybnoe Khoz-vo.* Urozhai, Kiev, no. 3.

Nesterenko, N.V. 1957. Opyt gibridizatsii Ural'skogo ripusa s chudskim sigom v prudovykh usloviyakh (Experiments on hybridization of Ural whitefish with Lake Chudskoe whitefish under pond conditions). *Izv. VNIORKh.* Leningrad, vol. 39.

Nesterenko, N.V. 1964. Vosproizvoditel'naya sposobnost' gibridov ripusa s chudskim sigom (Reproduction capacity of hybrids of Ural whitefish and Lake Chudskoe whitefish). *Trudy Uralsk. Otd. VNIORKh.* Leningrad, vol. 6.

Nikolyukin, N.I. 1952. Mezhvidovaya gibridizatsiya ryb (Interspecific Hybridization in Fish). *Saratov. Obl. Izd-vo.*

Nikolyukin, N.I. 1954. Skreshchivanie belugi so sterlyad'yu i vyrashchivanie gibridnoi molodi (Crossing beluga with sterlet and raising hybrid fry). *Trudy Saratovsk. Otd. Kasp Filiala, VNIRO.* Saratov, vol. 3.

Nikolyukin, N.I. 1965. Teoreticheskie predposylki primeneniya metoda gibridizatsii v rybovodstve (Theoretical considerations in using the method of hybridization in fish rearing). Sbornik *Teoreticheskie Osnovy Rybovodstva.* Nauka, Moscow.

Nikolyukin, N.I. 1968. Geterozis i ego ispol'zovanie v rybovodstve (Heterosis and its use in fish culture). Sbornik *Geterozis v Zhivotnovodstve.* Kolos, Moscow.

Nosal', A.D. 1967. Sozdanie gibridnykh form sigovykh dlya promyshlennogo vyrashchivaniya vo vnutrennykh vodoemakh Ukrainy (Creation of hybrid forms of whitefish for commercial raising in inland reservoirs of the Ukraine). Sbornik *Rybnoe Khoz-vo*. Urozhai, Kiev, no. 3.

Poliksenov, D.P. 1953. Opyt plememoi raboty s karpom v Belorusskikh rybkhozakh (Experiments in breeding work with carp in Belorussian fish farms). *Uch. Zap. Belorussk. Un-ta im. V.I. Lenina, Ser. Biol., Minsk,* no. 17.

Pulina, G.A. 1968. Vyrashchivanie inbrednykh i autbrednykh segoletkov karpa (Raising inbred and crossbred under-yearlings of carp). Sbornik *Nauch.-tekhn. Inform. VNIRO*. Moscow, no. 6.

Reshnie nauchnogo soveshchaniya po voprosam genetiki, selektsii i gibridizatsii ryb (Resolutions of the Scientific Council on the Problems of Genetics, Breeding, and Hybridization of Fish). Izd. VNIORKh, Leningrad.

Sevost'yanova, G.G. 1969. Massovyi otbor v forelovodstve (Mass selection in trout culture). *Genetika,* vol. 5, no. 7.

Sevost'yanova, G.G. 1969. Sravnitel'naya rybokhozyaistvennaya kharakteristika razlichnykh grupp raduzhnoi foreli (Comparative characterization of different groups of rainbow trout for fisheries). Sbornik *Genetika, Selektsiya i Gibridizatsiya Ryb*. Nauka, Moscow.

Shaskol'skii, D.V. 1954. O rodstvennom razvedenii karpa v promyshlennykh prudovykh khozyaistvakh (Pedigree breeding of carp in commercial fish farms). *Trudy VNIIPRKh,* vol. 7.

Turdakov, F.A. 1943. Vozrästnoi podbor (Age composition). *Zhurn. Obshch. Biol.,* vol. 4, no. 3.

Viktorovskii, R.M. 1966. Morfologicheskaya kharakteristika gibridov karpa (*Cuprinus carpio* L.) s linem (*Tinca tinca* L.) [Morphological characters of hybrids of carp (*Cuprinus carpio* L.) and tench (*Tinca tinca* L.)]. *Izv. GosNIORKh*. Leningrad, vol. 61.

Vinogradov, V.K. and L.V. Erokhina. 1966. Opyt gibridizatsii belogo i pestrogo tolstolobikov (Experiments in hybridization of silver carp and bigheads). *Rybokhozyaistvennoe Obvoenie Rastitel'noyadnykh Ryb*. Nauka, Moscow.

Vliyanie kachestva proizoditelei na potomstvo u ryb (Effect of the Quality of Spawners on Fish Progeny). 1965. Edited by V.I. Vladimirova.

Voropaev, N. 1970. Gibridy tolstolobikov (Hybrids of silver carp). *Rybovodstvo i Rybolovstvo,* no. 1.

Wolf, E. 1954. Development of disease-resistant strain of fish, *Trans. Amer. Fish. Soc.,* vol. 83.

Zobel, H. 1960. Steigerung der forellenproduction durch systematische zuchtauslese, *Dtsch. Fisch. Ztg.,* vol. 7, no. 2.

Chapter XVII

TRANSPORT OF LIVE FISH

Stock for one-year production farms, material for research in acclimatization, brood fish, and marketable fish are usually transported live. In addition to live fish of different ages, viable fertilized eggs are also transported. Transport may be undertaken with or without water. Live fish in water are transported by boat, by road—animal-drawn carts and motor trucks—and by rail. The fish are kept in special barrels, tarpaulin vats, polyethylene bags, cisterns, closed containers, fish boxes, slots, and other containers.

The following conditions must be observed in undertaking the transport of live fish:

1. Clearance from a veterinary inspector.

2. Maintenance of the fish for two to four hours in tanks with flowing clean water prior to their loading in transport carriers. The debris adhering to the body at the time of catching is thereby removed, the gills cleaned, and the guts freed of food and feces. Carp with empty guts withstand transport better than with filled guts. The requirement for oxygen is higher in fed fish than in hungry fish; hence feeding is stopped two days prior to fish transport. The rate of packing sazan, carp, tench, and Crucian carp in fish containers should not exceed more than 220 kg under a water temperature of 12°C, or 110 kg at a water temperature above 12°C.

3. Injured, weak, and especially fish suspected to be diseased are rejected.

4. Containers in which fish are carried should be clean, with a smooth inside devoid of sharp projections or nails, and filled with pure water devoid of harmful substances, including poisonous compounds, which is neutral or mildly alkaline.

5. The temperature of the water in the containers should be equal to that of the resource from which the fish are removed, and later equalized to the temperature of the water in the resource in which they are to be released (difference between water temperatures should not be more than 1 to 2°C at the initiation of transport and at its termination).

6. Loading the containers is done as quickly as possible (never over more than two hours) and very carefully to avoid injuring the fish.

7. Transportation is preferably done during the night, in the early morning hours, or two to three hours before darkness sets in.

8. En route, the water temperature should be checked regularly.

9. Sufficient ice should be stocked to cool warm-water en route at the rate of not less than 5 kg ice per 100 liters of water; the ice is crushed and placed not directly in the transporting container, but in a special box with a net bottom. Otherwise the pieces of ice may cause injury to the fish.

10. In case the water needs to be replaced en route, clean water from rivers, lakes, and ponds is used. Water from wells is poor in oxygen content and from city supply taps chlorinated; hence neither of these sources should be used.

11. In transporting fish long distances care is taken to avoid stops en route exceeding 30 minutes or under favorable conditions up to 45 minutes. In the case of an unavoidable longer halt the water in the containers is aerated in the same way as done in the case of an oxygen deficiency in the water en route.

12. Dead fish are removed from the water surface as well as from the bottom during prolonged transportation, especially in the warmer part of the year. Otherwise their decomposition impairs the oxygen regime of the water, which could lead to total mortality of the fish under transport.

13. A certificate is made on reaching destination indicating duration, conditions, and results (extent of losses, giving reasons), and signed by the consignor, those accompanying the fish during transport, and the consignee.

Primary Veterinary-Sanitary Requirements. Irrespective of the method and mode of transport it is essential to observe veterinary-sanitary rules strictly. Live fish and fertilized eggs may be transported only if the requisite veterinary certificate is issued. Veterinary inspection involves the following:

1. Only healthy active live fish may be transported—those without mechanical injury, tumors and molds on the body, skin, and eyes, with fins and scales intact and clean, and body covered with a thin layer of slime. Fertilized eggs are kept in special containers.

2. Live fish to be transported are treated with a 5% saline solution even in the absence of any infectious disease (see Chapter IV).

3. It is prohibited to transfer fish, crustaceans, and other invertebrates from fish farms to water resources prone to German measles, branchiomycosis, furunculosis, salmon vertigo, infectious anemia, and discocotylosis of trout. The transport of crustaceans and other aquatic invertebrates is also prohibited which might result in a large-scale spread of dangerous infectious and pathogenic diseases or mortality due to unestablished reasons. Fish with the following symptoms (even in isolated cases) may not be transported until a definite diagnosis has been made: bulging abdomen, scale erosion, blindness, exophthalmia, ulcers on the skin, complete or partial damage to the gills, white or gray coloration of the gills, occurrence of a large number of minute spots on the body surface, fins, or gills, blackening of the posterior one-third

of the body, black pigmented spots on the body, twisting of the spinal cord, or abnormal development of the skull.

4. Fish suffering from trichodiniasis, chilodonelliuasis, dactylogyrosis "A", dactylogyrosis "B", and gyrodactylosis are permitted for transport only after treatment with a salt or ammonia antiparasitic bath.

5. Within the limits of a district, territory, or autonomous republic, the question of transporting fish suffering from costiasis, coccidosis, ichthyophthyriasis, sanguinicoliasis, diplostomatosis, neascosis, caryophyllosis, bothriocephalosis, ligulosis, piscicolosis, lernaeosis, argulosis, and other diseases is decided by the veterinary department; within the limits of a republic this question is decided by the Head Veterinary Office; transportation between republics is decided by the Veterinary Department of the Ministry of Agriculture of the USSR. Here it is essential that prophylactic measures be taken against the entry of a pathogen into water resources which are free from the diseases listed above.

6. Exhausted, dull, and weak fish are removed from those to be transported.

7. Transport of fish in slots to water resources where infectious and pathogenic diseases are prevalent is not permitted.

8. Transport of fertilized eggs, fish, crustaceans, and other aquatic invertebrates beyond national boundaries is done in accordance with special instructions. A veterinary certificate is required asserting the absence of infectious and pathogenic diseases. A document prepared by the consignor must also be signed by the consignee stating that the fish received will be kept for at least one year in a special quarantine section of the farm. In the case of a successful quarantine procedure, further transport of exotic fish within the receiver country may be undertaken in accordance with the conditions laid down in the special instructions.

9. In the case of an outbreak of disease during transportation, on reaching their destination the fish are placed in quarantine ponds and proper corrective measures undertaken. If for some reason it is not possible to carry out the requisite measures, on completion of a veterinary inspection the fish are either utilized as food or animal feed or destroyed. In such cases the veterinary inspection office prepares the necessary documents in the presence of the consignor, consignee, and persons accompanying the fish during transport.

10. The water in which the fish are transported is not released in any pond.

Conscientious compliance with these procedures obviates the spread of diseases among fish at fish farms and other water resources, and the transmission of diseases from other countries. Furthermore, losses during transportation are notably reduced (or even eliminated). Finally, communicable diseases of man and animals deriving from fish can be totally prevented.

Transport of Live Fish in Water. In transporting fish in water the quality of the water is significant and its oxygen content the most important criterion. The oxygen content should be sufficient for normal respiration during transport. It depends on the intensity of oxygen consumption by the fish under transport as well as on the water temperature. The oxygen content is higher in relatively cooler water, the fish requirement for oxygen less, and the magnitude of the threshold (minimum) oxygen concentration increases. In other words, at low temperatures more favorable conditions are created.

Oxygen consumption differs in different fish species and in different age groups of the same species. In the case of fry it also depends on their size. Larger fry are closer to fish of older age groups in terms of respiration. The younger and more active the fish, the larger the quantity of oxygen consumed. The characteristic features of respiration of some fish species cultured in ponds are given in Table 72.

Table 72. Nature of respiration in some fish species (based on data from T.I. Privol'nev et al.)

Type of fish	Oxygen consumption/kg body weight at 10°C, mg/hr	Oxygen content at initial stage of suppressed respiration, mg/l	Minimum threshold O_2 content necessary for survival of fish, mg/l
Sterlet	68	7.0–7.5	3.5
Carp (sazan) weighing 500–700 g	45	2.0–2.5	0.3–0.5
Carp (fingerlings)	120	5.0–6.0	0.5–0.1 [*sic*]
Pike of average size	50	2.0–3.0	0.3–0.6
Pike of small size	70	—	—
Crucian carp	30	1.0–2.0	0.1
Rainbow trout, fingerlings	150–196	6.5–7.0	2.1–2.6
Rainbow trout weighing 202 g	211	—	—
Whitefishes:			
peled	100	3.5–4.0	1.0–1.5
Volkhov	125	4.5–5.0	—
Ladoga	135	—	1.5–2.0
Eel	25	2.0–2.5	0.2–0.3

A water temperature of 6 to 8°C is considered favorable for cold-loving fish and 10 to 12°C for warm-loving fish during transport in summer; these temperatures during spring and autumn are respectively 3 to 5°C and 5 to 6°C. During winter all fish are transported at a water temperature of 1 to 2°C.

Taking into account its significance, especially under prolonged transport, water temperature must be systematically checked to avoid notable fluctuations.

Depending on the volume of the transporting containers the water is aerated by different methods. In barrels and other small containers this is done by agitating the water a little below the surface with the help of a hand-operated pump. In motor and rail transport aeration is done from a rubber pipe with diffusers (15 to 20 needle-sized punctures in an area of 1 cm^2). Air bubbles approximately 0.10 mm in size pass through these punctures under pressure. Increasing the pressure increases the size of the air bubbles, which has a practical significance (the finer the air tension, the greater the quantity of oxygen passed into the water). Any method of aeration achieves another objective—freeing the water of carbon dioxide and chlorine (if chlorinated water, harmful for fish, has been used at railway stations). A chlorine concentration of about 0.5 mg/l is considered high and even a smaller quantity (0.2 mg/l) can drastically affect the respiration mechanism, ultimately leading to mortality. Chlorine mainly acts on the nervous system of fish. Its effect increases at low water temperatures (2 to 3°C). Water can be dechlorinated by passing air through it or by passing the water through a filter with activated charcoal. With intensive aeration the chlorine can be liberated from the water within 10 to 15 minutes. It is also possible to remove chlorine by adding such chemicals as hyposulphite (0.89 mg of pure hyposulphite per 1 mg of chlorine) or sulphite and ammonia. It should also be remembered that the water should contain no zinc.

The oxygen content in the water is affected by the pollution of organic substances excreted/secreted by fish during transportation. The mucus released by fish in water serves as a good substratum for putrefying bacteria, which also consume oxygen, thereby reducing its quantity. Such a reduction is especially dangerous in overcrowded conditions and during the summer when putrefying bacteria multiply rapidly. Hence it is recommended that under summer transportation the water in the containers be changed at least once a day and the temperature kept at a lower level to slow down the development of putrefying bacteria. Daily water replacement is generally not necessary during autumn-winter transportation of live fish.

The quantity of water required for the transport of fish depends on the biological peculiarities and age of the latter and the duration of transport. It is calculated on the basis of data on the respiration requirements of different fish species and the level of dissolved oxygen in water at different temperatures. For example, to estimate how much water would be required for transporting 250 kg peled at a water temperature of 5°C and duration of transport 4 hours, one need only look at the data presented in Table 72. It can be seen from the Table that peled at a water temperature of 10°C consume 100 mg of oxygen per hour per each kg of body weight. At a temperature of 5°C the consumption of oxygen is less, $100 \times 0.5 = 50$ mg/hr. Table 73 shows that the oxygen content per liter of water at 5°C is 12.73 mg. If 1.0 to 1.5 mg/l (or taking 1.5 mg/l) is deducted from this value, as the threshold

Table 73. Quantities of dissolved oxygen in water at different temperatures (from T.I. Privol'nev)

Water temperature, °C	Quantity of oxygen dissolved in water		Water temperature, °C	Quantity of oxygen dissolved in water	
	cm^3/l	mg/l		cm^3/l	mg/l
0	10.19	14.56	16	6.89	9.85
1	9.91	14.16	17	6.75	9.65
2	9.64	13.78	18	6.61	9.45
3	9.39	13.42	19	6.48	9.26
4	9.14	13.06	20	6.36	9.09
5	8.91	12.73	21	6.24	8.90
6	8.68	12.41	22	6.11	8.73
7	8.47	12.11	23	6.00	8.58
8	8.26	11.81	24	5.89	8.42
9	8.06	11.52	25	5.78	8.26
10	7.87	11.25	26	5.67	8.11
11	7.67	10.90	27	5.50	7.95
12	7.52	10.75	28	5.46	7.81
13	7.35	10.50	29	5.36	7.67
14	7.19	10.28	30	5.26	7.52
15	7.04	10.06			

(minimum) oxygen content or minimum reserve for survival of the fish, the remaining 11.23 mg/l of oxygen is used for respiration. For each kg of body weight of peled 50 mg of oxygen are required in one hour at a temperature of 5°C. Such a quantity of oxygen is present in $50:11.23=4.45$ l of water. To transport 250 kg peled over 4 hours without the addition of oxygen during the period of transport, a total of $4.45\times250\times4=4{,}450$ l of water would be required at the outset.

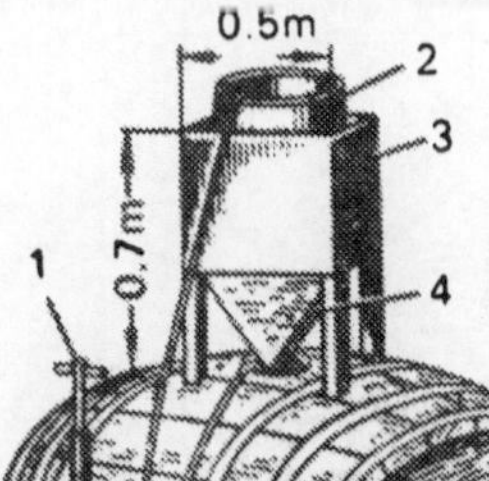

Figure 120. Oval barrel equipped with aerator, designed by Truten. *1*—pump; *2*—sieve; *3*—ice box; *4*—net.

CONTAINERS. When the quantity of fish to be transported is small and the distance short, animal-drawn carts and oval wooden barrels of 150 to 250-l capacity (Figure 120) are used. The barrels are best prepared from oak, birch, maple, aspen, spruce, or pine, using staves 1.5 to 2.0 cm thick. An opening of 2.5 cm $\times$ 30.0 cm in the upper end of the barrel permits a box with a net bottom to be inserted after the fish are loaded. The box is filled with clean ice which on melting keeps the temperature of the water in the barrel low.

New wooden barrels should not be used for transporting fish as the tannic acid exuded from the wood may have an adverse effect on them. New barrels must be steamed and soaked in clean water for several days. Any barrel used in transporting fish is thoroughly washed before usage. Barrels priorly used for pickling or salting of fruits and vegetables or for storing petrol, kerosene, and crude oil are absolutely taboo for the transportation of fish. In metal barrels the water tends to overheat necessitating covers of tarpaulin, moist straw, or grass.

To achieve better aeration of the water, prevent water spillage, and avoid injury and traumatization of the fish, the barrels are filled with water up to three-fourths of their volume. Barrels with a capacity of 250 l can be loaded with 1,200 to 1,600 standard carp yearlings at a water temperature of 10 to 12°C.

Barrels with fish are transported not only on animal-drawn carts (three barrels per cart), but also on motor trucks and even in goods wagons. Tarpaulin vats are also commonly used for holding fish, which are hung along the sides of the carrier (Figure 121) (body of the motor truck covered with tarpaulin). Common vehicles can also be covered with tarpaulin and used

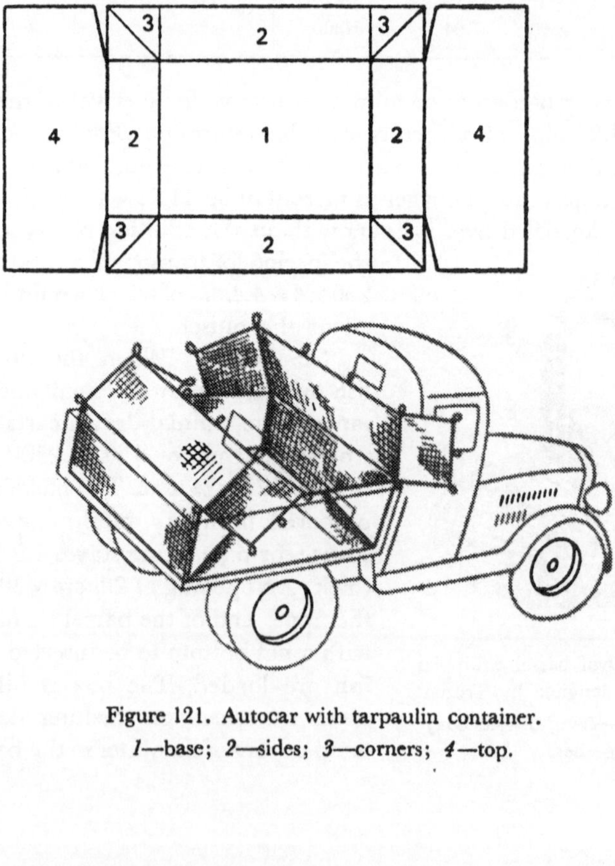

Figure 121. Autocar with tarpaulin container.
1—base; *2*—sides; *3*—corners; *4*—top.

for transport within a farm or over a short distance. The following minimum quantity of fish is loaded in tarpaulin vats on motor trucks (kg/m^3 of water): eel, Crucian carp, sterlet, tench, and carp—500; bream and pike—350 to 400; and salmon, trout, and whitefish—200 to 250.

Special live fish carriers with an aeration system have been designed on the chassis ZIS-150. Experience in Krasnodar has shown the usefulness of dismountable systems for transportation, which can be fixed on trucks of different makes. The cistern is equipped with two hatches and a system of air supply. It is loaded with 800 to 1,600 kg of marketable fish and 500 to 1,200 kg of stocking material. The loading norm for motor truck GAZ-51 fitted with a cistern therefore is 800 kg of marketable fish or 500 kg of stocking material. The loading norm for truck ZIS-150 is 1,200 and 900 kg respectively and a Ural-375—1,600 and 1,200 kg respectively.

Live fish are transported long distances in specially equipped rail wagons (Figure 122). The optimum temperature of water in fish tanks placed in rail wagons in spring and autumn is 3 to 5°C, and in warmer months 5 to 10°C lower than the water temperature of the pond from which the fish have been

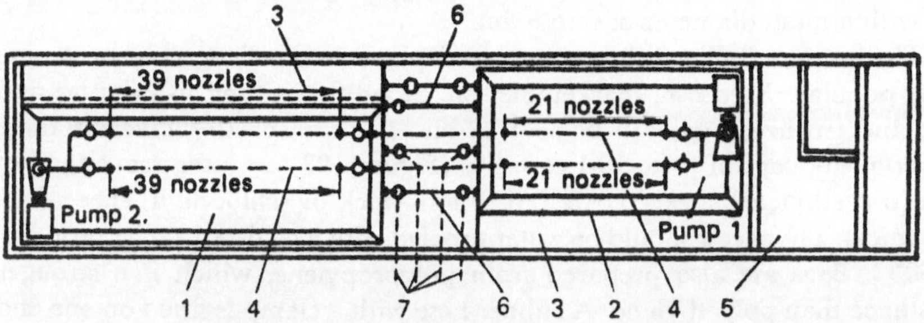

Figure 122. Rail wagon with aeration system.

1—large tank; *2*—small tank; *3*—passage near tanks; *4*—pipes of aeration system with nozzles; *5*—space for conductors where control panels are located; *6*—pipes connecting tanks; *7*—overspill.

taken, but not higher than 8°C. The temperature during winter should be 2 to 4°C, but not below 1.0 to 1.5°C. To avoid ice formation in the tanks containing fish when the atmospheric temperature is below 0°C, the space containing fish is heated. In warm months five to seven tons of ice are packed in the wagons before the fish are loaded and, if need be, the ice stock supplemented en route.

The norms of loading live fish in wagons which contain no special equipment for enforced aeration of water depend on the type and age of fish. The ratio of fish to water in such cases is generally 1:10; for transporting fish which are particularly demanding in oxygen (salmon and whitefish) the ratio is 1:15 to 1:25 and even more. The extent of losses should not be greater than 5% in transporting live fish by rail for three days and 7 to 8% over five days.

Fish transported by boat over considerable distances are carried in special slots. These consist of floating boxes with slits of about 2 cm on the sides. A container of this type is loaded on the average with one ton of live fish in a water volume of 10 m^3 (ratio 1:10). The fish load may be increased in the cooler parts of the year and reduced during summer. The slots loaded with fish are covered with mats of bast and reed to prevent the fish jumping out, especially fish of older age groups. The speed should be slow when transporting live fish (not more than 3 to 4 km/hr) and halts made at places with a mild water current (less than 0.5 m/sec). Water areas en route polluted with industrial wastes are particularly hazardous since the exposure of fish for a long time (sometimes even just a short period) to polluted zones can lead to total mortality. The norms of loading fish in slots during summer are as follows (kg/m^3 of water): Crucian carp, sterlet, sazan, and tench—70 to 90; bream, pike, roach, perch, and ide—50 to 60; and salmon, trout, and whitefish—30 to 40.

To prevent fish escapes through the slits the slots are lined with capron net with a mesh diameter of 6 to 8 mm.

The transport of live fish in water by airplanes and helicopters is becoming quite popular. Seed fish, marketable fish, brood fish, young fish (larvae and fry), and fertilized eggs can be sent by air. For transporting long distances polyethylene bags of 6- to 10-l capacity (Figure 123) are used, consisting of thin (polyethylene) sheets 0.048 to 0.10 mm thick or (chlorvinyl) sheets 0.15 mm thick. The sheets should be water proofed with a cold-resistance property of 60°C. Bags are also prepared from polypropylene, which is a stronger substance than polyethylene. A rubber hose with a clamp is fitted on one end

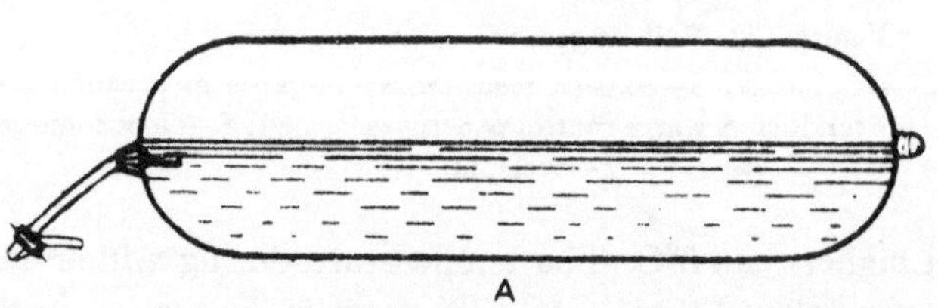

A

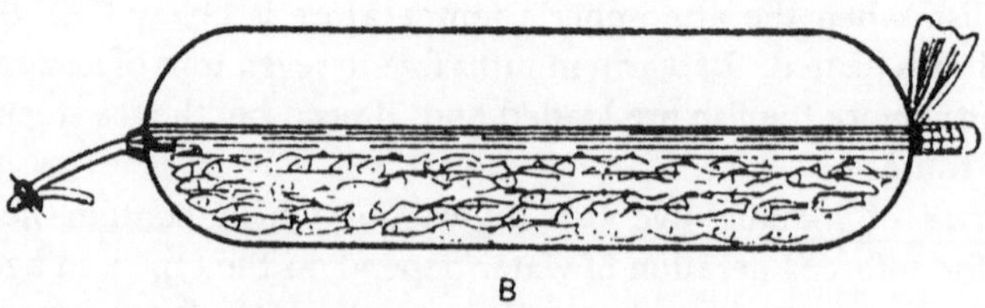

B

Figure 123. Polyethylene bag.
A—Welded at one end; *B*—Filled with fish and tied at one end.

of the bag for collecting water samples to determine the oxygen content; 10 cc of water suffice for analysis. A metal or glass tube is inserted into the rubber pipe. The bag is first filled with water, the fish released in it, and the pipe connected through the tube to an oxygen cylinder or balloon then fitted. The bag, squeezed shut by hand at the level of the water surface, is filled with oxygen by pressing the balloon or releasing oxygen from the cylinder (Figure 124). When the bag has been filled with oxygen it is squeezed shut by hand farther down, the upper part folded over, and the open end tied with a

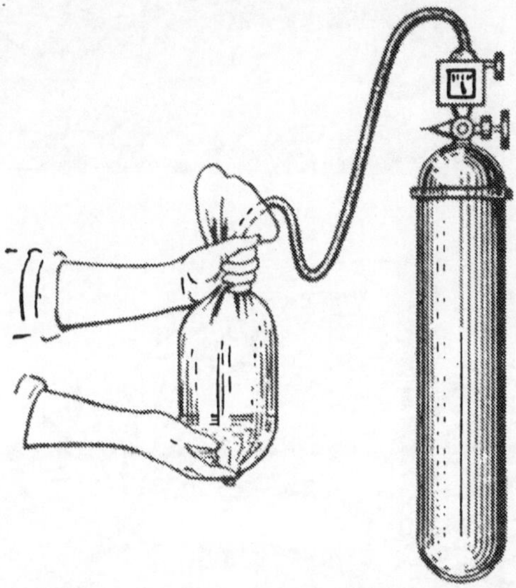

Figure 124. Filling of polyethylene bag with oxygen.

thick cotton twine. Capron twine or twine made of soft material is not suitable since the former cuts through the polyethylene and the latter is difficult to tie sufficiently tight to prevent oxygen leakage, and once so tied difficult to undo.

The readied bags, containing water, fish, and oxygen, are placed in a cardboard carton with the sleeves directed upward. The carton is lined with a polyethylene or chlorvinyl sheet (Figure 125) and ice inserted between the bags, which are arranged in one or two rows. For prolonged transportation it is better to arrange the bags in a single row since the pressure of the upper row may cause an oxygen leakage in the lower one. The thicker the polyethylene sheets, the longer the life of the bag. Bags made of sheets 0.047 mm thick are suitable for use once or twice, while those made of sheets 0.1 mm thick can be used three or four times.

In transporting carp in bags in autumn-winter the most favorable water temperature is 5°C and during summer 15°C. A reduction in water temper-

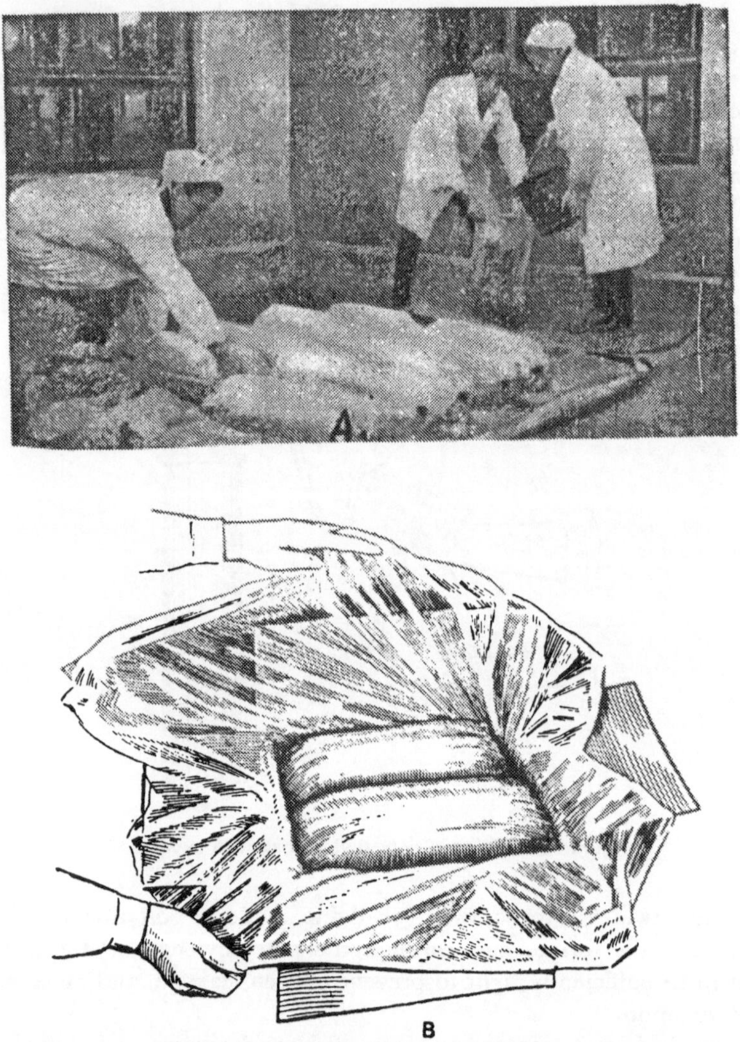

B

Figure 125.
A—Filling polyethylene bags with water, fish, and oxygen for transport; *B*—Packing bags in cardboard cartons.

ature during summer transport to 5 to 10°C is permissible only under a favorable oxygen regime. The ratio between water and fingerlings during transport up to 5 hours may be taken as 1.5:1 (by weight) and for 15 to 20 hours 2:1; for transporting fry this ratio should not be less than 4:1.

Fry are transported in metal vats, milk cans made of white metal, barrels, etc. by any mode of transport. The holding norms for larvae, fry and fingerlings per one liter of water (not taking into account water aeration en route or entry of oxygen from the air) are given in Table 74.

Table 74. Holding norms for larvae, fry, and fingerlings for transportation (from T.I. Privol'nev)

Type of young fish	Water temperature, °C	Duration of transport	Norms of loading fishes/l water	Remarks
Larvae of salmon	10	1	100–200	1. In case of availability of equipment for aeration of water, holding density can be significantly increased.
Larvae of whitefish	10	1	3,000–5,000	2. Young fish with accessory respiration through skin (carp, sazan, Crucian carp, and eel) can be transported without water in humid environment (see text).
Fingerlings of carp weighing approx. 1 to 3 g	10–15	—	100–300	

Phytophagous fish (grass carp, silver carp, and bighead) are transported in wagons and motor trucks, but more often in polyethylene bags carried by automobile or airplanes. Other methods of transport are also used, primarily those mentioned above for carp. In transporting live grass and silver carp in wagons over a period of up to 12 days, one wagon is loaded with 1,500 kg of two-year and three-year grass carp, and up to 800 kg of three-year and four-year silver carp. Silver carp withstand transportation less well than grass carp and their losses are generally significantly higher in transit.

Fish of older age groups and those weighing up to 5 g are transported in motor vehicles. The loading norm per vehicle is 30,000 fry. Under the conditions of transportation up to 10 hours, losses do not exceed 5%. Fingerlings around 15 g in body weight are transported in trucks at a water temperature of 5 to 10°C with a norm of 5,000 fish. Constant aeration with the help of a compressor is essential en route. Given these norms and proper conditions fry can be transported with very little loss. Fish of higher weight (500 to 700 g) with a duration of transportation of 25 hours are carried at a temperature of 5 to 8°C and a loading norm of 300 to 400 fish per carrier.

Polyethylene bags 42 to 44 cm wide (thickness of sheet 0.10 to 0.14 mm) are commonly used for transporting phytophagous fish. Brood fish and large fish are generally transported in bags of 50 to 80-l capacity, although for particularly large fish larger bags are used. In transporting fry in vats by motor trucks polyethylene bags of 40-l capacity are used (length of sleeve 88 cm). The length of the bag is increased to 1.35 to 1.55 m for packing fish of older age groups. The bags vary in number of polyethylene layers in accordance with the age of the fish and conditions of transportation: for larvae shipped by air or on motor trucks two-layered bags are used, and for fish of the older age groups and brood fish three-layered bags. Such bags are considerably more durable and the size of the packed bag at least 5 cm greater.

Brood fish are starved for several days before transportation, washed in frequently changed water to remove dirt and slime, and only then introduced into the bags. The entire procedure should be executed in the shortest time possible. If the bags are arranged in a hard container and the water does not cover the head of the fish completely, the bags should be tilted in such a way that the head and gills are covered with water. In the process of transporting large brood fish hard containers are used which are smaller in width than the diameter of the bags to ensure raising the water level and keeping the body in a horizontal position with the back upward. To avoid injury to the fish against the bottom of the hard container, a soft material is placed under the bags—grass, hay, straw, etc. The bags are transported in a slanted position, at an angle of 45°. The recommended loading norms are as follows: on motor truck GAZ-51—50 to 60 fish; on truck ZIL-150—70 to 80 fish; in helicopter MI-4—25 to 30 fish; in airplane AN-2—up to 30 fish; and in airplane IL-14—up to 100 fish.

It is better to transport brood fish in autumn at a temperature of 5 to 7°C; their transportation during spring is highly undesirable.

Larvae can be transported two to four days after hatching or soon after an air bubble is visible in the swim bladder. No injury to the fish during their introduction into the bags is permissible. Weak and frail larvae are carefully screened out since they die shortly after introduction into the bags and start to decompose. The toxic substances released from dead larvae soil the water for the remaining larvae and cause their mortality.

Advanced larvae are kept for 12 to 24 hours in enclosures (15,000 to 20,000 in each) for the evacuation of food from the guts before being packed in polyethylene bags. The enclosures, 70 cm × 70 cm × 45 cm in size, are prepared from capron net Nos. 10 to 12, fixed on wooden frames, and floated in a canal in running water.

The water temperature in the polyethylene bags is regularly checked during transportation of larvae, and not allowed to drop below 15°C. Significant differences in temperature at the time of releasing the larvae into the pond after transportation are not permissible, and the duration of transport should not exceed 24 to 30 hours. A prolonged period of transportation leads to significant mortality among larvae not only in the bags but also at the time of their release into ponds. The process of preparing plastic bags and packing them with fish is carried out in a well-illuminated place or shed to ensure round-the-clock working conditions.

The loading norms of larvae in polyethylene bags of 40-l volume (water one-third, oxygen two-thirds) are as follows: in the case of transportation of up to 5 hours—100,000 fish, and more than 5 hours—50,000 fish. The losses should not exceed 3 to 5% during transportation up to 10 hours nor 10 to 15% for 10 to 24 hours.

In addition to bags, canisters are also used for transporting larvae of phytophagous fish. Two tubes are inserted in the lid through which oxygen is supplied in the same way as in the case of bags. Experience has shown that canisters are more durable and the loading of larvae in them twice as rapid compared to loading bags. Disinfection is essential after every use.

Advanced larvae and fry may be transported in vats, motor carriers, and polyethylene bags. A plastic bag may contain from 5,000 to 25,000 larvae. At a water temperature below 20°C, with larvae of small size and a duration of transportation not exceeding 24 hours, the loading norms may be increased to 25,000 larvae per bag. Losses should not exceed 3 to 5% during the period of transport.

The bags are prepared in the same way as described above—filled with water up to one-third their volume, fishes released in them, and a rubber pipe inserted through which oxygen is fed to replace the air which has been removed. The oxygen is taken either from an aviation cylinder of the KB-1 and KB-2 type at a pressure of 30 atm or under stationary conditions from the usual hexagonal cylinder provided with a reducing valve (pressure 150 atm). The end of the bag is wrapped with water-insulation tape (50 cm per bag) and to make it airtight fastened with a screw-type metal clip. The bags are kept on a tarpaulin for 15 to 20 minutes before loading in the carrier in order to check them for air-tightness.

The bags are then arranged in cardboard cartons 33 cm × 60 cm × 33 cm or 40 cm × 65 cm × 40 cm previously lined with chlorvinyl sheets, thick paper, or cardboard pieces. These are tied with a rope not less than 1 cm in diameter. The weight of one carton is 20 kg.

During long-distance transportation and in high temperatures the water in the bags is apt to become highly polluted with excreta; it must be totally replaced and fresh oxygen added.

Carp larvae produced on a commercial scale are transported over short distances (up to two hours) in milk cans or polyethylene bags with oxygen. The density of larvae in each can or bag may be 200,000 to 300,000.

For prolonged transportation (more than two hours) motor trucks equipped with compressors are used, and for even greater distances, airplanes. Carp are packed in polyethylene bags in the same way as described for phytophagous fish undergoing a transportation period of not more than 20 hours.

The transportation of sander, cisco, and broad whitefish is assuming greater proportions each year. Experience in Karelia has shown that sander at the age of three to seven years is more tolerant of transportation under unfavorable conditions of the external environment; sexually mature fish withstand transportation less well and are particularly sensitive to jolting and oxygen deficiency during prespawning and spawning periods, as well as to pollution of the water with excreta. Young sander up to one year of age cannot withstand transportation and fish one to three years of age can only be transported under a temperature of 14 to 16°C. The following loading norms are recommended for motor carriers under a duration of transport not exceeding more than 12 hours (Saltup, 1967):

Average weight, kg	Loading norms, no. of fish	Weight, kg
1.5	150	230
1.0	200	200
0.5	300	150
0.3	400	120
0.2	300	60

Sanders of lower weight can withstand transportation better in wagons than large-sized fish. The loading norms for the transportation of brood fish and young sander in specially equipped slots in the conditions of the Volga delta are as follows: (a) brood fish—800 to 1,000 fish; and (b) young fish (yearlings) to foraging ponds—20,000 to 30,000 fish.* The sander in this case should have no body injury and it is recommended that the water be sufficiently saturated with oxygen, not less than 60% saturation at a temperature of 7 to 9°C.

Motor trucks equipped with tarpaulin vats or cans of 30- to 36-l capacity are used for transportation of fingerlings of whitefishes, as well as motor carriers fitted with cisterns. Losses are about 0.4% while carrying 3,000 yearlings, with a duration of transportation of 11.5 hours, and a temperature of 10 to 11°C. Cisco are transported at a water temperature of about 4°C, but not higher than 6 to 8°C. If sufficient care is not taken, cisco easily lose their scales and suffer injuries. Such fish should not be transported. The fingerlings of cisco can be transported long distances for a period of 5 to 6 hours at a temperature of 1 to 2°C with a density in the usual tarpaulin vats of 5 to 7 fish/l. The density can be increased to 10 to 15 fish/l at the same temperature if the transportation period is only 1.0 to 1.5 hours.

Transportation of Live Fish Without Water. Fish are usually transported without water in airplanes or helicopters. The material is arranged in crates of plywood (veneer) or tin boxes and still better of duralu-

*Original reads 200,000 to 300,000 yearlings, which would seem to be a misprint—General Editor.

minium with transverse planks; the crates are 60 cm × 75 cm × 10 cm or 55 cm × 55 cm × 10 cm in size (Figure 126). They are stacked one on top of the other, with seven to eight crates in a stack. The corners are raised on the upper side of each crate to prevent the one lying above from slipping into the underlying one. Holes are made at the bottom of each crate for drainage of water and a drain pipe fitted into the lower-most crate in a stack. The bottom of the crates is covered with moss, muslin pieces, or mats made of bast and reed. A water tank in which pieces of ice are floated is fitted above the stack of crates. Through levers the pilot can release cold water on the fish in the course of the flight. Between 300 to 400 carp yearlings of standard weight are packed in each crate.

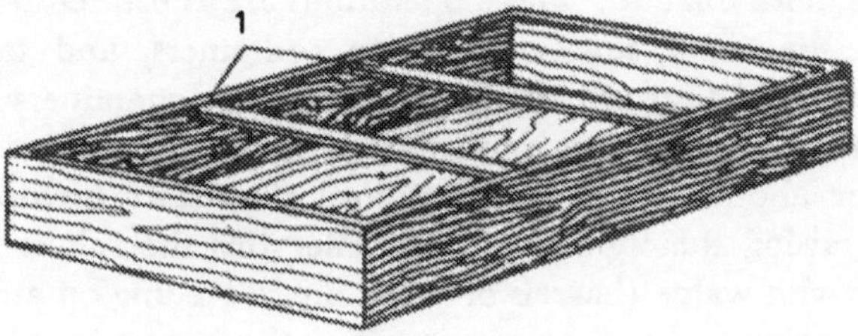

Figure 126. Crate for transporting live fish by airplane.
1—transverse planks.

Fish can also be transported by motor truck in crates without water. In this case too the crates are stacked on the truck, and covered with a tarpaulin tied in place with a rope to prevent the fish from drying. A tank or box containing water and ice is fitted above the stacks of crates as a means of keeping the fish moist. Brood fish are transported in boxes with removable wooden slats and a thin tarpaulin or oil cloth covering. The fish are placed between the slats and their heads covered with a wet cloth to prevent desiccation of the gills.

The duration of transportation of carp of different age groups by air without water should not exceed (under the necessary temperature at a given time of the year) the periods indicated in Table 75.

The transport of fish by any mode is preferably done in the cooler part of the day. It is better to release the transported fish in ponds either early morning or two to three hours before darkness.

If the temperature is higher than 15°C, the fish must be carried in water by air, keeping the ratio of water and fish at 10 : 1. If the flight is longer than two to three hours, the water should be aerated even if only with a simple pump.

Fish transported without water are prepared in the same way as for transport in water. Loading and unloading of the fish should be done as

Table 75. Duration of transportation of carp by airplane without water (hours)

Age group	During spring at different air temperatures			During autumn at different air temperatures		
	1–5°C	5–10°C	10–15°C	1–5°C	5–10°C	10–15°C
Under-yearlings	—	—	—	2.5	2.0	1.5
Yearlings	2.0	1.5	1.0	—	—	—
Replenishing stock	2.5	2.0	1.5	3.0	2.0	1.5
Brood fish	3.0	2.5	1.5	4.0	3.0	2.0

quickly as possible: loading should take no more than 15 to 20 minutes and unloading not more than 10. These operations are expedited if not only the fish but also clean water, stock of fresh ice, containers, and tarpaulin are ready by the time of despatch of the transport. The containers or boxes, on reaching their destination, are immediately unloaded and the fish flooded with water for about five minutes or until they start breathing quietly, rhythmically raising their gill opercula. After this the fish are transferred to a container with water (barrels or tarpaulin vats slung on automobiles or carriers) and transported to the pond in which they are to be released. If the difference between the water temperature in the container and that of the pond water is not more than 1 to 2°C the fish are released immediately into the ponds. If, however, this difference is greater, the water temperature in the containers must first be equalized by gradually adding pond water to them; only then are the fish transferred to the pond. Transference is done very carefully in different shallow areas of the pond irrespective of the mode of transportation (with or without water).

Transportation of Fertilized Eggs. Fertilized eggs are usually transported in special isothermic containers in which a humid atmosphere can be maintained. In the warmer part of the year the atmosphere inside the containers in which the eggs are stored is cooled to slow down egg development and reduce the intensity of gas exchange. However, excessive cooling of the eggs may lead to their mortality. Fish eggs at different stages of development are characterized by a different tolerance to temperature as well as to mechanical injury and other external treatments. It is better to transport fertilized eggs after the appearance of pigmentation in the eyes of motile embryos. Like carp, the eggs of salmon and whitefish are also characterized by higher resistance in the initial period of development (up to four to seven days), which subsequently reduces until pigmented eyes appear in the motile embryos. Losses are much higher if the eggs are transported during the period prior to the hatching of larvae. The transportation of fertilized eggs of coregonids cannot continue for more than seven days from the time of fertilization, and of salmon eggs for no more than two to three days. In the first two to three days after fertilization eggs may be transported over a distance not exceeding a few hours. It is better to transport eggs over long distances by air and over short distances by rail or motor trucks. Polyethylene bags are often used as containers but for the eggs of autumn spawning fish, isothermic containers.

In the USA and Canada cuvettes made of styrofoam are used for transporting salmon eggs on a commercial scale. Styrofoam is light weight and has high insulation properties which prevent sharp fluctuations in air temperature (Figure 127). Each cuvette weighs 158 to 160 g. There are 49 holes in 7 rows at the bottom of the cuvette and the diameter of each hole is 4 mm. Generally, 10,000 eggs are arranged on a moist muslin sheet in six to eight layers in each cuvette. The cuvettes are then stacked in such a way that a space of 3.5 mm is left between them and one container does not press on the other. Air mainly reaches the eggs through openings in the bottom of the cuvettes. Five containers are packed into a cardboard or plywood box and the total weight of each carton is about 16 kg. In 1965, 100,000 fertilized eggs of steelhead salmon were carried in such cuvettes from the USA to the USSR with no loss during a transportation period of 55 hours.

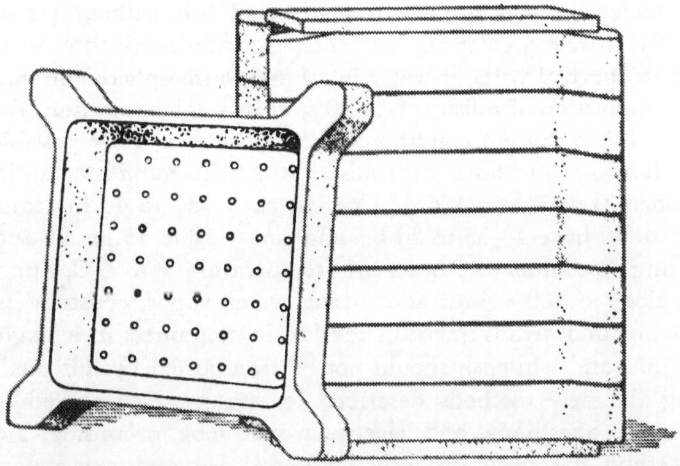

Figure 127. Cuvette made of styrofoam for transporting eggs.

Live fish can be transported in humid conditions without water. Such transportation is based on carrying live fish up to a state of anabiosis. Some fish species important for pond culture (carp, tench, Crucian carp) are quite hardy and can survive for a fairly long time without water. Respiration occurs partially through the skin; the duration of such respiration depends on the age and fish species. The longest survival (up to 100 hr) of carp fingerlings in humid atmosphere was recorded at 4°C; at 1°C they survived for 28 hr and at 15°C for 33 hr. While the fingerlings of carp survived at 4°C for 100 hr, two-year carp at this temperature had a maximum survival period of 36 hr. The season of the year is quite important. Two-year carp from December to June can live without water from 28 to 34 hr but in September–October not more than 8 hr. The difference can probably be attributed to the adaptation of the metabolic process to the winter period.

The maximum survival time for some fish species under a humid atmosphere at a temperature of 4°C is summarized below.

Species	Weight, g	Time, hr	Range, hr
Crucian carp	100–213	288	36–288
Tench	100–500	79	24– 79
Catfish	263–1,300	56	24– 56
Pike	400–800	20	9– 20
Sander	550–986	2	—

At a higher temperature (15°C) fry live for a much shorter period in humid air, especially juvenile pike—from 1 hr 20 min to 5 hr; survival at a temperature of 24°C is only 30 to 60 min.

Temperature, humidity, partial pressures of oxygen and carbon dioxide, etc.—all influence the duration of survival of fish without water (E.A. Veselov, 1947). The greater the percentage of skin respiration in the fish, the longer its survival without water in a humid atmosphere and vice versa. If the skin respiration of a fish is 8 to 10% of its total respiration, the transportation of fish in humid conditions without water is quite possible.

The following durations of transportation in humid conditions are recommended (Privol'nev, 1964): Crucian carp—30 to 48 hr, catfish and tench—24 to 30 hr, eel—15 to 20 hr, and carp—12 to 15 hr. In such cases the following conditions are observed: temperature 1 to 4°C, atmospheric humidity close to 100% saturation using water vapor, constant inflow of oxygen or air, and arrangement of the fish in containers in a single layer. Sander, trout, and whitefish should not be transported by this method.

Among the new methods described for transporting chilled live fish without water anesthesia and electronarcosis look promising. However, neither method has been properly developed yet and hence cannot be recommended at this time.

References

Danchenko, A. 1967. Tristerny dlya perevozki ryby (Tanks for transporting fish). *Rybovodstvo i Rybolovstvo,* no. 2.

Golovkov, G. 1969. Perevozka chira (Transporting broad whitefish). *Izv. VNIORKh.* Leningrad, vol. 63.

Instruktsiya po veterinarnomu nadzoru za perevozkoi zhivoi ryby, prednazna chennoi dlya ryborazvedeniya i akklimatizatsii (Instructions on Veterinary Inspection of Transportation of Live Fish Meant for Fish Rearing and Acclimatization). Izd. MSKh SSSR, Upravlenie Veterinarii, Moscow, 1962.

Instruktsiya po perevozke zhivoi ryby dlya akklimatizatsii zarbleniya vodoemov (Instructions for Transport of Live Fish for Acclimatization and Releasing Fish in Reservoirs). Izd. VNIORKh, Leningrad, 1964.

Kamburova, St. 1967. Rybov"dstvo (Fish Farming). Sofia.

Karpanin, D.P. and R.Ya. Kosyreva. 1956. Biotekhnika razvedeniya sudaka v del'te Volgi (Biotechnique of sander farming in the Volga delta). *Rybnoe Khoz-vo,* no. 1.

Kosyreva, R.Ya., N.L. Nechaeva and A.P. Ivanov. 1966. Novaya tara dlya oplodotvorennoi ikry lososya (A new container for the fertilized eggs of salmon). *Sbornik Nauchno-tekhn. Inform. VNIRO.* Moscow, no. 5.

Kruzhalina, E.I. and Yu.I. Orlov. 1960. Perovozka karpovykh ryb v polietilenovykh paketakh (Transportation of carp in polyethylene bags). *Izd. VNIRO.* Moscow.

Orlov, Yu.I. and E.I. Kruzhalina. 1959. Primenenie paketov iz plastmassovykh plenok dlya perevozki zhivoi ryby i drugikh vodnykh zhivotnykh (Use of plastic bags for transporting live fish and other aquatic animals). *Izd. VNIRO.* Moscow.

Privol'nev, T.I. 1959. Perevozka i khranenie zhivoi ryby (Transportation and Storage of Live Fish). Pishchepromizdat, Moscow.

Privol'nev, T.I. 1964. Vyzhivanie i perevozka zhivoi ryby bez vody vo vlazhnoi atmosfere (Survival and waterless transportation of live fish in humid weather). *Izv. GosNIORKh.* Leningrad, vol. 58.

Saltup, B.N. 1967. Zaryblenie Karelskikh ozer sudakom (Stocking of sander in Karelian lakes). *Rybnoe Khoz-vo,* no. 1.

Stel'tsova, S.V. 1953. Deistvie khlorirovannoi vody na ryb (Effect of chlorinated water on fish). *Izv. VNIORKh.* Leningrad, vol. 33.

LIST OF BASIC REFERENCES

Cherfas, B.I. 1956. Rybovodstvo v estestvennykh vodoemakh (Fish Culture in Natural Reservoirs). Pischchepromizdat, Moscow.

Direktivy XXIV s"ezda KPSS po pyatiletnemu planu razvitiya nardonogo khozyaistva SSSR na 1971-1975 gg (Directives of the XXIV Congress of the Communist Party of the Soviet Union on the Five-Year Plan for Development of the National Economy of the USSR for 1971-1975). Gospolitizdat, Moscow, 1971.

Eleonskii, A.N. 1946. Prudovoe rybovodstvo (Pond Fisheries). Pishchepromizdat, Moscow.

Ioganzen, B. and G. Krivoshchekov. 1970. Sel'skokhozyaistvennoe rybovodstvo Sibiri (Fish Farming in Siberia). Novosibirsk.

Kamburova, St. 1967. Rybov"dstvo (Fish Farming). Sofia.

Karzinkin, G.S. 1952. Osnovy biologicheskoi produktivnosti vodoemov (Fundamentals of the Biological Productivity of Reservoirs). Pishchepromizdat, Moscow.

Lipin, A.N. 1950. Presnye vody i ikh zhizn' (Sweet Waters and the Life in Them). Uchpedgiz, Moscow.

Martyshev, F.G. 1968. Bol'she ostorozhnosti (Be more careful!). *Rybovodstvo i Rybolovstvo,* no. 6.

Martyshev, F.G. 1968. Dostizheniya prudovogo rybovodstva za 50 let Sovetskoi vlasti (Progress in pond fisheries during 50 years of Soviet rule). *Doklady TSKhA*, no. 139.

Nikol'skii, G.V. 1963. E'kologiya ryb (Fish Ecology). Vysshaya Shkola, Moscow.

Nikol'skii, G.V. 1971. Chastnaya ikhtiologiya (Special Ichthyology). Vysshaya Shkola, Moscow.

Nikolyukin, N.I. 1972. Otdalennaya gibridizatsiya ryb (Distant Hybridization of Fish). Pishchevaya Promyshlennost', Moscow.

Pegel', V.A. 1959. Fiziologiya pishchevareniya ryb (Physiology of Digestion in Fish). Tomsk.

Pryanishnikov, D.N. 1946. Agrokhimiya (Agricultural Chemistry). Sel'khozgiz, Moscow.

Rybovodstvo i bolezni ryb (Fish Farming and the Diseases of Fish). Sbornik

Izd. VASKhNIL. Moscow, 1969.

Shpet, G.I. 1968. Biologicheskaya produktivnost' ryb i drugikh zhivotnykh (Biological Productivity of Fish and Other Animals). Urozhai, Kiev.

Stroganov, N.S. 1962. E'kologicheskaya fiziologiya ryb (Ecological Physiology of Fish). Izd-vo MGU.

Stroganov, N.S. and N.S. Buzinova. 1969. Gidrokhimiya (Hydrochemistry). Izd-vo MGU.

Sudakevich, F. 1869. Obzor iskusstvennogo rybovodstva za granitsei i v Rossii (Review of Artificial Fish Culture Abroad and in Russia). St. Petersburg.

Zernov, S.A. 1949. Obshchaya gidrobiologiya (General Hydrobiology). Izd-vo Akad. Nauk SSSR, Moscow.

Printed in India